STOCHASTIC CALCULUS

A Practical Introduction

PROBABILITY AND STOCHASTICS SERIES
Edited by Richard Durrett and Mark Pinsky

Linear Stochastic Control Systems, Guanrong Chen, Goong Chen, and Shia-Hsun Hsu

Advances in Queueing: Theory, Methods, and Open Problems, Jewgeni H. Dshalalow

Stochastics Calculus: A Practical Introduction, Richard Durrette

Chaos Expansion, Multiple Weiner-Ito Integrals and Applications, Christian Houdre and Victor Perez-Abreu

White Noise Distribution Theory, Hui-Hsiung Kuo

Topics in Contemporary Probability, J. Laurie Snell

Probability
and Stochastics Series

STOCHASTIC CALCULUS

A Practical Introduction

Richard Durrett

CRC Press
Boca Raton   New York   London   Tokyo

Tim Pletscher: Acquiring Editor
Dawn Boyd: Cover Designer
Susie Carlisle: Marketing Manager
Arline Massey: Associate Marketing Manager
Paul Gottehrer: Assistant Managing Editor, EDP
Kevin Luong: Pre-Press
Sheri Schwartz: Manufacturing Assistant

Library of Congress Cataloging-in-Publication Data

Catalog record is available from the Library of Congress

This book contains information obtained from authentic and highly regarded sources. Reprinted material is quoted with permission, and sources are indicated. A wide variety of references are listed. Reasonable efforts have been made to publish reliable data and information, but the author and the publisher cannot assume responsibility for the validity of all materials or for the consequences of their use.

Neither this book nor any part may be reproduced or transmitted in any form or by any means, electronic or mechanical, including photocopying, microfilming, and recording, or by any information storage or retrieval system, without prior permission in writing from the publisher.

CRC Press, Inc.'s consent does not extend to copying for general distribution, for promotion, for creating new works, or for resale. Specific permission must be obtained in writing from CRC Press for such copying.

Direct all inquiries to CRC Press, Inc., 2000 Corporate Blvd., N.W., Boca Raton, Florida 33431.

© 1996 by CRC Press, Inc.

No claim to original U.S. Government works
International Standard Book Number 0-8493-8071-5
Printed in the United States of America 1 2 3 4 5 6 7 8 9 0
Printed on acid-free paper

Preface

This book is the re-incarnation of my first book *Brownian Motion and Martingales in Analysis*, which was published by Wadsworth in 1984. For more than a decade I have used Chapters 1, 2, 8, and 9 of that book to give "reading courses" to graduate students who have completed the first year graduate probability course and were interested in learning more about processes that move continuously in space and time. Taking the advice from biology that "form follows function" I have taken that material on stochastic integration, stochastic differential equations, Brownian motion and its relation to partial differential equations to be the core of this book (Chapters 1-5). To this I have added other practically important topics: one dimensional diffusions, semigroups and generators, Harris chains, and weak convergence. I have struggled with this material for almost twenty years. I now think that I understand most of it, so to help you master it in less time, I have tried to explain it as simply and clearly as I can.

My students' motivations for learning this material have been diverse: some have wanted to apply ideas from probability to analysis or differential geometry, others have gone on to do research on diffusion processes or stochastic partial differential equations, some have been interested in applications of these ideas to finance, or to problems in operations research. My motivation for writing this book, like that for *Probability Theory and Examples*, was to simplify my life as a teacher by bringing together in one place useful material that is scattered in a variety of sources.

An old joke says that "if you copy from one book that is plagiarism, but if you copy from ten books that is scholarship." From that viewpoint this is a scholarly book. Its main contributors for the various subjects are (a) stochastic integration and differential equations: Chung and Williams (1990), Ikeda and Watanabe (1981), Karatzas and Shreve (1991), Protter (1990), Revuz and Yor (1991), Rogers and Williams (1987), Stroock and Varadhan (1979); (b) partial differential equations: Folland (1976), Friedman (1964), (1975), Port and Stone (1978), Chung and Zhao (1995); (c) one dimensional diffusions: Karlin and Taylor (1981); (d) semi-groups and generators: Dynkin (1965), Revuz and Yor (1991); (e) weak convergence: Billingsley (1968), Ethier and Kurtz (1986), Stroock and Varadhan (1979). If you bought all those books you would spend more than $1000 but for a fraction of that cost you can have this book, the intellectual equivalent of the ginzu knife.

Preface

Shutting off the laugh-track and turning on the violins, the road from this book's first publication in 1984 to its rebirth in 1996 has been a long and winding one. In the second half of the 80's I accumulated an embarrassingly long list of typos from the first edition. Some time at the beginning of the 90's I talked to the editor who brought my first three books into the world, John Kimmel, about preparing a second edition. However, after the work was done, the second edition was personally killed by Bill Roberts, the President of Brooks/Cole. At the end of 1992 I entered into a contract with Wayne Yuhasz at CRC Press to produce this book. In the first few months of 1993, June Meyerman typed most of the book into TeX. In the Fall Semester of 1993 I taught a course from this material and began to organize it into the current form. By the summer of 1994 I thought I was almost done. At this point I had the good (and bad) fortune of having Nora Guertler, a student from Lyon, visit for two months. When she was through making an average of six corrections per page, it was clear that the book was far from finished.

During the 1994-95 academic year most of my time was devoted to preparing the second edition of my first year graduate textbook *Probability: Theory and Examples*. After that experience my brain cells could not bear to work on another book for another several months, but toward the end of 1995 they decided "it is now or never." The delightful Cornell tradition of a long winter break, which for me stretched from early December to late January, provided just enough time to finally finish the book.

I am grateful to my students who have read various versions of this book and also made numerous comments: Don Allers, Hassan Allouba, Robert Battig, Marty Hill, Min-jeong Kang, Susan Lee, Gang Ma, and Nikhil Shah. Earlier in the process, before I started writing, Heike Dengler, David Lando and I spent a semester reading Protter (1990) and Jacod and Shiryaev (1987), an enterprise which contributed greatly to my education.

The ancient history of the revision process has unfortunately been lost. At the time of the proposed second edition, I transferred a number of lists of typos to my copy of the book, but I have no record of the people who supplied the lists. I remember getting a number of corrections from Mike Brennan and Ruth Williams, and it is impossible to forget the story of Robin Pemantle who took *Brownian Motion, Martingales and Analysis* as his only math book for a year-long trek through the South Seas and later showed me his fully annotated copy. However, I must apologize to others whose contributions were recorded but whose names were lost. Flame me at rtd1@cornell.edu and I'll have something to say about you in the next edition.

Rick Durrett

About the Author

Rick Durrett received his Ph.D. in Operations research from Stanford University in 1976. He taught in the Mathematics Department at UCLA for nine years before becoming a Professor of Mathematics at Cornell University. He was a Sloan Fellow 1981–83, Guggenheim Fellow 1988–89, and spoke at the International Congress of Math in Kyoto 1990.

Durrett is the author of a graduate textbook, *Probability: Theory and Examples*, and an undergraduate one, *The Essentials of Probability*. He has written almost 100 papers with a total of 38 co-authors and seen 19 students complete their Ph.D.'s under his direction. His recent research focuses on the applications of stochastic spatial models to various problems in biology.

Stochastic Calculus: A Practical Introduction

1. Brownian Motion
   - 1.1 Definition and Construction 1
   - 1.2 Markov Property, Blumenthal's 0-1 Law 7
   - 1.3 Stopping Times, Strong Markov Property 18
   - 1.4 First Formulas 26

2. Stochastic Integration
   - 2.1 Integrands: Predictable Processes 33
   - 2.2 Integrators: Continuous Local Martingales 37
   - 2.3 Variance and Covariance Processes 42
   - 2.4 Integration w.r.t. Bounded Martingales 52
   - 2.5 The Kunita-Watanabe Inequality 59
   - 2.6 Integration w.r.t. Local Martingales 63
   - 2.7 Change of Variables, Itô's Formula 68
   - 2.8 Integration w.r.t. Semimartingales 70
   - 2.9 Associative Law 74
   - 2.10 Functions of Several Semimartingales 76
     Chapter Summary 79
   - 2.11 Meyer-Tanaka Formula, Local Time 82
   - 2.12 Girsanov's Formula 90

3. Brownian Motion, II
   - 3.1 Recurrence and Transience 95
   - 3.2 Occupation Times 100
   - 3.3 Exit Times 105
   - 3.4 Change of Time, Lévy's Theorem 111
   - 3.5 Burkholder Davis Gundy Inequalities 116
   - 3.6 Martingales Adapted to Brownian Filtrations 119

4. Partial Differential Equations
   - **A. Parabolic Equations**
   - 4.1 The Heat Equation 126
   - 4.2 The Inhomogeneous Equation 130
   - 4.3 The Feynman-Kac Formula 137
   - **B. Elliptic Equations**
   - 4.4 The Dirichlet Problem 143
   - 4.5 Poisson's Equation 151
   - 4.6 The Schrödinger Equation 156
   - **C. Applications to Brownian Motion**
   - 4.7 Exit Distributions for the Ball 164
   - 4.8 Occupation Times for the Ball 167
   - 4.9 Laplace Transforms, Arcsine Law 170

5. Stochastic Differential Equations

- 5.1 Examples 177
- 5.2 Itô's Approach 183
- 5.3 Extension 190
- 5.4 Weak Solutions 196
- 5.5 Change of Measure 202
- 5.6 Change of Time 207

6. One Dimensional Diffusions

- 6.1 Construction 211
- 6.2 Feller's Test 214
- 6.3 Recurrence and Transience 219
- 6.4 Green's Functions 222
- 6.5 Boundary Behavior 229
- 6.6 Applications to Higher Dimensions 234

7. Diffusions as Markov Processes

- 7.1 Semigroups and Generators 245
- 7.2 Examples 250
- 7.3 Transition Probabilities 255
- 7.4 Harris Chains 258
- 7.5 Convergence Theorems 268

8. Weak Convergence

- 8.1 In Metric Spaces 271
- 8.2 Prokhorov's Theorems 276
- 8.3 The Space C 282
- 8.4 Skorohod's Existence Theorem for SDE 285
- 8.5 Donsker's Theorem 287
- 8.6 The Space D 293
- 8.7 Convergence to Diffusions 296
- 8.8 Examples 305

Solutions to Exercises 311

References 335

Index 339

1 Brownian Motion

1.1. Definition and Construction

In this section we will define Brownian motion and construct it. This event, like the birth of a child, is messy and painful, but after a while we will be able to have fun with our new arrival. We begin by reducing the definition of a d-dimensional Brownian motion with a general starting point to that of a one dimensional Brownian motion starting at 0. The first two statements, (1.1) and (1.2), are part of our definition.

(1.1) **Translation invariance.** $\{B_t - B_0, t \geq 0\}$ is independent of B_0 and has the same distribution as a Brownian motion with $B_0 = 0$.

(1.2) **Independence of coordinates.** If $B_0 = 0$ then $\{B_t^1, t \geq 0\} \ldots, \{B_t^d, t \geq 0\}$ are independent one dimensional Brownian motions starting at 0.

Now we define a one dimensional **Brownian motion** starting at 0 to be a process B_t, $t \geq 0$ taking values in $\mathbf{R}$ that has the following properties:

(a) If $t_0 < t_1 < \ldots < t_n$ then $B(t_0), B(t_1) - B(t_0), \ldots B(t_n) - B(t_{n-1})$ are independent.

(b) If $s, t \geq 0$ then

$$P(B(s+t) - B(s) \in A) = \int_A (2\pi t)^{-1/2} \exp(-x^2/2t)\, dx$$

(c) With probability one, $B_0 = 0$ and $t \to B_t$ is continuous.

(a) says that B_t has independent increments. (b) says that the increment $B(s+t) - B(s)$ has a normal distribution with mean vector 0 and variance t. (c) is self-explanatory. The reader should note that above we have sometimes written B_s and sometimes written $B(s)$, a practice we will continue in what follows.

An immediate consequence of the definition that will be useful many times is:

(1.3) Scaling relation. If $B_0 = 0$ then for any $t > 0$,

$$\{B_{st}, s \geq 0\} \stackrel{d}{=} \{t^{1/2} B_s, s \geq 0\}$$

To be precise, the two families of random variables have the same finite dimensional distributions, i.e., if $s_1 < \ldots < s_n$ then

$$(B_{s_1 t}, \ldots, B_{s_n t}) \stackrel{d}{=} (t^{1/2} B_{s_1}, \ldots, t^{1/2} B_{s_n})$$

In view of (1.2) it suffices to prove this for a one dimensional Brownian motion. To check this when $n = 1$, we note that $t^{1/2}$ times a normal with mean 0 and variance s is a normal with mean 0 and variance st. The result for $n > 1$ follows from independent increments.

A second equivalent definition of one dimensional Brownian motion starting from $B_0 = 0$, which we will occasionally find useful, is that B_t, $t \geq 0$, is a real valued process satisfying

(a') B_t is a **Gaussian process** (i.e., all its finite dimensional distributions are multivariate normal),

(b') $EB_s = 0$, $EB_s B_t = s \wedge t = \min\{s,t\}$,

(c) With probability one, $t \to B_t$ is continuous.

It is easy to see that (a) and (b) imply (a'). To get (b') from (a) and (b) suppose $s < t$ and write
$$EB_s B_t = E(B_s^2) + E(B_s(B_t - B_s))$$
$$= s + EB_s E(B_t - B_s) = s$$

The converse is even easier. (a') and (b') specify the finite dimensional distributions of B_t, which by the last calculation must agree with the ones defined in (a) and (b).

The first question that must be addressed in any treatment of Brownian motion is, "Is there a process with these properties?" The answer is "Yes," of course, or this book would not exist. For pedagogical reasons we will pursue an approach that leads to a dead end and then retreat a little to rectify the difficulty. In view of our definition we can restrict our attention to a one dimensional Brownian motion starting from a fixed $x \in \mathbf{R}$. We could take $x = 0$ but will not for reasons that become clear in the remark after (1.5).

Section 1.1 Definition and Construction   3

For each $0 < t_1 < \ldots < t_n$ define a measure on $\mathbf{R}^n$ by

$$\mu_{x,t_1,\ldots t_n}(A_1 \times \cdots \times A_n) = \int_{A_1} dx_1 \ldots \int_{A_n} dx_n \prod_{m=1}^{n} p_{t_m - t_{m-1}}(x_{m-1}, x_m)$$

where $x_0 = x$, $t_0 = 0$,

$$p_t(a, b) = (2\pi t)^{-1/2} \exp(-(b-a)^2/2t)$$

and $A_i \in \mathcal{R}$ the Borel subsets of $\mathbf{R}$. From the formula above it is easy to see that for fixed x the family μ is a consistent set of finite dimensional distributions (f.d.d.'s), that is, if $\{s_1, \ldots s_{n-1}\} \subset \{t_1, \ldots t_n\}$ and $t_j \notin \{s_1, \ldots s_{n-1}\}$ then

$$\mu_{x,s_1,\ldots s_{n-1}}(A_1 \times \cdots \times A_{n-1}) = \mu_{x,t_1,\ldots t_n}(A_1 \times \cdots \times A_{j-1} \times \mathbf{R} \times A_j \times \cdots \times A_{n-1})$$

This is clear when $j = n$. To check the equality when $1 \leq j < n$, it is enough to show that

$$\int p_{t_j - t_{j-1}}(x, y) p_{t_{j+1} - t_j}(y, z) \, dy = p_{t_{j+1} - t_{j-1}}(x, z)$$

By translation invariance, we can without loss of generality assume $x = 0$, but all this says in that case is the sum of independent normals with mean 0 and variances $t_j - t_{j-1}$ and $t_{j+1} - t_j$ has a normal distribution with mean 0 and variance $t_{j+1} - t_{j-1}$. With the consistency of f.d.d.'s verified we get our first construction of Brownian motion:

(1.4) **Theorem.** Let $\Omega_o = \{\text{functions } \omega : [0, \infty) \to \mathbf{R}\}$ and $\mathcal{F}_o$ be the σ-field generated by the **finite dimensional sets** $\{\omega : \omega(t_i) \in A_i \text{ for } 1 \leq i \leq n\}$ where $A_i \in \mathcal{R}$. For each $x \in \mathbf{R}$, there is a unique probability measure ν_x on $(\Omega_o, \mathcal{F}_o)$ so that $\nu_x\{\omega : \omega(0) = x\} = 1$ and when $0 < t_1 \cdots < t_n$

(∗) $$\nu_x(\{\omega : \omega(t_i) \in A_i\}) = \mu_{x,t_1,\ldots t_n}(A_1 \times \cdots \times A_n)$$

This follows from a generalization of Kolmogorov's extension theorem. We will not bother with the details since at this point we are at the dead end referred to above. If $C = \{\omega : t \to \omega(t) \text{ is continuous}\}$ then $C \notin \mathcal{F}_o$, that is, C is not a measurable set. The easiest way of proving $C \notin \mathcal{F}_o$ is to do

Exercise 1.1. $A \in \mathcal{F}_o$ if and only if there is a sequence of times $t_1, t_2, \ldots \in [0, \infty)$ and a $B \in \mathcal{R}^{\{1,2,\ldots\}}$ (the infinite product σ-field $\mathcal{R} \times \mathcal{R} \times \cdots$) so that $A = \{\omega : (\omega(t_1), \omega(t_2), \ldots) \in B\}$. In words, all events in $\mathcal{F}_o$ depend on only countably many coordinates.

The above problem is easy to solve. Let $\mathbf{Q}_2 = \{m2^{-n} : m, n \geq 0\}$ be the **dyadic rationals**. If $\Omega_q = \{\omega : \mathbf{Q}_2 \to \mathbf{R}\}$ and $\mathcal{F}_q$ is the σ-field generated by the finite dimensional sets, then enumerating the rationals $q_1, q_2, \ldots$ and applying Kolmogorov's extension theorem shows that we can construct a probability ν_x on $(\Omega_q, \mathcal{F}_q)$ so that $\nu_x\{\omega : \omega(0) = x\} = 1$ and $(*)$ in (1.4) holds when the $t_i \in \mathbf{Q}_2$. To extend the definition from $\mathbf{Q}_2$ to $[0, \infty)$ we will show:

(1.5) Theorem. Let $T < \infty$ and $x \in \mathbf{R}$. ν_x assigns probability one to paths $\omega : \mathbf{Q}_2 \to \mathbf{R}$ that are uniformly continuous on $\mathbf{Q}_2 \cap [0, T]$.

Remark. It will take quite a bit of work to prove (1.5). Before taking on that task, we will attend to the last measure theoretic detail: we tidy things up by moving our probability measures to $(C, \mathcal{C})$ where $C = \{\text{continuous } \omega : [0, \infty) \to \mathbf{R}\}$ and $\mathcal{C}$ is the σ-field generated by the coordinate maps $t \to \omega(t)$. To do this, we observe that the map ψ that takes a uniformly continuous point in Ω_q to its extension in C is measurable, and we set

$$P_x = \nu_x \circ \psi^{-1}.$$

Our construction guarantees that $B_t(\omega) = \omega_t$ has the right finite dimensional distributions for $t \in \mathbf{Q}_2$. Continuity of paths and a simple limiting argument shows that this is true when $t \in [0, \infty)$.

As mentioned earlier the generalization to $d > 1$ is straightforward since the coordinates are independent. In this generality $C = \{\text{continuous } \omega : [0, \infty) \to \mathbf{R}^d\}$ and $\mathcal{C}$ is the σ-field generated by the coordinate maps $t \to \omega(t)$. The reader should note that the result of our construction is one set of random variables $B_t(\omega) = \omega(t)$, and a family of probability measures P_x, $x \in \mathbf{R}^d$, so that under P_x, B_t is a Brownian motion with $P_x(B_0 = x) = 1$. It is enough to construct the Brownian motion starting from an initial point x, since if we want a Brownian motion starting from an initial measure μ (i.e., have $P_\mu(B_0 \in A) = \mu(A)$) we simply set

$$P_\mu(A) = \int \mu(dx) P_x(A)$$

Proof of (1.5) By (1.1) and (1.3), we can without loss of generality suppose $B_0 = 0$ and prove the result for $T = 1$. In this case, part (b) of the definition and the scaling relation (1.3) imply

$$E_0(|B_t - B_s|^4) = E_0|B_{t-s}|^4 = C(t-s)^2$$

where $C = E|B_1|^4 < \infty$. From the last observation we get the desired uniform continuity by using a result due to Kolmogorov. In this proof, we do not use the independent increments property of Brownian motion; the only thing we use is the moment condition. In Section 2.11 we will need this result when the X_t take values in a space S with metric ρ; so, we will go ahead and prove it in that generality.

(1.6) Theorem. Suppose that $E\rho(X_s, X_t)^\beta \leq K|t-s|^{1+\alpha}$ where $\alpha, \beta > 0$. If $\gamma < \alpha/\beta$ then with probability one there is a contant $C(\omega)$ so that

$$\rho(X_q, X_r) \leq C|q-r|^\gamma \quad \text{for all } q, r \in \mathbf{Q}_2 \cap [0,1]$$

Proof Let $\gamma < \alpha/\beta$, $\eta > 0$, $I_n = \{(i,j) : 0 \leq i \leq j \leq 2^n, 0 < j - i \leq 2^{n\eta}\}$ and $G_n = \{\rho(X(j2^{-n}), X(i2^{-n})) \leq ((j-i)2^{-n})^\gamma \text{ for all } (i,j) \in I_n\}$. Since $a^\beta P(|Y| > a) \leq E|Y|^\beta$ we have

$$P(G_n^c) \leq \sum_{(i,j) \in I_n} ((j-i)2^{-n})^{-\beta\gamma} E\rho(X(j2^{-n}), X(i2^{-n}))^\beta$$

$$\leq K \sum_{(i,j) \in I_n} ((j-i)2^{-n})^{-\beta\gamma+1+\alpha}$$

by our assumption. Now the number of $(i,j) \in I_n$ is $\leq 2^n 2^{n\eta}$, so

$$P(G_n^c) \leq K \cdot 2^n 2^{n\eta} \cdot (2^{n\eta} 2^{-n})^{-\beta\gamma+1+\alpha} = K 2^{-n\lambda}$$

where $\lambda = (1-\eta)(1+\alpha-\beta\gamma) - (1+\eta)$. Since $\gamma < \alpha/\beta$, we can pick η small enough so that $\lambda > 0$. To complete the proof now we will show

(1.7) Lemma. Let $A = 3 \cdot 2^{(1-\eta)\gamma}/(1-2^{-\gamma})$. On $H_N = \cap_{n=N}^\infty G_n$ we have

$$\rho(X_q, X_r) \leq A|q-r|^\gamma \quad \text{for } q, r \in \mathbf{Q}_2 \cap [0,1]$$
$$\text{with } |q-r| \leq 2^{-(1-\eta)N}$$

(1.6) follows easily from (1.7):

$$P(H_N^c) \leq \sum_{n=N}^\infty P(G_n^c) \leq K \sum_{n=N}^\infty 2^{-n\lambda} = \frac{K 2^{-N\lambda}}{1-2^{-\lambda}}$$

This shows $\rho(X_q, X_r) \leq A|q-r|^\gamma$ for $|q-r| \leq \delta(\omega)$ and implies that we have $\rho(X_q, X_r) \leq C(\omega)|q-r|^\gamma$ for $q, r \in [0,1]$.

Proof of (1.7) Let $q, r \in \mathbf{Q}_2 \cap [0,1]$ with $0 < r - q < 2^{-(1-\eta)N}$. Pick $m \geq N$ so that

$$2^{-(m+1)(1-\eta)} \leq r - q < 2^{-m(1-\eta)}$$

and write
$$r = j2^{-m} + 2^{-r(1)} + \cdots + 2^{-r(\ell)}$$
$$q = i2^{-m} - 2^{-q(1)} - \cdots - 2^{-q(k)}$$
where $m < r(1) < \cdots < r(\ell)$ and $m < q(1) < \cdots < q(k)$. Now $0 < r - q < 2^{-m(1-\eta)}$, so $(j - i) < 2^{m\eta}$ and it follows that on H_N

(a) $$\rho(X(i2^{-m}), X(j2^{-m})) \leq ((2^{m\eta})2^{-m})^{\gamma}$$

On H_N it follows from the triangle inequality that

(b) $$\rho(X_q, X(i2^{-m})) \leq \sum_{h=1}^{k}(2^{-q(h)})^{\gamma} \leq \sum_{h=m}^{\infty}(2^{-\gamma})^h \leq C_\gamma 2^{-\gamma m}$$

where $C_\gamma = 1/(1 - 2^{-\gamma}) > 1$. Repeating the last computation shows

(c) $$\rho(X_r, X(j2^{-m})) \leq C_\gamma 2^{-\gamma m}$$

Combining (a)–(c) gives

$$\rho(X_q, X_r) \leq 3C_\gamma 2^{-\gamma m(1-\eta)} \leq 3C_\gamma 2^{(1-\eta)\gamma}|r - q|^\gamma$$

since $2^{-m(1-\eta)} \leq 2^{1-\eta}|r - q|$. This completes the proof of (1.7) and hence of (1.6) and (1.5) □

The scaling relation (1.2) implies
$$E|B_t - B_s|^{2m} = C_m|t - s|^m \text{ where } C_m = E|B_1|^{2m}$$

So using (1.6) with $\beta = 2m$ and $\alpha = m - 1$ and then letting $m \to \infty$ gives:

(1.8) Theorem. Brownian paths are Hölder continuous with exponent γ for any $\gamma < 1/2$.

It is easy to show:

(1.9) Theorem. With probability one, Brownian paths are not Lipschitz continuous (and hence not differentiable) at any point.

Proof Let $A_n = \{\omega :$ there is an $s \in [0, 1]$ so that $|B_t - B_s| \leq C|t - s|$ when $|t - s| \leq 3/n\}$. For $1 \leq k \leq n - 2$ let

$$Y_{k,n} = \max\left\{\left|B\left(\frac{k+j}{n}\right) - B\left(\frac{k+j-1}{n}\right)\right| : j = 0, 1, 2\right\}$$

$G_n = \{$ at least one $Y_{k,n}$ is $\leq 5C/n\}$

We claim that $A_n \subset G_n$. To prove this we consider $s = 1$, which we claim is the worst possibility. In this case we want to conclude that $Y_{n-2,n} \leq 5C/n$. For this we observe that $j = 0$ is the worst case, but even then

$$\left|B_{(n-3)/n} - B_{(n-2)/n}\right| \leq \left|B_{(n-3)/n} - B_1\right| + \left|B_1 - B_{(n-2)/n}\right| \leq C\left(\frac{3}{n} + \frac{2}{n}\right)$$

Using $A_n \subset G_n$ and the scaling relation (1.2) gives

$$P(A_n) \leq P(G_n) \leq n\, P\left(|B_{1/n}| \leq \frac{5C}{n}\right)^3$$

$$= n\, P\left(|B_1| \leq \frac{5C}{n^{1/2}}\right)^3 \leq n \cdot \left\{\frac{10C}{n^{1/2}} \cdot (2\pi)^{-1/2}\right\}^3$$

since $\exp(-x^2/2) \leq 1$. Letting $n \to \infty$ shows $P(A_n) \to 0$. Noticing $n \to A_n$ is increasing shows $P(A_n) = 0$ for all n and completes the proof. □

Exercise 1.2. Show by considering k increments instead of 3 that if $\gamma > 1/2 + 1/k$ then with probability 1, Brownian paths are not Hölder continuous with exponent γ at any point of $[0,1]$.

The next result is more evidence that $B_t - B_s \approx \sqrt{t-s}$.

Exercise 1.3. Let $\Delta_{m,n} = B(tm2^{-n}) - B(t(m-1)2^{-n})$. Compute

$$E\left(\sum_{m \leq 2^n} \Delta_{m,n}^2 - t\right)^2$$

and use the Borel-Cantelli lemma to conclude that $\sum_{m \leq 2^n} \Delta_{m,n}^2 \to t$ a.s. as $n \to \infty$.

Remark. The last result is true if we consider a sequence of partitions $\Pi_1 \subset \Pi_2 \subset \ldots$ with mesh $\to 0$. See Freedman (1970) p.42-46. The true quadratic variation, defined as the sup over all partitions, is ∞ for Brownian motion.

1.2. Markov Property, Blumenthal's 0-1 Law

Intuitively the Markov property says

> "given the present state, B_s, any other information about what happened before time s is irrelevant for predicting what happens after time s."

Since Brownian motion is translation invariant, the Markov property can be more simply stated as:

"if $s \geq 0$ then $B_{t+s} - B_s$, $t \geq 0$ is a Brownian motion that is independent of what happened before time s."

This should not be surprising: the independent increments property ((a) in the definition) implies that if $s_1 \leq s_2 \ldots \leq s_m = s$ and $0 < t_1 \ldots < t_n$ then

$$(B_{t_1+s} - B_s, \ldots B_{t_n+s} - B_s) \text{ is independent of } (B_{s_1}, \ldots B_{s_n})$$

The major obstacle in proving the Markov property for Brownian motion then is to introduce the measure theory necessary to state and prove the result. The first step in doing this is to explain what we mean by "what happened before time s." The technical name for what we are about to define is a **filtration**, a fancy term for an increasing collection of σ-fields, $\mathcal{F}_s$, i.e., if $s \leq t$ then $\mathcal{F}_s \subset \mathcal{F}_t$. Since we want $B_s \in \mathcal{F}_s$, i.e., B_s is measurable with respect to $\mathcal{F}_s$, the first thing that comes to mind is

$$\mathcal{F}_s^o = \sigma(B_r : r \leq s)$$

For technical reasons, it is convenient to replace $\mathcal{F}_s^o$ by

$$\mathcal{F}_s^+ = \cap_{t > s} \mathcal{F}_t^o$$

The fields $\mathcal{F}_s^+$ are nicer because they are **right continuous**. That is,

$$\cap_{t>s} \mathcal{F}_t^+ = \cap_{t>s} (\cap_{u>t} \mathcal{F}_u^o) = \cap_{u>s} \mathcal{F}_u^o = \mathcal{F}_s^+$$

In words the $\mathcal{F}_s^+$ allow us an "infinitesimal peek at the future," i.e., $A \in \mathcal{F}_s^+$ if it is in $\mathcal{F}_{s+\epsilon}^o$ for any $\epsilon > 0$. If B_t is a Brownian motion and f is any measurable function with $f(u) > 0$ when $u > 0$ the random variable

$$\limsup_{t \downarrow s} (B_t - B_s)/f(t-s)$$

is measurable with respect to $\mathcal{F}_s^+$ but not $\mathcal{F}_s^o$. Exercises 2.9 and 2.10 consider what happens when we take $f(u) = \sqrt{u}$ and $f(u) = \sqrt{u \log \log(1/u)}$. However, as we will see in (2.6), there are no interesting examples of sets that are in $\mathcal{F}_s^+$ but not in $\mathcal{F}_s^o$. The two σ-fields are the same (up to null sets).

To state the Markov property we need some notation. First, recall that we have a family of measures P_x, $x \in \mathbf{R}^d$, on $(C, \mathcal{C})$ so that under P_x, $B_t(\omega) = \omega(t)$ is a Brownian motion with $B_0 = x$. In order to define "what happens after time

Section 1.2 Markov Property, Blumenthal's 0-1 Law

s," it is convenient to define the **shift transformations** $\theta_s : C \to C$, for $s \geq 0$ by
$$(\theta_s \omega)(t) = \omega(s+t) \quad \text{for } t \geq 0$$

In words, we cut off the part of the path before time s and then shift the path so that time s becomes time 0. To prepare for the next result, we note that if $Y : C \to \mathbf{R}$ is $\mathcal{C}$ measurable then $Y \circ \theta_s$ is a function of the future after time s. To see this, consider the simple example $Y(\omega) = f(\omega(t))$. In this case

$$Y \circ \theta_s = f(\theta_s \omega(t)) = f(\omega(s+t)) = f(B_{s+t})$$

Likewise, if $Y(\omega) = f(\omega_{t_1}, \ldots \omega_{t_n})$ then

$$Y \circ \theta_s = f(B_{s+t_1}, \ldots B_{s+t_n})$$

(2.1) The Markov property. If $s \geq 0$ and Y is bounded and $\mathcal{C}$ measurable then for all $x \in \mathbf{R}^d$

$$E_x(Y \circ \theta_s | \mathcal{F}_s^+) = E_{B_s} Y$$

where the right hand side is $\varphi(y) = E_y Y$ evaluated at $y = B(s)$.

Explanation. In words, this says that the conditional expectation of $Y \circ \theta_s$ given $\mathcal{F}_s^+$ is just the expected value of Y for a Brownian motion starting at B_s. To explain why this implies "given the present state, B_s, any other information about what happened before time s is irrelevant for predicting what happens after time s," we begin by recalling (see (1.1) in Chapter 5 of Durrett (1995)) that if $\mathcal{G} \subset \mathcal{F}$ and $E(Z|\mathcal{F}) \in \mathcal{G}$ then $E(Z|\mathcal{F}) = E(Z|\mathcal{G})$. Applying this with $\mathcal{F} = \mathcal{F}_s^+$ and $\mathcal{G} = \sigma(B_s)$ we have

$$E_x(Y \circ \theta_s | \mathcal{F}_s^+) = E_x(Y \circ \theta_s | B_s)$$

If we recall that $X = E(Z|\mathcal{F})$ is our best guess at Z given the information in $\mathcal{F}$ (in the sense of minimizing $E(Z - X)^2$ over $X \in \mathcal{F}_s$, see (1.4) in Chapter 4 of Durrett (1995)) then we see that our best guess at $Y \circ \theta_s$ given B_s is the same as our best guess given $\mathcal{F}_s^+$, i.e., any other information in $\mathcal{F}_s^+$ is irrelevant for predicting $Y \circ \theta_s$.

Proof By the definition of conditional expectation, what we need to show is

(MP) $\qquad E_x(Y \circ \theta_s ; A) = E_x(E_{B_s} Y ; A) \quad \text{for all } A \in \mathcal{F}_s^+$

We will begin by proving the result for a special class of Y's and a special class of A's. Then we will use the $\pi - \lambda$ theorem and the monotone class theorem to extend to the general case. Suppose

$$Y(\omega) = \prod_{1 \le m \le n} f_m(\omega(t_m))$$

where $0 < t_1 < \ldots < t_n$ and the $f_m : \mathbf{R}^d \to \mathbf{R}$ are bounded and measurable. Let $0 < h < t_1$, let $0 < s_1 \ldots < s_k \le s+h$, and let $A = \{\omega : \omega(s_j) \in A_j, 1 \le j \le k\}$ where $A_j \in \mathcal{R}^d$ for $1 \le j \le k$. We will call these A's the **finite dimensional** or **f.d. sets** in $\mathcal{F}^o_{s+h}$.

From the definition of Brownian motion it follows that if $0 = u_0 < u_1 < \ldots u_\ell$ then the joint density of $(B_{u_1}, \ldots B_{u_\ell})$ is given by

$$P_x(B_{u_1} = y_1, \ldots B_{u_\ell} = y_\ell) = \prod_{i=1}^{\ell} p_{u_i - u_{i-1}}(y_{i-1}, y_i)$$

where $y_0 = x$. From this it follows that

$$E_x \left(\prod_{i=1}^{\ell} g_i(B_{u_i}) \right) = \int dy_1 \, p_{u_1 - u_0}(y_0, y_1) g_1(y_1)$$

$$\ldots \int dy_\ell \, p_{u_\ell - u_{\ell-1}}(y_{\ell-1}, y_\ell) g_\ell(y_\ell)$$

Applying this result with the u_i given by $s_1, \ldots, s_k, s+h, s+t_1, \ldots s+t_n$ and the obvious choices for the g_i we have

$$E_x(Y \circ \theta_s; A) = E_x \left(\prod_{j=1}^{k} 1_{A_j}(B_{s_j}) \cdot 1_{\mathbf{R}^d}(B_{s+h}) \cdot \prod_{m=1}^{n} f_m(B_{s+t_m}) \right)$$

$$= \int_{A_1} dx_1 \, p_{s_1}(x, x_1) \ldots \int_{A_k} dx_k \, p_{s_k - s_{k-1}}(x_{k-1}; x_k)$$

$$\cdot \int_{\mathbf{R}^d} dy \, p_{s+h-s_k}(x_k, y) \, \varphi(y, h)$$

where

$$\varphi(y, h) = \int dy_1 \, p_{t_1 - h}(y, y_1) f_1(y_1) \ldots \int dy_n \, p_{t_n - t_{n-1}}(y_{n-1}, y_n) f_n(y_n)$$

Using the formula for $E_x \prod_{i=1}^{k} g_i(B_{u_i})$ again we have

Section 1.2 Markov Property, Blumenthal's 0-1 Law

(MP-1) $E_x(Y \circ \theta_s; A) = E_x(\varphi(B_{s+h}, h); A)$ for all f.d. sets A in $\mathcal{F}^o_{s+h}$

To extend the class of A's to all of $\mathcal{F}^o_{s+h}$, we use

(2.2) The $\pi - \lambda$ theorem. Let $\mathcal{A}$ be a collection of subsets of Ω that contains Ω and is closed under intersection (i.e., if $A, B \in \mathcal{A}$ then $A \cap B \in \mathcal{A}$). Let $\mathcal{G}$ be a collection of sets that satisfy

(i) if $A, B \in \mathcal{G}$ and $A \supset B$ then $A - B \in \mathcal{G}$

(ii) If $A_n \in \mathcal{G}$ and $A_n \uparrow A$, then $A \in \mathcal{G}$.

If $\mathcal{A} \subset \mathcal{G}$ then the σ-field generated by $\mathcal{A}$, $\sigma(\mathcal{A}) \subset \mathcal{G}$.

Proof See (2.1) in the Appendix of Durrett (1995).

(MP-2) $E_x(Y \circ \theta_s; A) = E_x(\varphi(B_{s+h}, h); A)$ for all $A \in \mathcal{F}^o_{s+h}$

Proof Fix Y and let $\mathcal{G}$ be the collection of sets A for which the desired equality is true. A simple subtraction shows that (i) in (2.2) holds, while the monotone convergence theorem shows that (ii) holds. If we let $\mathcal{A}$ be the collection of finite dimensional sets in $\mathcal{F}^o_{s+h}$ then we have shown $\mathcal{A} \subset \mathcal{G}$ so $\mathcal{F}^o_{s+h} = \sigma(\mathcal{A}) \subset \mathcal{G}$ which is the desired conclusion. □

Our next step is

(MP-3) $E_x(Y \circ \theta_s; A) = E_x(\varphi(B_s, 0); A)$ for all $A \in \mathcal{F}^+_s$

Proof Since $\mathcal{F}^+_s \subset \mathcal{F}^o_{s+h}$ for any $h > 0$, all we need to do is let $h \to 0$ in (MP-2). It is easy to see that

$$\psi(y_1) = f_1(y_1) \int dy_2\, p_{t_2 - t_1}(y_1, y_2) f_2(y_2)$$

$$\cdots \int dy_n\, p_{t_n - t_{n-1}}(y_{n-1}, y_n) f_n(y_n)$$

is bounded and measurable. Using the dominated convergence theorem shows that if $h \to 0$ and $x_h \to x$ then

$$\varphi(x_h, h) = \int dy_1\, p_{t_1 - h}(x_h, y_1) \psi(y_1) \to \varphi(x, 0)$$

Using (MP-2) and the bounded convergence theorem now gives the desired result. □

12   Chapter 1   Brownian Motion

(MP-3) shows that (MP) holds for $Y = \prod_{1 \leq m \leq n} f_m(\omega(t_m))$ when the f_m are bounded and measurable. To extend to general bounded measurable Y we will use

(2.3) Monotone class theorem. Let $\mathcal{A}$ be a collection of subsets of Ω that contains Ω and is closed under intersection (i.e., if $A, B \in \mathcal{A}$ then $A \cap B \in \mathcal{A}$). Let $\mathcal{H}$ be a vector space of real valued functions on Ω satisfying

(i) If $A \in \mathcal{A}$, $1_A \in \mathcal{H}$.

(ii) If $0 \leq f_n \in \mathcal{H}$ and $f_n \uparrow f$, a bounded function, then $f \in \mathcal{H}$.

Then $\mathcal{H}$ contains all the bounded functions on Ω that are measurable with respect to $\sigma(\mathcal{A})$.

Proof   See (1.4) of Chapter 5 of Durrett (1995).

To get (2.1) from (2.3), fix an $A \in \mathcal{F}_s^+$ and let $\mathcal{H} =$ the collection of bounded functions Y for which (MP) holds. $\mathcal{H}$ clearly is a vector space satisfying (ii). Let $\mathcal{A}$ be the collection of sets of the form $\{\omega : \omega(t_j) \in A_j, 1 \leq j \leq n\}$ where $A_j \in \mathcal{R}^d$. The special case treated above shows that if $A \in \mathcal{A}$ then $1_A \in \mathcal{H}$. This shows (i) holds and the desired conclusion follows from (2.3). □

The next seven exercises give typical applications of the Markov property. Perhaps the simplest example is

Exercise 2.1. Let $0 < s < t$. If $f : \mathbf{R}^d \to \mathbf{R}$ is bounded and measurable

$$E_x(f(B_t)|\mathcal{F}_s) = E_{B_s} f(B_{t-s})$$

Exercise 2.2. Take $f(x) = x_i$ and $f(x) = x_i x_j$ with $i \neq j$ in Exercise 2.1 to conclude that B_t^i and $B_t^i B_t^j$ are martingales if $i \neq j$.

The next two exercises prepare for calculations in Section 1.4.

Exercise 2.3. Let $T_0 = \inf\{s > 0 : B_s = 0\}$ and let $R = \inf\{t > 1 : B_t = 0\}$. R is for right or return. Use the Markov property at time 1 to get

$$(2.4) \qquad P_x(R > 1 + t) = \int p_1(x,y) P_y(T_0 > t)\, dy$$

Exercise 2.4. Let $T_0 = \inf\{s > 0 : B_s = 0\}$ and let $L = \sup\{t \leq 1 : B_t = 0\}$. L is for left or last. Use the Markov property at time $0 < t < 1$ to conclude

$$(2.5) \qquad P_0(L \leq t) = \int p_t(0,y) P_y(T_0 > 1 - t)\, dy$$

Exercise 2.5. Let G be an open set and let $T = \inf\{t : B_t \notin G\}$. Let K be a closed subset of G and suppose that for all $x \in K$ we have $P_x(T > 1, B_t \in K) \geq \alpha$, then for all integers $n \geq 1$ and $x \in K$ we have $P_x(T > n, B_t \in K) \geq \alpha^n$.

The next two exercises prepare for calculations in Chapter 4.

Exercise 2.6. Let $0 < s < t$. If $h : \mathbf{R} \times \mathbf{R}^d \to \mathbf{R}$ is bounded and measurable

$$E_x\left(\int_0^t h(r, B_r)\,dr \bigg| \mathcal{F}_s\right) = \int_0^s h(r, B_r)\,dr + E_{B(s)} \int_0^{t-s} h(s+u, B_u)\,du$$

Exercise 2.7. Let $0 < s < t$. If $f : \mathbf{R}^d \to \mathbf{R}$ and $h : \mathbf{R} \times \mathbf{R}^d \to \mathbf{R}$ are bounded and measurable then

$$E_x\left(f(B_t)\exp\left(\int_0^t h(r, B_r)\,dr\right)\bigg|\mathcal{F}_s\right)$$
$$= \exp\left(\int_0^s h(r, B_r)\,dr\right) E_{B_s}\left\{f(B_{t-s})\exp\left(\int_0^{t-s} h(s+u, B_u)\,du\right)\right\}$$

The reader will see many other applications of the Markov property below, so we turn our attention now to a "triviality" that has surprising consequences. Since

$$E_x(Y \circ \theta_s | \mathcal{F}_s^+) = E_{B_s} Y \in \mathcal{F}_s^o$$

it follows (see, e.g., (1.1) in Chapter 5 of Durrett (1995)) that

$$E_x(Y \circ \theta_s | \mathcal{F}_s^+) = E_x(Y \circ \theta_s | \mathcal{F}_s^o)$$

From the last equation it is a short step to

(2.6) Theorem. If $Z \in \mathcal{C}$ is bounded then for all $s \geq 0$ and $x \in \mathbf{R}^d$,

$$E_x(Z|\mathcal{F}_s^+) = E_x(Z|\mathcal{F}_s^o).$$

Proof By the monotone class theorem, (2.3), it suffices to prove the result when

$$Z = \prod_{m=1}^n f_m(B(t_m))$$

and the f_m are bounded and measurable. In this case $Z = X(Y \circ \theta_s)$ where $X \in \mathcal{F}_s^o$ and $Y \in \mathcal{C}$, so using a property of conditional expectation (see, e.g., (1.3) in Chapter 4 of Durrett (1995)) and the Markov property (2.1) gives

$$E_x(Z|\mathcal{F}_s^+) = XE_x(Y \circ \theta_s|\mathcal{F}_s^+) = XE_{B_s}Y \in \mathcal{F}_s^o$$

and the proof is complete. □

If we let $Z \in \mathcal{F}_s^+$ then (2.6) implies $Z = E_x(Z|\mathcal{F}_s^o) \in \mathcal{F}_s^o$, so the two σ-fields are the same up to null sets. At first glance, this conclusion is not exciting. The fun starts when we take $s = 0$ in (2.6) to get

(2.7) **Blumenthal's 0-1 law.** If $A \in \mathcal{F}_0^+$ then for all $x \in \mathbf{R}^d$,

$$P_x(A) \in \{0, 1\}$$

Proof Using (i) the fact that $A \in \mathcal{F}_0^+$, (ii) (2.6), (iii) $\mathcal{F}_0^o = \sigma(B_0)$ is trivial under P_x, and (iv) if $\mathcal{G}$ is trivial $E(X|\mathcal{G}) = EX$ gives

$$1_A = E_x(1_A|\mathcal{F}_0^+) = E_x(1_A|\mathcal{F}_0^o) = P_x(A) \qquad P_x \text{ a.s.}$$

This shows that the indicator function 1_A is a.s. equal to the number $P_x(A)$ and it follows that $P_x(A) \in \{0, 1\}$. □

In words, the last result says that the **germ field**, $\mathcal{F}_0^+$, is trivial. This result is very useful in studying the local behavior of Brownian paths. Until further notice we will restrict our attention to one dimensional Brownian motion.

(2.8) **Theorem.** If $\tau = \inf\{t \geq 0 : B_t > 0\}$ then $P_0(\tau = 0) = 1$.

Proof $P_0(\tau \leq t) \geq P_0(B_t > 0) = 1/2$ since the normal distribution is symmetric about 0. Letting $t \downarrow 0$ we conclude

$$P_0(\tau = 0) = \lim_{t \downarrow 0} P_0(\tau \leq t) \geq 1/2$$

so it follows from (2.7) that $P_0(\tau = 0) = 1$. □

Once Brownian motion must hit $(0, \infty)$ immediately starting from 0, it must also hit $(-\infty, 0)$ immediately. Since $t \to B_t$ is continuous, this forces:

(2.9) **Theorem.** If $T_0 = \inf\{t > 0 : B_t = 0\}$ then $P_0(T_0 = 0) = 1$.

Section 1.2 Markov Property, Blumenthal's 0-1 Law   15

Combining (2.8) and (2.9) with the Markov property you can prove

Exercise 2.8. If $a < b$ then with probability one there is a local maximum of B_t in (a, b). Since with probability one this holds for all rational $a < b$, the local maxima of Brownian motion are dense.

Another typical application of (2.7) is

Exercise 2.9. Let $f(t)$ be a function with $f(t) > 0$ for all $t > 0$. Use (2.7) to conclude that $\limsup_{t \downarrow 0} B(t)/f(t) = c$ P_0 a.s. where $c \in [0, \infty]$ is a constant.

In the next exercise we will see that $c = \infty$ when $f(t) = t^{1/2}$. The law of the iterated logarithm (see Section 7.9 of Durrett (1995)) shows that $c = 2^{1/2}$ when $f(t) = (t \log \log(1/t))^{1/2}$.

Exercise 2.10. Show that $\limsup_{t \downarrow 0} B(t)/t^{1/2} = \infty$ P_0 a.s., so with probability one Brownian paths are not Hölder continuous of order $1/2$ at 0.

Remark. Let $\mathcal{H}_\gamma(\omega)$ be the set of times at which the path $\omega \in C$ is Hölder continuous of order γ. (1.6) shows that $P(\mathcal{H}_\gamma = [0, \infty)) = 1$ for $\gamma < 1/2$. Exercise 1.2 shows that $P(\mathcal{H}_\gamma = \emptyset) = 1$ for $\gamma > 1/2$. The last exercise shows $P(t \in \mathcal{H}_{1/2}) = 0$ for each t, but B. Davis (1983) has shown $P(\mathcal{H}_{1/2} \neq \emptyset) = 1$.

Comic Relief. There is a wonderful way of expressing the complexity of Brownian paths that I learned from Wilfrid Kendall.

"If you run Brownian motion in two dimensions for a positive amount of time, it will write your name."

Of course, on top of your name it will write everybody else's name, as well as all the works of Shakespeare, several pornographic novels, and a lot of nonsense. Thinking of the function g as our signature we can make a precise statement as follows:

(2.10) Theorem. Let $g : [0, 1] \to \mathbf{R}^d$ be a continuous function with $g(0) = 0$, let $\epsilon > 0$ and let $t_n \downarrow 0$. Then P_0 almost surely,

$$\sup_{0 \leq \theta \leq 1} \left| \frac{B(\theta t_n)}{\sqrt{t_n}} - g(\theta) \right| < \epsilon \quad \text{for infinitely many } n$$

Proof In view of Blumenthal's 0-1 law, (2.7), and the scaling relation (1.3), we can prove this if we can show that

$$(\star) \qquad P_0 \left(\sup_{0 \leq \theta \leq 1} |B(\theta) - g(\theta)| < \epsilon \right) > 0 \quad \text{for any } \epsilon > 0$$

This was easy for me to believe, but not so easy for me to prove when students asked me to fill in the details. The first step is to treat the dead man's signature $g(x) \equiv 0$.

Exercise 2.11. Show that if $\epsilon > 0$ and $t < \infty$ then

$$P_0\left(\sup_i \sup_{0 \le s \le t} |B_s^i| < \epsilon\right) > 0$$

In doing this you may find Exercise 2.5 helpful. In (5.4) of Chapter 5 we will get the general result from this one by change of measure. Can the reader find a simple direct proof of $(\star)$?

With our discussion of Blumenthal's 0-1 law complete, the distinction between $\mathcal{F}_s^+$ and $\mathcal{F}_s^o$ is no longer important, so we will make one final improvement in our σ-fields and remove the superscripts. Let

$$\mathcal{N}_x = \{A : A \subset B \text{ with } P_x(B) = 0\}$$
$$\mathcal{F}_s^x = \sigma(\mathcal{F}_s^+ \cup \mathcal{N}_x)$$
$$\mathcal{F}_s = \cap_x \mathcal{F}_s^x$$

$\mathcal{N}_x$ are the **null sets** and $\mathcal{F}_s^x$ are the completed σ-fields for P_x. Since we do not want the filtration to depend on the initial state we take the intersection of all the completed σ-fields. This technicality will be mentioned at one point in the next section but can otherwise be ignored.

(2.7) concerns the behavior of B_t as $t \to 0$. By using a trick we can use this result to get information about the behavior as $t \to \infty$.

(2.11) Theorem. If B_t is a Brownian motion starting at 0 then so is the process defined by $X_0 = 0$ and $X_t = t\, B(1/t)$ for $t > 0$.

Proof By (1.2) it suffices to prove the result in one dimension. We begin by observing that the strong law of large numbers implies $X_t \to 0$ as $t \to 0$, so X has continuous paths and we only have to check that X has the right f.d.d.'s. By the second definition of Brownian motion, it suffices to show that (i) if $0 < t_1 < \ldots < t_n$ then $(X(t_1), \ldots X(t_n))$ has a multivariate normal distribution with mean 0 (which is obvious) and (ii) if $s < t$ then

$$E(X_s X_t) = st E(B(1/s)B(1/t)) = s \qquad \square$$

Section 1.2 Markov Property, Blumenthal's 0-1 Law   17

(2.11) allows us to relate the behavior of B_t as $t \to \infty$ to the behavior as $t \to 0$. Combining this idea with Blumenthal's 0-1 law leads to a very useful result. Let
$$\mathcal{F}'_t = \sigma(B_s : s \geq t) = \text{ the future after time } t$$
$$\mathcal{T} = \cap_{t \geq 0} \mathcal{F}'_t = \text{ the tail } \sigma\text{-field}$$

(2.12) Theorem. If $A \in \mathcal{T}$ then either $P_x(A) \equiv 0$ or $P_x(A) \equiv 1$.

Remark. Notice that this is stronger than the conclusion of Blumenthal's 0-1 law (2.7). The examples $A = \{\omega : \omega(0) \in B\}$ show that for A in the germ σ-field $\mathcal{F}_0^+$ the value of $P_x(A)$ may depend on x.

Proof Since the tail σ-field of B is the same as the germ σ-field for X, it follows that $P_0(A) \in \{0, 1\}$. To improve this to the conclusion given observe that $A \in \mathcal{F}'_1$, so 1_A can be written as $1_B \circ \theta_1$. Applying the Markov property, (2.1), gives

$$P_x(A) = E_x(1_B \circ \theta_1) = E_x(E_x(1_B \circ \theta_1 | \mathcal{F}_1)) = E_x(E_{B_1} 1_B)$$
$$= \int (2\pi)^{-d/2} \exp(-|y-x|^2/2) P_y(B) \, dy$$

Taking $x = 0$ we see that if $P_0(A) = 0$ then $P_y(B) = 0$ for a.e. y with respect to Lebesgue measure, and using the formula again shows $P_x(A) = 0$ for all x. To handle the case $P_0(A) = 1$ observe that $A^c \in \mathcal{T}$ and $P_0(A^c) = 0$, so the last result implies $P_x(A^c) = 0$ for all x. □

The next result is a typical application of (2.12). The argument here is a close relative of the one for (2.8).

(2.13) Theorem. Let B_t be a one dimensional Brownian motion and let $A = \cap_n \{B_t = 0 \text{ for some } t \geq n\}$. Then $P_x(A) = 1$ for all x.

In words, one dimensional Brownian motion is recurrent. It will return to 0 "infinitely often," i.e., there is a sequence of times $t_n \uparrow \infty$ so that $B_{t_n} = 0$. We have to be careful with the interpretation of the phrase in quotes since starting from 0, B_t will return to 0 infinitely many times by time $\epsilon > 0$.

Proof We begin by noting that under P_x, $B_t/\sqrt{t}$ has a normal distribution with mean $x/\sqrt{t}$ and variance 1, so if we use χ to denote a standard normal,

$$P_x(B_t < 0) = P_x(B_t/\sqrt{t} < 0) = P(\chi < -x/\sqrt{t})$$

and $\lim_{t\to\infty} P_x(B_t < 0) = 1/2$. If we let $T_0 = \inf\{t : B_t = 0\}$ then the last result and the fact that Brownian paths are continuous implies that for all $x > 0$

$$P_x(T_0 < \infty) \geq 1/2$$

Symmetry implies that the last result holds for $x < 0$, while (2.9) (or the Markov property) covers the last case $x = 0$.

Combining the fact that $P_x(T_0 < \infty) \geq 1/2$ for all x with the Markov property shows that

$$P_x(B_t = 0 \text{ for some } t \geq n) = E_x(P_{B_n}(T_0 < \infty)) \geq 1/2$$

Letting $n \to \infty$ it follows that $P_x(A) \geq 1/2$ but $A \in \mathcal{T}$ so (2.12) implies $P_x(B_t = 0 \text{ i.o.}) \equiv 1$. □

1.3. Stopping Times, Strong Markov Property

We call a random variable S taking values in $[0, \infty]$ a **stopping time** if for all $t \geq 0$, $\{S < t\} \in \mathcal{F}_t$. To bring this definition to life think of B_t as giving the price of a stock and S as the time we choose to sell it. Then the decision to sell before time t should be measurable with respect to the information known at time t.

In the last definition we have made a choice between $\{S < t\}$ and $\{S \leq t\}$. This makes a big difference in discrete time but none in continuous time (for a right continuous filtration $\mathcal{F}_t$):

If $\{S \leq t\} \in \mathcal{F}_t$ then $\{S < t\} = \cup_n \{S \leq t - 1/n\} \in \mathcal{F}_t$.

If $\{S < t\} \in \mathcal{F}_t$ then $\{S \leq t\} = \cap_n \{S < t + 1/n\} \in \mathcal{F}_t$.

The first conclusion requires only that $t \to \mathcal{F}_t$ is increasing. The second relies on the fact that $t \to \mathcal{F}_t$ is right continuous. (3.2) and (3.3) below show that when checking something is a stopping time it is nice to know that the two definitions are equivalent.

(3.1) Theorem. If G is an open set and $T = \inf\{t \geq 0 : B_t \in G\}$ then T is a stopping time.

Proof Since G is open and $t \to B_t$ is continuous $\{T < t\} = \cup_{q<t}\{B_q \in G\}$ where the union is over all rational q, so $\{T < t\} \in \mathcal{F}_t$. Here, we need to use the rationals so we end up with a countable union. □

(3.2) Theorem. If T_n is a sequence of stopping times and $T_n \downarrow T$ then T is a stopping time.

Section 1.3 Stopping Times, Strong Markov Property

Proof $\{T < t\} = \cup_n \{T_n < t\}$. □

(3.3) Theorem. If T_n is a sequence of stopping times and $T_n \uparrow T$ then T is a stopping time.

Proof $\{T \le t\} = \cap_n \{T_n \le t\}$. □

(3.4) Theorem. If K is a closed set and $T = \inf\{t \ge 0 : B_t \in K\}$ then T is a stopping time.

Proof Let $D(x,r) = \{y : |x-y| < r\}$, let $G_n = \cup\{D(x, 1/n) : x \in K\}$, and let $T_n = \inf\{t \ge 0 : B_t \in G_n\}$. Since G_n is open, it follows from (3.1) that T is a stopping time. I claim that as $n \uparrow \infty$, $T_n \uparrow T$. To prove this notice that $T \ge T_n$ for all n, so $\lim T_n \le T$. To prove $T \le \lim T_n$ we can suppose that $T_n \uparrow t < \infty$. Since $B(T_n) \in \bar{G}_n$ for all n and $B(T_n) \to B(t)$, it follows that $B(t) \in K$ and $T \le t$. □

Remark. As the reader might guess the hitting time of a Borel set A, $T_A = \inf\{t : B_t \in A\}$, is a stopping time. However, this turns out to be a difficult result to prove and is not true unless the filtration is completed as we did at the end of the last section. Hunt was the first to prove this. The reader can find a discussion of this result in Section 10 of Chapter 1 of Blumenthal and Getoor (1968) or in Chapter 3 of Dellacherie and Meyer (1978). We will not worry about that result here since (3.1) and (3.4), or in a pinch the next result, will be adequate for all the hitting times we will consider.

Exercise 3.1. Suppose A is an F_σ, i.e., a countable union of closed sets. Show that $T_A = \inf\{t : B_t \in A\}$ is a stopping time.

Exercise 3.2. Let S be a stopping time and let $S_n = ([2^n S] + 1)/2^n$ where $[x] = $ the largest integer $\le x$. That is,

$$S_n = (m+1)2^{-n} \text{ if } m2^{-n} \le S < (m+1)2^{-n}$$

In words, we stop at the first time of the form $k2^{-n}$ after S (i.e., $> S$). From the verbal description it should be clear that S_n is a stopping time. Prove that it is.

Exercise 3.3. If S and T are stopping times, then $S \wedge T = \min\{S,T\}$, $S \vee T = \max\{S,T\}$, and $S+T$ are also stopping times. In particular, if $t \ge 0$, then $S \wedge t$, $S \vee t$, and $S+t$ are stopping times.

Exercise 3.4. Let T_n be a sequence of stopping times. Show that

$$\sup_n T_n, \quad \inf_n T_n, \quad \limsup_n T_n, \quad \liminf_n T_n \quad \text{are stopping times}$$

Our next goal is to state and prove the strong Markov property. To do this, we need to generalize two definitions from Section 1.2. Given a nonnegative random variable $S(\omega)$ we define the random shift θ_S which "cuts off the part of ω before $S(\omega)$ and then shifts the path so that time $S(\omega)$ becomes time 0."

$$(\theta_S \omega)(t) = \begin{cases} \omega(S(\omega)+t) & \text{on } \{S < \infty\} \\ \Delta & \text{on } \{S = \infty\} \end{cases}$$

Here Δ is an extra point we add to C to cover the case in which the whole path gets shifted away. Some authors like to adopt the convention that all functions have $f(\Delta) = 0$ to take care of the second case. However, we will usually explicitly restrict our attention to $\{S < \infty\}$ so that the second half of the definition will not come into play.

The second quantity $\mathcal{F}_S$, "the information known at time S," is a little more subtle. We could have defined

$$\mathcal{F}_s = \cap_{\epsilon>0} \sigma(B_{t \wedge (s+\epsilon)}, t \geq 0)$$

so by analogy we could set

$$\mathcal{F}_S = \cap_{\epsilon>0} \sigma(B_{t \wedge (S+\epsilon)}, t \geq 0)$$

The definition we will now give is less transparent but easier to work with.

$$\mathcal{F}_S = \{A : A \cap \{S \leq t\} \in \mathcal{F}_t \text{ for all } t \geq 0\}$$

In words, this makes the reasonable demand that the part of A that lies in $\{S \leq t\}$ should be measurable with respect to the information available at time t. Again we have made a choice between $\leq t$ and $< t$ but as in the case of stopping times, this makes no difference and it is useful to know that the two definitions are equivalent.

Exercise 3.5. When $\mathcal{F}_t$ is right continuous, the definition of $\mathcal{F}_S$ is unchanged if we replace $\{S \leq t\}$ by $\{S < t\}$.

For practice with the definition of $\mathcal{F}_S$ do

Section 1.3 Stopping Times, Strong Markov Property

Exercise 3.6. Let S be a stopping time and let $A \in \mathcal{F}_S$. Show that

$$R = \begin{cases} S & \text{on } A \\ \infty & \text{on } A^c \end{cases} \quad \text{is a stopping time}$$

Exercise 3.7. Let S and T be stopping times.
(i) $\{S < t\}$, $\{S > t\}$, $\{S = t\}$ are in $\mathcal{F}_S$.
(ii) $\{S < T\}$, $\{S > T\}$, and $\{S = T\}$ are in $\mathcal{F}_S$ (and in $\mathcal{F}_T$).

Two properties of $\mathcal{F}_S$ that will be useful below are:

(3.5) Theorem. If $S \leq T$ are stopping times then $\mathcal{F}_S \subset \mathcal{F}_T$.

Proof If $A \in \mathcal{F}_S$ then $A \cap \{T \leq t\} = (A \cap \{S \leq t\}) \cap \{T \leq t\} \in \mathcal{F}_t$. □

(3.6) Theorem. If $T_n \downarrow T$ are stopping times then $\mathcal{F}_T = \cap_n \mathcal{F}(T_n)$.

Proof (3.5) implies $\mathcal{F}(T_n) \supset \mathcal{F}_T$ for all n. To prove the other inclusion, let $A \in \cap \mathcal{F}(T_n)$. Since $A \cap \{T_n < t\} \in \mathcal{F}_t$ and $T_n \downarrow T$, it follows that $A \cap \{T < t\} \in \mathcal{F}_t$, so $A \in \mathcal{F}_T$. □

The last result and Exercises 3.2 and 3.7 allow us to prove something that is obvious from the verbal definition.

Exercise 3.8. $B_S \in \mathcal{F}_S$, i.e., the value of B_S is measurable with respect to the information known at time S! To prove this let $S_n = ([2^n S] + 1)/2^n$ be the stopping times defined in Exercise 3.2. Show $B(S_n) \in \mathcal{F}_{S_n}$ then let $n \to \infty$ and use (3.6).

The next result goes in the opposite direction from (3.6). Here $\mathcal{G}_n \uparrow \mathcal{G}$ means $n \to \mathcal{G}_n$ is increasing and $\mathcal{G} = \sigma(\mathcal{G}_n)$.

Exercise 3.9. Let $S < \infty$ and T_n be stopping times and suppose that $T_n \uparrow \infty$ as $n \uparrow \infty$. Show that $\mathcal{F}_{S \wedge T_n} \uparrow \mathcal{F}_S$ as $n \uparrow \infty$.

We are now ready to state the strong Markov property, which says that the Markov property holds at stopping times.

(3.7) Strong Markov property. Let $(s, \omega) \to Y(s, \omega)$ be bounded and $\mathcal{R} \times \mathcal{C}$ measurable. If S is a stopping time then for all $x \in \mathbf{R}^d$

$$E_x(Y_S \circ \theta_S | \mathcal{F}_S) = E_{B(S)} Y_S \quad \text{on } \{S < \infty\}$$

where the right-hand side is $\varphi(y,t) = E_y Y_t$ evaluated at $y = B(S)$, $t = S$.

Remark. In most applications the function that we apply to the shifted path will not depend on s but this flexibility is important in Example 3.3. The verbal description of this equation is much like that of the ordinary Markov property:

"the conditional expectation of $Y \circ \theta_S$ given $\mathcal{F}_S^+$ is just the expected value of Y_S for a Brownian motion starting at B_S."

Proof We first prove the result under the assumption that there is a sequence of times $t_n \uparrow \infty$, so that $P_x(S < \infty) = \sum P_x(S = t_n)$. In this case we simply break things down according to the value of S, apply the Markov property and put the pieces back together. If we let $Z_n = Y_{t_n}(\omega)$ and $A \in \mathcal{F}_S$ then

$$E_x(Y_S \circ \theta_S; A \cap \{S < \infty\}) = \sum_{n=1}^{\infty} E_x(Z_n \circ \theta_{t_n}; A \cap \{S = t_n\})$$

Now if $A \in \mathcal{F}_S$, $A \cap \{S = t_n\} = (A \cap \{S \le t_n\}) - (A \cap \{S \le t_{n-1}\}) \in \mathcal{F}(t_n)$, so it follows from the Markov property that the above sum is

$$= \sum_{n=1}^{\infty} E_x(E_{B(t_n)} Z_n; A \cap \{S = t_n\}) = E_x(E_{B(S)} Y_S; A \cap \{S < \infty\})$$

To prove the result in general we let $S_n = ([2^n S] + 1)/2^n$ where $[x] = $ the largest integer $\le x$. In Exercise 3.2 you showed that S_n is a stopping time. To be able to let $n \to \infty$ we restrict our attention to Y's of the form

(*) $$Y_s(\omega) = f_0(s) \prod_{m=1}^{n} f_m(\omega(t_m))$$

where $0 < t_1 < \ldots < t_n$ and $f_0, \ldots, f_n$ are real valued, bounded and continuous. If f is bounded and continuous then the dominated convergence theorem implies that

$$x \to \int dy \, p_t(x,y) f(y)$$

is continuous. From this and induction it follows that

$$\varphi(x,s) = E_x Y_s = f_0(s) \int dy_1 \, p_{t_1}(x, y_1) f(y_1)$$
$$\ldots \int dy_n \, p_{t_n - t_{n-1}}(y_{n-1}, y_n) f(y_n)$$

is bounded and continuous.

Having assembled the necessary ingredients we can now complete the proof. Let $A \in \mathcal{F}_S$. Since $S \leq S_n$, (3.5) implies $A \in \mathcal{F}(S_n)$. Applying the special case of (3.7) proved above to S_n and observing that $\{S_n < \infty\} = \{S < \infty\}$ gives

$$E_x(Y_{S_n} \circ \theta_{S_n}; A \cap \{S < \infty\}) = E_x(\varphi(B(S_n), S_n) \ ; \ A \cap \{S < \infty\})$$

Now as $n \to \infty$, $S_n \downarrow S$, $B(S_n) \to B(S)$, $\varphi(B(S_n), S_n) \to \varphi(B(S), S)$ and

$$Y_{S_n} \circ \theta_{S_n} \to Y_S \circ \theta_S,$$

so the bounded convergence theorem implies that (3.7) holds when Y has the form given in (*).

To complete the proof now we use the monotone class theorem, (2.3). Let $\mathcal{H}$ be the collection of bounded functions for which (3.7) holds. Clearly $\mathcal{H}$ is a vector space that satisfies (ii) if $Y_n \in \mathcal{H}$ are nonnegative and increase to a bounded Y, then $Y \in \mathcal{H}$. To check (i) now, let $\mathcal{A}$ be the collection of sets of the form $\{\omega : \omega(t_j) \in G_j\}$ where G_j is an open set. If G is open the function 1_G is a decreasing limit of the continuous functions $f_k(x) = (1 - k \operatorname{dist}(x, G))^+$, where $\operatorname{dist}(x, G)$ is the distance from x to G, so if $A \in \mathcal{A}$ then $1_A \in \mathcal{H}$. This shows (i) holds and the desired conclusion follows from (2.3). □

Example 3.1. Zeros of Brownian motion. Consider one dimensional Brownian motion, let $R_t = \inf\{u > t : B_u = 0\}$ and let $T_0 = \inf\{u > 0 : B_u = 0\}$. Now (2.13) implies $P_x(R_t < \infty) = 1$, so $B(R_t) = 0$ and the strong Markov property and (2.9) imply

$$P_x(T_0 \circ \theta_{R_t} > 0 | \mathcal{F}_{R_t}) = P_0(T_0 > 0) = 0$$

Taking the expected value of the last equation we see that

$$P_x(T_0 \circ \theta_{R_t} > 0 \text{ for some rational } t) = 0$$

From this it follows that with probability one, if a point $u \in \mathcal{Z}(\omega) \equiv \{t : B_t(\omega) = 0\}$ is isolated on the left (i.e., there is a rational $t < u$ so that $(t, u) \cap \mathcal{Z}(\omega) = \emptyset$) then it is a decreasing limit of points in $\mathcal{Z}(\omega)$. This shows that the closed set $\mathcal{Z}(\omega)$ has no isolated points and hence must be uncountable. For the last step see Hewitt and Stromberg (1969), page 72.

If we let $|\mathcal{Z}(\omega)|$ denote the Lebesgue measure of $\mathcal{Z}(\omega)$ then Fubini's theorem implies

$$E_x(|\mathcal{Z}(\omega) \cap [0, T]|) = \int_0^T P_x(B_t = 0) \, dt = 0$$

So $\mathcal{Z}(\omega)$ is a set of measure zero. $\square$

Example 3.2. Let G be an open set, $x \in G$, let $T = \inf\{t : B_t \notin G\}$, and suppose $P_x(T < \infty) = 1$. Let $A \subset \partial G$, the boundary of G, and let $u(x) = P_x(B_T \in A)$. I claim that if we let $\delta > 0$ be chosen so that $D(x, \delta) = \{y : |y - x| < \delta\} \subset G$ and let $S = \inf\{t \geq 0 : B_t \notin D(x, \delta)\}$, then

$$u(x) = E_x u(B_S)$$

Intuition Since $D(x, \delta) \subset G$, B_t cannot exit G without first exiting $D(x, \delta)$ at B_S. When $B_S = y$, the probability of exiting G in A is $u(y)$ independent of how B_t got to y.

Proof To prove the desired formula, we will apply the strong Markov property, (3.7), to

$$Y = 1_{(B_T \in A)}$$

To check that this leads to the right formula, we observe that since $D(x, \delta) \subset G$, we have $B_T \circ \theta_S = B_T$ and $1_{(B_T \in A)} \circ \theta_S = 1_{(B_T \in A)}$. In words, ω and the shifted path $\theta_S \omega$ must exit G at the same place.

Since $S \leq T$ and we have supposed $P_x(T < \infty) = 1$, it follows that $P_x(S < \infty) = 1$. Using the strong Markov property now gives

$$E_x(1_{(B_T \in A)} \circ \theta_S | \mathcal{F}_S) = E_{B(S)} 1_{(B_T \in A)} = u(B_S)$$

Using the definition of u, $1_{(B_T \in A)} \circ \theta_S = 1_{(B_T \in A)}$, and taking expected value of the previous display, we have

$$\begin{aligned} u(x) = E_x 1_{(B_T \in A)} &= E_x \left(1_{(B_T \in A)} \circ \theta_S\right) \\ &= E_x E_x \left(1_{(B_T \in A)} \circ \theta_S | \mathcal{F}_S\right) = E_x u(B_S) \end{aligned} \qquad \square$$

Exercise 3.10. Let G, T, $D(x, \delta)$ and S be as above, but now suppose that $E_y T < \infty$ for all $y \in G$. Let g be a bounded function and let $u(y) = E_y(\int_0^T g(B_s)\, ds)$. Show that for $x \in G$

$$u(x) = E_x \left(\int_0^S g(B_s)\, ds + u(B_S) \right)$$

Our third application shows why we want to allow the function Y that we apply to the shifted path to depend on the stopping time S.

Example 3.3. Reflection principle. Let B_t be a one dimensional Brownian motion, let $a > 0$ and let $T_a = \inf\{t : B_t = a\}$. Then

(3.8) $$P_0(T_a < t) = 2P_0(B_t > a)$$

Intuitive proof We observe that if B_s hits a at some time $s < t$ then the strong Markov property implies that $B_t - B(T_a)$ is independent of what happened before time T_a. The symmetry of the normal distribution and $P_0(B_t = a) = 0$ then imply

(3.9) $$P_0(T_a < t, B_t > a) = \frac{1}{2}P_0(T_a < t)$$

Multiplying by 2, then using $\{B_t > a\} \subset \{T_a < t\}$ we have

$$P_0(T_a < t) = 2P_0(T_a < t, B_t > a) = 2P_0(B_t > a)$$

Proof To make the intuitive proof rigorous we only have to prove (3.9). To extract this from the strong Markov property (3.7), we let

$$Y_s(\omega) = \begin{cases} 1 & \text{if } s < t, \omega(t-s) > a \\ 0 & \text{otherwise} \end{cases}$$

We do this so that if we let $S = \inf\{s < t : B_s = a\}$ with $\inf \emptyset = \infty$ then

$$Y_S(\theta_S \omega) = 1_{(B_t > a)} \quad \text{on } \{S < \infty\} = \{T_a < t\}$$

If we let $\varphi(x, s) = E_x Y_s$ the strong Markov property implies

$$E_0(Y_S \circ \theta_S | \mathcal{F}_S) = \varphi(B_S, S) \quad \text{on } \{S < \infty\} = \{T_a < t\}$$

Now $B_S = a$ on $\{S < \infty\}$ and $\varphi(a, s) = 1/2$ if $s < t$, so

$$P_0(T_a < t, B_t > a) = E_0(1/2 \,;\, T_a < t)$$

which proves (3.9).

Exercise 3.11. Generalize the proof of (3.9) to conclude that if $u < v \leq a$ then

(3.10) $$P_0(T_a < t, u < B_t < v) = P_0(2a - v < B_t < 2a - u)$$

26 Chapter 1 Brownian Motion

To explain our interest in (3.10), let $M_t = \max_{0 \le s \le t} B_s$ to rewrite it as
$$P_0(M_t > a, u < B_t < v) = P_0(2a - v < B_t < 2a - u)$$
Letting the interval (u, v) shrink to x we see that
$$P_0(M_t > a, B_t = x) = P_0(B_t = 2a - x) = \frac{1}{\sqrt{2\pi t}} e^{-(2a-x)^2/2t}$$
Differentiating with respect to a now we get the joint density

(3.11) $$P_0(M_t = a, B_t = x) = \frac{2(2a - x)}{\sqrt{2\pi t^3}} e^{-(2a-x)^2/2t}$$

1.4. First Formulas

We have two interrelated aims in this section: the first to understand the behavior of the hitting times $T_a = \inf\{t : B_t = a\}$ for a one dimensional Brownian motion B_t; the second to study the behavior of Brownian motion in the upper half space $H = \{x \in \mathbf{R}^d : x_d > 0\}$. We begin with T_a and observe that the reflection principle (3.8) implies
$$P_0(T_a < t) = 2P_0(B_t > a) = 2\int_a^\infty (2\pi t)^{-1/2} \exp(-x^2/2t)\, dx$$
Here, and until further notice, we are dealing with a one dimensional Brownian motion. To find the probability density of T_a, we change variables $x = t^{1/2}a/s^{1/2}$, $dx = -t^{1/2}a/2s^{3/2}ds$ to get

(4.1) $$P_0(T_a < t) = 2\int_t^0 (2\pi t)^{-1/2} \exp(-a^2/2s)\left(-t^{1/2}a/2s^{3/2}\right) ds$$
$$= \int_0^t (2\pi s^3)^{-1/2} a \exp(-a^2/2s)\, ds$$

Using the last formula we can compute the distribution of $L = \sup\{t \le 1 : B_t = 0\}$ and $R = \inf\{t \ge 1 : B_t = 0\}$, completing work we started in Exercises 2.3 and 2.4. By (2.5) if $0 < s < 1$ then
$$P_0(L \le s) = \int_{-\infty}^\infty p_s(0, x) P_x(T_0 > 1 - s)\, dx$$
$$= 2\int_0^\infty (2\pi s)^{-1/2} \exp(-x^2/2s) \int_{1-s}^\infty (2\pi r^3)^{-1/2} x \exp(-x^2/2r)\, dr\, dx$$
$$= \frac{1}{\pi} \int_{1-s}^\infty (sr^3)^{-1/2} \int_0^\infty x \exp(-x^2(r+s)/2rs)\, dx\, dr$$
$$= \frac{1}{\pi} \int_{1-s}^\infty (sr^3)^{-1/2} rs/(r+s)\, dr$$

Our next step is to let $t = s/(r+s)$ to convert the integral over $r \in [1-s, \infty)$ into one over $t \in [0, s]$. $dt = -s/(r+s)^2 dr$, so to make the calculations easier we first rewrite the integral as

$$\frac{1}{\pi} \int_{1-s}^{\infty} \left(\frac{(r+s)^2}{rs} \right)^{1/2} \frac{s}{(r+s)^2} dr$$

Changing variables as indicated above and then again with $t = u^2$ to get

(4.2)
$$P_0(L \leq s) = \frac{1}{\pi} \int_0^s (t(1-t))^{-1/2} dt$$
$$= \frac{2}{\pi} \int_0^{\sqrt{s}} (1-u^2)^{-1/2} du = \frac{2}{\pi} \arcsin(\sqrt{s})$$

The reader should note that, contrary to intuition, the density function of $L =$ the last 0 before time 1,

$$P_0(L = t) = \frac{1}{\pi} \int_0^s (t(1-t))^{-1/2} \quad \text{for } 0 < t < 1$$

is symmetric about $1/2$ and blows up near 0 and 1. This is one of two arcsine laws associated with Brownian motion. We will encounter the other one in Section 4.9.

The computation for R is much easier and is left to the reader.

Exercise 4.1. Show that the probability density for R is given by

$$P_0(R = 1+t) = 1/(\pi t^{1/2}(1+t)) \quad \text{for } t \geq 0$$

Notation. In the last two displays and in what follows we will often write $P(T = t) = f(t)$ as short hand for T has density function $f(t)$.

As our next application of (4.1) we will compute the distribution of B_τ where $\tau = \inf\{t : B_t \notin H\}$ and $H = \{z : z_d > 0\}$ and, of course, B_t is a d dimensional Brownian motion. Since the exit time depends only on the last coordinate, it is independent of the first $d-1$ coordinates and we can compute the distribution of B_τ by writing for $x, \theta \in \mathbf{R}^{d-1}$, $y \in \mathbf{R}$

$$P_{(x,y)}(B_\tau = (\theta, 0)) = \int_0^\infty ds\, P_{(x,y)}(\tau = s)(2\pi s)^{-(d-1)/2} e^{-|x-\theta|^2/2s}$$
$$= \frac{y}{(2\pi)^{d/2}} \int_0^\infty ds\, s^{-(d+2)/2} e^{-(|x-\theta|^2+y^2)/2s}$$

by (4.1) with $a = y$. Changing variables $s = (|x - \theta|^2 + y^2)/2t$ gives

$$\frac{y}{(2\pi)^{d/2}} \int_\infty^0 \frac{-(|x-\theta|^2 + y^2)}{2t^2} \, dt \left(\frac{2t}{|x-\theta|^2 + y^2}\right)^{(d+2)/2} e^{-t}$$

so we have

(4.3) $$P_{(x,y)}(B_\tau = (\theta, 0)) = \frac{y}{(|x-\theta|^2 + y^2)^{d/2}} \frac{\Gamma(d/2)}{\pi^{d/2}}$$

where $\Gamma(\alpha) = \int_0^\infty y^{\alpha-1} e^{-y} \, dy$ is the usual gamma function.

When $d = 2$ and $x = 0$, probabilists should recognize this as a Cauchy distribution. At first glance, the fact that B_τ has a Cauchy distribution might be surprising, but a moment's thought reveals that this must be true. To explain this, we begin by looking at the behavior of the hitting times $\{T_a, a \geq 0\}$ as the level a varies.

(4.4) Theorem. Under P_0, $\{T_a, a \geq 0\}$ has stationary independent increments.

Proof The first step is to notice that if $0 < a < b$ then

$$T_b \circ \theta_{T_a} = T_b - T_a,$$

so if f is bounded and measurable, the strong Markov property (3.7) and translation invariance imply

$$E_0\left(f(T_b - T_a) | \mathcal{F}_{T_a}\right) = E_0\left(f(T_b) \circ \theta_{T_a} | \mathcal{F}_{T_a}\right)$$
$$= E_a f(T_b) = E_0 f(T_{b-a})$$

The desired result now follows from the next lemma which will be useful later.

(4.5) Lemma. Suppose Z_t is adapted to $\mathcal{G}_t$ and that for each bounded measurable function f

$$E(f(Z_t - Z_s) | \mathcal{G}_s) = E f(Z_{t-s})$$

Then Z_t has stationary independent increments.

Proof Let $t_0 < t_1 \ldots < t_n$, and let f_i, $1 \leq i \leq n$ be bounded and measurable. Conditioning on $\mathcal{F}_{t_{n-1}}$ and using the hypothesis we have

$$E\left(\prod_{i=1}^n f_i(Z_{t_i} - Z_{t_{i-1}})\right)$$
$$= E\left(\prod_{i=1}^{n-1} f_i(Z_{t_i} - Z_{t_{i-1}}) \cdot E(f_n(Z_{t_n} - Z_{t_{n-1}}) | \mathcal{F}_{t_{n-1}})\right)$$
$$= E\left(\prod_{i=1}^{n-1} f_i(Z_{t_i} - Z_{t_{i-1}})\right) E f(Z_{t_n - t_{n-1}})$$

By induction it follows that

$$E\left(\prod_{i=1}^n f_i(Z_{t_i} - Z_{t_{i-1}})\right) = \prod_{i=1}^n Ef_i(Z_{t_i - t_{i-1}})$$

which implies the desired conclusion. □

The scaling relation (1.2) implies

(4.6) $$T_a \stackrel{d}{=} a^2 T_1$$

So consulting Section 2.7 of Durrett (1995) we see that T_a has a stable law with index $\alpha = 1/2$ and skewness parameter $\kappa = 1$ (since $T_a \geq 0$).

To explain the appearance of the Cauchy distribution in the hitting locations, let $\tau_a = \inf\{t : B_t^2 = a\}$ (where B_t^2 is the second component of a two dimensional Brownian motion) and observe that another application of the strong Markov property implies

(4.7) **Theorem.** Under P_0, $\{C_s = B(\tau_s), s \geq 0\}$ has stationary independent increments.

Exercise 4.2. Prove (4.7).

The scaling relation (1.3) and an obvious symmetry imply

(4.8) $$C_s \stackrel{d}{=} sC_1 \qquad C_s \stackrel{d}{=} -C_s$$

so again consulting Section 2.7 of Durrett (1995) we can conclude that C_s has the symmetric stable distribution with index $\alpha = 1$ and hence must be Cauchy.

To give a direct derivation of the last fact let $\varphi_\theta(s) = E(\exp(i\theta C_s))$. Then (4.7) and (4.8) imply

$$\varphi_\theta(s)\varphi_\theta(t) = \varphi_\theta(s+t), \quad \varphi_\theta(s) = \varphi_{\theta s}(1), \quad \varphi_\theta(s) = \varphi_{-\theta}(s)$$

The fact that $C_s \stackrel{d}{=} -C_s$ implies that $\varphi_\theta(s)$ is real. Since $\theta \to \varphi_\theta(1)$ is continuous, the second equation implies $s \to \varphi_\theta(s) = \varphi_{\theta s}(1)$ is continuous, and a simple argument (see Exercise 4.3) shows that for each θ, $\varphi_\theta(s) = \exp(c_\theta s)$. The last two equations imply that $c_s = sc_1$ and $c_{-\theta} = c_\theta$ so $c_\theta = -\kappa|\theta|$ for some κ, so C_s has a Cauchy distribution. Since the arguments above apply equally well to (B_t^1, cB_t^2), where c is any constant, we cannot determine κ with this argument.

Exercise 4.3. Suppose $\varphi(s)$ is real valued continuous and satisfies $\varphi(s)\varphi(t) = \varphi(s+t)$, $\varphi(0) = 1$. Let $\psi(s) = \ln(\varphi(s))$ to get the additive equation: $\psi(s) + \psi(t) = \psi(s+t)$. Use the equation to conclude that $\psi(m2^{-n}) = m2^{-n}\psi(1)$ for all integers $m, n \geq 0$, and then use continuity to extend this to $\psi(t) = t\psi(1)$.

Exercise 4.4. Adapt the argument for $\varphi_\theta(s) = E(\exp(i\theta C_s))$ to show

$$E_0(\exp(-\lambda T_a)) = \exp(-a\kappa\sqrt{\lambda})$$

The representation of the Cauchy process given above allows us to see that its sample paths $a \to T_a$ and $s \to C_s$ are very bad.

Exercise 4.5. If $u < v$ then $P_0(a \to T_a$ is discontinuous in $(u, v)) = 1$.

Exercise 4.6. If $u < v$ then $P_0(s \to C_s$ is discontinuous in $(u, v)) = 1$.

Hint. By independent increments the probabilities in Exercises 4.5 and 4.6 only depend on $v - u$ but then scaling implies that they do not depend on the size of the interval.

The discussion above has focused on how B_t leaves H. The rest of the section is devoted to studying where B_t goes before it leaves H. We begin with the case $d = 1$.

(4.9) Theorem. If $x, y > 0$, then $P_x(B_t = y, T_0 > t) = p_t(x, y) - p_t(x, -y)$ where

$$p_t(x, y) = (2\pi t)^{-1/2} e^{-(y-x)^2/2t}$$

Proof The proof is a simple extension of the argument we used in Section 1.3 to prove that $P_0(T_a \leq t) = 2P_0(B_t \geq a)$. Let $f \geq 0$ with $f(x) = 0$ when $x \leq 0$. Clearly

$$E_x(f(B_t); T_0 > t) = E_x f(B_t) - E_x(f(B_t); T_0 \leq t)$$

If we let $\bar{f}(x) = f(-x)$, then it follows from the strong Markov property and symmetry of Brownian motion that

$$\begin{aligned} E_x(f(B_t); T_0 \leq t) &= E_x[E_0 f(B_{t-T_0}); T_0 \leq t] \\ &= E_x[E_0 \bar{f}(B_{t-T_0}); T_0 \leq t] \\ &= E_x[\bar{f}(B_t); T_0 \leq t] = E_x(\bar{f}(B_t)) \end{aligned}$$

since $\bar{f}(y) = 0$ for $y \geq 0$. Combining this with the first equality shows

$$E_x(f(B_t); T_0 > t) = E_x f(B_t) - E_x \bar{f}(B_t)$$
$$= \int (p_t(x,y) - p_t(x,-y))f(y)dy \qquad \square$$

The last formula generalizes easily to $d \geq 2$.

(4.10) Theorem. Let $\tau = \inf\{t : B_t^d = 0\}$. If $x, y \in H$,

$$P_x(B_t = y, \tau > t) = p_t(x,y) - p_t(x,\bar{y})$$

where $\bar{y} = (y_1, \ldots, y_{d-1}, -y_d)$.

Exercise 4.7. Prove (4.10).

2 Stochastic Integration

In this chapter we will define our stochastic integral $I_t = \int_0^t H_s dX_s$. To motivate the developments, think of X_s as being the price of a stock at time s and H_s as the number of shares we hold, which may be negative (selling short). The integral I_t then represents the net profits at time t, relative to our wealth at time 0. To check this note that the infinitesimal rate of change of the integral $dI_t = H_t \, dX_t$ = the rate of change of the stock times the number of shares we hold.

In the first section we will introduce the integrands H_s, the "predictable processes," a mathematical version of the notion that the number of shares held must be based on the past behavior of the stock and not on the future performance. In the second section, we will introduce our integrators X_s, the "continuous local martingales." Intuitively, martingales are fair games, while the "local" refers to the fact that we reduce the integrability requirements to admit a wider class of examples. We restrict our attention to the case of martingales with continuous paths $t \to X_t$ to have a simpler theory.

2.1. Integrands: Predictable Processes

To motivate the class of integrands we consider, we will discuss integration w.r.t. discrete time martingales. Here, we will assume that the reader is familiar with the basics of martingale theory, as taught for example in Chapter 4 of Durrett (1995). However, we will occasionally present results whose proofs can be found there.

Let $X_n, n \geq 0$, be a martingale w.r.t. $\mathcal{F}_n$. If $H_n, n \geq 1$, is any process, we can define

$$(H \cdot X)_n = \sum_{m=1}^{n} H_m(X_m - X_{m-1})$$

To motivate the last formula and the restriction we are about to place on the H_m, we will consider a concrete example. Let $\xi_1, \xi_2, \ldots$ be independent with $P(\xi_i = 1) = P(\xi_i = -1) = 1/2$, and let $X_n = \xi_1 + \cdots + \xi_n$. X_n is the **symmetric simple random walk** and is a martingale with respect to $\mathcal{F}_n = \sigma(\xi_1, \ldots, \xi_n)$.

If we consider a person flipping a fair coin and betting $1 on heads each time then X_n gives their net winnings at time n. Suppose now that the person bets an amount H_m on heads at time m (with $H_m < 0$ interpreted as a bet of $-H_m$ on tails). I claim that $(H \cdot X)_n$ gives her net winnings at time n. To check this note that if $H_m > 0$ our gambler wins her bet at time m and increases her fortune by H_m if and only if $X_m - X_{m-1} = 1$.

The gambling interpretation of the stochastic integral suggests that it is natural to let the amount bet at time n depend on the outcomes of the first $n-1$ flips but not on the flip we are betting on, or on later flips. A process H_n that has $H_n \in \mathcal{F}_{n-1}$ for all $n \geq 1$ (here $\mathcal{F}_0 = \{\emptyset, \Omega\}$, the trivial σ-field) is said to be **predictable** since its value at time n can be predicted (with certainty) at time $n-1$. The next result shows that we cannot make money by gambling on a fair game.

(1.1) Theorem. Let X_n be a martingale. If H_n is predictable and each H_n is bounded, then $(H \cdot X)_n$ is a martingale.

Proof It is easy to check that $(H \cdot X)_n \in \mathcal{F}_n$. The boundedness of the H_n implies $E|(H \cdot X)_n| < \infty$ for each n. With this established, we can compute conditional expectations to conclude

$$E((H \cdot X)_{n+1}|\mathcal{F}_n) = (H \cdot X)_n + E(H_{n+1}(X_{n+1} - X_n)|\mathcal{F}_n)$$
$$= (HX)_n + H_{n+1}E(X_{n+1} - X_n|\mathcal{F}_n) = (H \cdot X)_n$$

since $H_{n+1} \in \mathcal{F}_n$ and $E(X_{n+1} - X_n|\mathcal{F}_n) = 0$. □

The last theorem can be interpreted as: you can't make money by gambling on a fair game. This conclusion does not hold if we only assume that H_n is **optional**, that is, $H_n \in \mathcal{F}_n$, since then we can base our bet on the outcome of the coin we are betting on.

Example 1.1. If X_n is the symmetric simple random walk considered above and $H_n = \xi_n$ then

$$(H \cdot X)_n = \sum_{m=1}^{n} \xi_m \cdot \xi_m = n$$

since $\xi_m^2 = 1$. □

In continuous time, we still want the metatheorem "you can't make money gambling on a fair game" to hold, i.e., we want our integrals to be martingales. However, since the present (t) and past ($< t$) are not separated, the definition of the class of allowable integrands is more subtle. We will begin by considering a simple example that indicates one problem that must be dealt with.

Section 2.1 Integrands: Predictable Processes   35

Example 1.2. Let $(\Omega, \mathcal{F}, P)$ be a probability space on which there is defined a random variable T with $P(T \le t) = t$ for $0 \le t \le 1$ and an independent random variable ξ with $P(\xi = 1) = P(\xi = -1) = 1/2$. Let

$$X_t = \begin{cases} 0 & t < T \\ \xi & t \ge T \end{cases}$$

and let $\mathcal{F}_t = \sigma(X_s : s \le t)$. In words we wait until time T and then flip a coin.

X_t is a martingale with respect to $\mathcal{F}_t$. However, if we define the stochastic integral $I_t = \int_0^t X_s dX_s$ to be the ordinary Lebesgue-Stieltjes integral then

$$Y_1 = \int_0^1 X_s \, dX_s = X_T \cdot \xi = \xi^2 = 1$$

To check this note that the measure dX_s corresponds to a mass of size ξ at T and hence the integral is ξ times the value there. Noting now that $Y_0 = 0$ while $Y_1 = 1$ we see that I_t is not a martingale. □

The problem with the last example is the same as the problem with the one in discrete time — our bet can depend on the outcome of the event we are betting on. Again there is a gambling interpretation that illustrates what is wrong. Consider the game of roulette. After the wheel is spun and the ball is rolled, people can bet at any time before ($<$) the ball comes to rest but not after ($\ge$). One way of guaranteeing that our bet be made strictly before T, i.e., a sufficient condition, is to require that the amount of money we have bet at time t is left continuous. This implies, for instance, that we cannot react instantaneously to take advantage of a jump in the process we are betting on.

The simplest left continuous integrand we can imagine is made by picking $a < b$, $C \in \mathcal{F}_a$, and setting

(∗) $$H(s, \omega) = C(\omega) 1_{(a,b]}(t)$$

In words, we buy $C(\omega)$ shares of stock at time a based on our knowledge then, i.e., $C \in \mathcal{F}_a$. We hold them to time b and then sell them all. Clearly, the examples in (∗) should be allowable integrands; one should be able to add two or more of these, and take limits. To encompass the possibilities in the previous sentence, we let Π be the smallest σ-field containing all sets of the form $(a, b] \times A$ where $A \in \mathcal{F}_a$. The Π we have just defined is called the **predictable σ-field**. We will demand that our integrands H are measurable w.r.t. Π.

In the previous paragraph, we took the "bottom-up" approach to the definition of Π. That is, we started with some simple examples and then extended the class of integrands by taking sums and limits. For the rest of the section, we will take a "top-down" approach. We will start with some natural requirements and then add more until we arrive at Π. The descending path is more confusing since it involves four definitions that only differ in subtle ways. However, the reader need not study this material in detail. We will only use the predictable σ-field. The other definitions are included only to allow the reader to make the connection with other treatments and to indirectly make the point that the measure theoretic questions associated with continuous time processes can be quite difficult.

Let $\mathcal{F}_t$ be a right continuous filtration. The first and most intuitive concept of "depending on the past behavior of the stock and not on the future performance" is the following:

$H(s,\omega)$ is said to be **adapted** if for each t we have $H_t \in \mathcal{F}_t$.

We encountered this notion in our discussion of the Markov property in Chapter 1. In words, it says that the value at time t can be determined from the information we have at time t. The definition above, while intuitive, is not strong enough. We are dealing with a function defined on a product space $[0,\infty) \times \Omega$, so we need to worry about measurability as a function of the two variables.

H is said to be **progressively measurable** if for each t the mapping $(s,\omega) \to H(s,\omega)$ from $[0,t]$ to $\mathbf{R}$ is $\mathcal{R} \times \mathcal{F}_t$ measurable.

This is a reasonable definition. However, the "modern" approach is to use a slightly different definition that gives us a slightly smaller σ-field.

Let Λ be the σ-field of subsets of $[0,\infty) \times \Omega$ that is generated by the adapted processes that are right continuous and have left limits, i.e., the smallest σ-field which makes all of these processes measurable. A process H is said to be **optional** if $H(s,\omega)$ is measurable w.r.t. Λ.

According to Dellacherie and Meyer (1978) page 122, the optional σ-field is contained in the progressive σ-field and in the case of the natural filtration of Brownian motion the inclusion is strict. The subtle distinction between the last two σ-fields is not important for us. The only purpose here for the last two definitions is to prepare for the next one.

Let Π' be the σ-field of subsets of $[0,\infty) \times \Omega$ that is generated by the left continuous adapted processes. A process H is said to be **predictable** if $H(s,\omega) \in \Pi'$.

As we will now show, the new definition of predictable is the same as the old one. We have added the $'$ only to make the next statement and proof possible.

(1.2) Theorem. $\Pi = \Pi'$.

Proof Since all the processes used to define Π are left continuous, we have $\Pi \subset \Pi'$. To argue the other inclusion, let $H(s,\omega)$ be adapted and left continuous and let $H^n(s,\omega) = H(m2^{-n},\omega)$ for $m2^{-n} < s \leq (m+1)2^{-n}$. Clearly $H^n \in \Pi'$. Further, since H is left continuous, $H^n(s,\omega) \to H(s,\omega)$ as $n \to \infty$. □

The distinction between the optional and predictable σ-fields is not important for Brownian motion since in that case the two σ-fields coincide. Our last fact is that, in general, $\Pi \subset \Lambda$.

Exercise 1.1. Show that if $H(s,\omega) = 1_{(a,b]}(s)1_A(\omega)$ where $A \in \mathcal{F}_a$, then H is the limit of a sequence of optional processes; therefore, H is optional and $\Pi \subset \Lambda$.

2.2. Integrators: Continuous Local Martingales

In Section 2.1 we described the class of integrands that we will consider: the predictable processes. In this section, we will describe our integrators: continuous local martingales. Continuous, of course, means that for almost every ω, the sample path $s \to X_s(\omega)$ is continuous. To define local martingale we need some notation. If T is a nonnegative random variable and Y_t is any process we define

$$Y_t^T = \begin{cases} Y_{T \wedge t} & \text{on } \{T > 0\} \\ 0 & \text{on } \{T = 0\} \end{cases}$$

(2.1) Definition. X_t is said to be a **local martingale** (w.r.t. $\{\mathcal{F}_t, t \geq 0\}$) if there are stopping times $T_n \uparrow \infty$ so that $X_t^{T_n}$ is a martingale (w.r.t. $\{\mathcal{F}_{t \wedge T_n} : t \geq 0\}$). The stopping times T_n are said to **reduce** X.

We need to set $X_t^T \equiv 0$ on $\{T = 0\}$ to deal with the fact that X_0 need not be integrable. In most of our concrete examples X_0 is a constant and we can take $T_1 > 0$ a.s. However, the more general definition is convenient in a number of situations: for example (i) below and the definition of the variance process in (3.1).

In the same way we can define local submartingale, locally bounded, locally of bounded variation, etc. In general, we say that a process Y is **locally A** if there is a sequence of stopping time $T_n \uparrow \infty$ so that the stopped process Y_t^T has property A. (Of course, strictly speaking this means we should say locally a submartingale but we will continue to use the other term.)

Why local martingales?

There are several reasons for working with local martingales rather than with martingales.

(i) This frees us from worrying about integrability. For example, let X_t be a martingale with continuous paths, and let φ be a convex function. Then $\varphi(X_t)$ is always a local submartingale (see Exercise 2.3). However, we can conclude that $\varphi(X_t)$ is a submartingale only if we know $E|\varphi(X_t)| < \infty$ for each t, a fact that may be either difficult to check or false in some cases.

(ii) Often we will deal with processes defined on a random time interval $[0, \tau)$. If $\tau < \infty$, then the concept of martingale is meaningless, since for large t X_t is not defined on the whole space. However, it is trivial to define a local martingale: there are stopping times $T_n \uparrow \tau$ so that ...

(iii) Since most of our theorems will be proved by introducing stopping times T_n to reduce the problem to a question about nice martingales, the proofs are no harder for local martingales defined on a random time interval than for ordinary martingales.

Reason (iii) is more than just a feeling. There is a construction that makes it almost a theorem. Let X_t be a local martingale defined on $[0, \tau)$ and let $T_n \uparrow \tau$ be a sequence of stopping times that reduces X. Let $T_0 = 0$, suppose $T_1 > 0$ a.s., and for $k \geq 1$ let

$$\gamma(t) = \begin{cases} t - (k-1) & T_{k-1} + (k-1) \leq t \leq T_k + (k-1) \\ T_k & T_k + (k-1) \leq t \leq T_k + k \end{cases}$$

To understand this definition it is useful to write it out for $k = 1, 2, 3$:

$$\gamma(t) = \begin{cases} t & [0, T_1] \\ T_1 & [T_1, T_1 + 1] \\ t - 1 & [T_1 + 1, T_2 + 1] \\ T_2 & [T_2 + 1, T_2 + 2] \\ t - 2 & [T_2 + 2, T_3 + 2] \\ T_3 & [T_3 + 2, T_3 + 3] \end{cases}$$

In words, the time change expands $[0, \tau)$ onto $[0, \infty)$ by waiting one unit of time each time a T_n is encountered. Of course, strictly speaking γ compresses $[0, \infty)$ onto $[0, \tau)$ and this is what allows $X_{\gamma(t)}$ to be defined for all $t \geq 0$. The reason for our fascination with the time change can be explained by:

Section 2.2 Integrators: Continuous Local Martingales 39

(2.2) Theorem. $\{X_{\gamma(t)}, \mathcal{F}_{\gamma(t)}, t \geq 0\}$ is a martingale.

First we need to show:

(2.3) The Optional Stopping Theorem. Let X be a continuous local martingale. If $S \leq T$ are stopping times and $X_{T \wedge t}$ is a uniformly integrable martingale then $E(X_T|\mathcal{F}_S) = X_S$.

Proof (7.4) in Chapter 4 of Durrett (1995) shows that if $L \leq M$ are stopping times and $Y_{M \wedge n}$ is a uniformly integrable martingale w.r.t. $\mathcal{G}_n$ then

$$E(Y_M|\mathcal{G}_L) = Y_L$$

To extend the result from discrete to continuous time let $S_n = ([2^n S] + 1)/2^n$. Applying the discrete time result to the uniformly integrable martingale $Y_m = X_{T \wedge m2^{-n}}$ with $L = 2^n S_n$ and $M = \infty$ we see that

$$E(X_T|\mathcal{F}_{S_n}) = X_{S_n}$$

Letting $n \to \infty$ and using the dominated convergence theorem for conditional expectations ((5.9) in Chapter 4 of Durrett (1995)) the result follows. □

Proof of (2.2) Let $n = [t] + 1$. Since $\gamma(t) \leq T_n \wedge n$, using the optional stopping theorem, (2.2), gives $X_{\gamma(t)} = E(X_{T_n \wedge n}|\mathcal{F}_{\gamma(t)})$. Taking conditional expectation with respect to $\mathcal{F}_{\gamma(s)}$ we get

$$E(X_{\gamma(t)}|\mathcal{F}_{\gamma(s)}) = E(X_{T_n \wedge n}|\mathcal{F}_{\gamma(s)}) = X_{\gamma(s)}$$

proving the desired result. □

The next result is an example of the simplifications that come from assuming local martingales are continuous.

(2.4) Theorem. If X is a continuous local martingale, we can always take the sequence which reduces X to be $T_n = \inf\{t : |X_t| > n\}$ or any other sequence $T'_n \leq T_n$ that has $T'_n \uparrow \infty$ as $n \uparrow \infty$.

Proof Let S_n be a sequence that reduces X. If $s < t$, then applying the optional stopping theorem to $X_r^{S_n}$ at times $r = s \wedge T'_m$ and $t \wedge T'_m$ gives

$$E(X(t \wedge T'_m \wedge S_n) 1_{(S_n > 0)}|\mathcal{F}(s \wedge T'_m \wedge S_n)) = X(s \wedge T'_m \wedge S_n) 1_{(S_n > 0)}$$

Multiplying by $1_{(T'_m>0)} \in \mathcal{F}_0 \subset \mathcal{F}(s \wedge T'_m \wedge S_n)$ we get

$$E(X(t \wedge T'_m \wedge S_n)1_{(S_n>0,T'_m>0)}|\mathcal{F}(s \wedge T'_m \wedge S_n)) = X(s \wedge T'_m \wedge S_n)1_{(S_n>0,T'_m>0)}$$

As $n \uparrow \infty$, $\mathcal{F}(s \wedge T'_m \wedge S_n) \uparrow \mathcal{F}(s \wedge T'_m)$, and

$$X(r \wedge T'_m \wedge S_n)1_{(S_n>0,T'_m>0)} \to X(r \wedge T'_m)1_{(T'_m>0)}$$

for all $r \geq 0$, and $|X(r \wedge T'_m \wedge S_n)|1_{(S_n>0,T'_m>0)} \leq m$, so it follows from the dominated convergence theorem for conditional expectations (see (5.9) in Chapter 4 of Durrett (1995)) that

$$E(X(t \wedge T'_m)1_{(T'_m>0)}|\mathcal{F}(s \wedge T'_m)) = X(s \wedge T'_m)1_{(T'_m>0)}$$

proving the desired result. □

In our definition of local martingale in (2.1), we assumed that $X_t^{T_n}$ is a martingale (w.r.t. $\{\mathcal{F}_{t \wedge T_n}, t \geq 0\}$). We did this with the proof of (2.4) in mind. The next exercise shows that we get the same definition if we assume $X_t^{T_n}$ is a martingale (w.r.t. $\{\mathcal{F}_t, t \geq 0\}$).

Exercise 2.1. Let S be a stopping time. Then X_t^S is a martingale w.r.t. $\mathcal{F}_{t \wedge S}, t \geq 0$, if and only if it is a martingale w.r.t. $\mathcal{F}_t, t \geq 0$.

In the first version of this book I said

> "You should think of a local martingale as something that would be a martingale if it had $E|X_t| < \infty$."

As many people have pointed out to me, there is a famous example which shows that this is WRONG.

Example 2.1. Let B_t be a Brownian motion in dimension $d \geq 3$. In Section 3.1 we will see that $X_t = 1/|B_t|^{d-2}$ is a local martingale. Now

$$E_x|X_t|^p = \int (2\pi t)^{-d/2} e^{-|y-x|^2/2t} |y|^{-(d-2)p} \, dy < \infty$$

if and only if $p < d/(d-2)$ since (i) there is no trouble with the convergence for large y and (ii) ignoring $(2\pi t)^{-d/2} e^{-|y-x|^2/2t}$ which is bounded near 0 and changing to polar coordinates we have

$$\int_0^1 r^{-(d-2)p} r^{d-1} \, dr < \infty$$

if and only if $-(d-2)p + (d-1) > -1$, i.e., $p < d/(d-2)$.

To see that X_t is not a martingale we begin by noting that $(B_t - x) =_d t^{1/2}(B_1 - x)$ under P_x so as $t \to \infty$, $|B_t| \to \infty$ and $X_t \to 0$ in probability. To show that $E_x X_t \to 0$ we observe that for any $R < \infty$

$$E_x X_t \leq (2\pi t)^{-d/2} \int_{|y| \leq R} |y|^{-(d-2)} \, dy + R^{-(d-2)}$$

The integral is convergent so $\limsup_{t \to \infty} E_x X_t \leq R^{-(d-2)}$. Since R is arbitrary $E_x X_t \to 0$. Since $E_x X_0 > 0$ it follows that $E_x X_t$ is not constant and hence X_t is not a martingale. □

An even worse counterexample is provided by

Exercise 2.2. In Section 3.1 we will learn that if B_t is a two dimensional Brownian motion then $X_t = \log |B_t|$ is a local martingale. Show $E|X_t|^p < \infty$ for any $p < \infty$ but $\lim_{t \to \infty} E X_t = \infty$ so X is not a martingale.

The last example may leave the reader with the impression that no amount of integrability will guarantee that a local martingale is an honest martingale, but as the next very useful result shows, this is a false impression.

(2.5) Theorem. If X_t is a local submartingale and

$$E\left(\sup_{0 \leq s \leq t} |X_s|\right) < \infty$$

for each t then X_t is a submartingale.

Proof Clearly our assumption implies $E|X_t| < \infty$ for each t so all we have to check is that $E(X_t | \mathcal{F}_s) \geq X_s$. To do this, we note that if T_n is a sequence of stopping times that reduces X

$$E\left(X_t^{T_n} \Big| \mathcal{F}_{s \wedge T_n}\right) \geq X_s^{T_n}$$

then we let $n \to \infty$ and apply the dominated convergence theorem for conditional expectations. See (5.9) in Chapter 4 of Durrett (1995). □

As usual, multiplying by -1 shows that the last result holds for supermartingales, and once it is true for super- and sub- it is true for martingales (by applying the two other results). The following special case will be important:

(2.6) Corollary. A bounded local martingale is a martingale.

Here and in what follows when we say that a process X_t is **bounded** we mean there is a constant M so that with probability 1, $|X_t| \leq M$ for all $t \geq 0$.

In some circumstances we will need a convergence theorem for local martingales. The following simple result is sufficient for our needs.

(2.7) Theorem. Let X_t be a local martingale on $[0, \tau)$. If

$$E\left(\sup_{0 \leq s < \tau} |X_s|\right) < \infty$$

then $X_\tau = \lim_{t \uparrow \tau} X_t$ exists a.s. and $EX_0 = EX_\tau$.

Proof Let γ be the time scale of (2.2) and let $Y_t = X_{\gamma(t)}$. Y_t is a martingale. The upcrossing inequality and the proof of the martingale convergence theorem given in Section 4.2 of Durrett (1995) generalize in a straightforward way to continuous time and allow us to conclude that $Y_\infty = \lim_{t \uparrow \infty} Y_t$ exists a.s. and $X_\tau = \lim_{t \uparrow \tau} X_t$ exists a.s. To prove the last conclusion we note that $EY_0 = EY_t$ then use the a.s. convergence and the dominated convergence theorem to conclude $EX_0 = EY_0 = EY_\infty = EX_\tau$. □

Exercise 2.3. Let X_t be a continuous local martingale and let φ be a convex function. Then $\varphi(X_t)$ is a local submartingale.

Exercise 2.4. Let X be a continuous local martingale and let $R < \infty$ be a stopping time. Then $Y_s = X_{R+s}$ is a local martingale with respect to $\mathcal{G}_s = \mathcal{F}_{R+s}$.

Exercise 2.5. Show that if X is a continuous local martingale with $X_t \geq 0$ and $EX_0 < \infty$ then X_t is a supermartingale.

2.3. Variance and Covariance Processes

The next definition may look mysterious but it is very important.

(3.1) Theorem. If X_t is a continuous local martingale, then we define the **variance process** $\langle X \rangle_t$ to be the unique continuous predictable increasing process A_t that has $A_0 = 0$ and makes $X_t^2 - A_t$ a local martingale.

To warm up for the proof of (3.1) we will prove a result in discrete time.

(3.2) Theorem. Suppose X_n is a martingale with $EX_n^2 < \infty$ for all n. Then there is a unique predictable process A_n with $A_0 = 0$ so that $X_n^2 - A_n$ is a martingale. Furthermore, $n \to A_n$ is increasing.

Proof Let $A_0 = 0$ and define for $n \geq 1$

$$A_n = A_{n-1} + E(X_n^2|\mathcal{F}_{n-1}) - X_{n-1}^2$$

From the definition, it is immediate that $A_n \in \mathcal{F}_{n-1}$, A_n is increasing (since X_n^2 is a submartingale), and

$$E(X_n^2 - A_n|\mathcal{F}_{n-1}) = E(X_n^2|\mathcal{F}_{n-1}) - A_n$$
$$= X_{n-1}^2 - A_{n-1}$$

so A has the desired properties. To see that A is unique, observe that if B is another process with desired properties, then $A_n - B_n$ is a martingale and $A_n - B_n \in \mathcal{F}_{n-1}$. Therefore

$$A_n - B_n = E(A_n - B_n|\mathcal{F}_{n-1}) = A_{n-1} - B_{n-1}$$

and it follows by induction that $A_n - B_n = A_0 - B_0 = 0$. □

The key to the uniqueness may be summarized as "any predictable discrete time martingale is constant." Since Brownian motion is a predictable martingale, the last statement fails in continuous time and we need another assumption.

(3.3) Theorem. Any continuous local martingale X_t that is predictable and locally of bounded variation is constant (in time).

Proof By subtracting X_0, we may suppose that $X_0 = 0$ and prove that X is identically 0. Let V_t be the variation of X on $[0,t]$. We leave it to the reader to check

Exercise 3.1. If X is continuous and locally of bounded variation then $t \to V_t$ is continuous.

Define $S = \inf\{s : V_s \geq K\}$. Since $t \leq S$ implies $|X_t| \leq K$, (2.4) implies that the stopped process $M_t = X(t \wedge S)$ is a bounded martingale. Now if $s < t$ using well known properties of conditional expectation (see e.g., (1.3) and other results in Chapter 4 of Durrett (1995)) we have

$$(3.4) \quad E((M_t - M_s)^2|\mathcal{F}_s) = E(M_t^2|\mathcal{F}_s) - 2M_s E(M_t|\mathcal{F}_s) + M_s^2$$
$$= E(M_t^2|\mathcal{F}_s) - M_s^2 = E(M_t^2 - M_s^2|\mathcal{F}_s)$$

an equation that we will refer to as the **orthogonality of martingale increments** since the key to its validity is the fact that $E(M_s(M_t - M_s)|\mathcal{F}_s) = 0$ (and hence $E(M_s(M_t - M_s)) = 0$).

44 Chapter 2 Stochastic Integration

Letting $0 = t_0 < t_1 \ldots < t_n = t$ be a subdivision of $[0,t]$, the orthogonality of martingale increments, the inequality $\sum_i a_i^2 \leq (\sum_i |a_i|) \sup_j |a_j|$, and the fact that $V_{t \wedge S} \leq K$ imply

$$EM_t^2 = E\left(\sum_{m=1}^n M_{t_m}^2 - M_{t_{m-1}}^2\right)$$

$$= E\sum_{m=1}^n (M_{t_m} - M_{t_{m-1}})^2$$

$$\leq E\left(V_{t \wedge S} \sup_m |M_{t_m} - M_{t_{m-1}}|\right)$$

$$\leq K\, E \sup_m |M_{t_m} - M_{t_{m-1}}|$$

If we take a sequence of partitions $\Delta_n = \{0 = t_0^n < t_1^n \ldots < t_{k(n)}^n = t\}$ in which the **mesh** $|\Delta_n| = \sup_m |t_m^n - t_{m-1}^n| \to 0$ then continuity implies $\sup_m |M_{t_m^n} - M_{t_{m-1}^n}| \to 0$. Since the $\sup_m \leq 2K$ the bounded convergence theorem implies $E \sup_m |M_{t_m^n} - M_{t_{m-1}^n}| \to 0$. This shows $EM_t^2 = 0$ so $M_t = 0$ a.s. In the last conclusion t is arbitrary, so with probability one we have $M_t = 0$ for all rational t. The desired result now follows from continuity. □

With (3.3) established we can do the

Proof of uniqueness in (3.1) Suppose A_t and A'_t are two processes with the desired properties. Then $A_t - A'_t$ is a continuous local martingale that is locally of bounded variation and hence must be constant. Since $A_0 - A'_0 = 0$ it follows that $A_t = A'_t$ for all t. □

Proof of existence in (3.1) when X_t is a bounded martingale The existence of A is considerably more complicated than its uniqueness. To inspire the reader for the battle to come, we would like to point out that (i) we are about to prove the special case that we need of the celebrated Doob-Meyer decomposition and (ii) the construction will provide some useful information, e.g., (3.6). Though (3.1) and (3.8) are quite important, the details of their proofs are not, so the squeamish reader can skip ahead to the boldfaced "End of Existence Proof" five pages ahead.

Given a partition $\Delta = \{0 = t_0 < t_1 < t_2 \ldots\}$ with $\lim_n t_n = \infty$, we let $k(t) = \sup\{k : t_k < t\}$ be the index of the last point before time t. (To avoid confusion later, we emphasize now that $k(t)$ is a number not a random variable.) We next define for a process X, an approximate quadratic variation by

$$Q_t^\Delta(X) = \sum_{k=1}^{k(t)} (X_{t_k} - X_{t_{k-1}})^2 + (X_t - X_{t_{k(t)}})^2$$

Section 2.3 Variance and Covariance Processes

If the reader recalls what we are trying to define (see (3.1)) then the reason for the definition is explained by

(a) **Lemma.** If X_t is a bounded continuous martingale then $X_t^2 - Q_t^\Delta(X)$ is a martingale.

Proof To prove this, we first note that reasoning as in the proof of (3.4) one can conclude that if $r < s < t$ then

$$E\left((X_t - X_r)^2|\mathcal{F}_s\right) = E\left((X_t - X_s)^2|\mathcal{F}_s\right)$$
$$- 2(X_s - X_r)E(X_t - X_s|\mathcal{F}_s) + (X_s - X_r)^2$$
$$= E\left((X_t - X_s)^2|\mathcal{F}_s\right) + (X_s - X_r)^2$$

since $E(X_t - X_s|\mathcal{F}_s) = 0$.

Writing $Q_t^\Delta = Q_s^\Delta + (Q_t^\Delta - Q_s^\Delta)$ and working out the difference,

$$Q_t^\Delta - Q_s^\Delta = \sum_{k=1}^{k(t)} (X_{t_k} - X_{t_{k-1}})^2 + (X_t - X_{t_{k(t)}})^2$$
$$- \sum_{k=1}^{k(s)} (X_{t_k} - X_{t_{k-1}})^2 + (X_t - X_{t_{k(s)}})^2$$
$$= (X_{t_{k(s)+1}} - X_{t_{k(s)}})^2 - (X_s - X_{t_{k(s)}})^2$$
$$+ \sum_{k=k(s)+2}^{k(t)} (X_{t_k} - X_{t_{k-1}})^2 + (X_t - X_{t_{k(t)}})^2$$

The next step is to take conditional expectation with respect to $\mathcal{F}_s$ and use the first formula with $r = t_{k(s)}$ and $t = t_{k(s)+1}$. To neaten up the result, define a sequence u_n, $k(s) - 1 \le n \le k(t) + 1$ by letting $u_{k(s)-1} = s$, $u_i = t_i$ for $k(s) \le i \le k(t)$, and $u_{k(t)+1} = t$.

$$E(X_t^2 - Q_t^\Delta(X)|\mathcal{F}_s)$$
$$= E(X_t^2|\mathcal{F}_s) - Q_s^\Delta(X) - E\left(\sum_{i=k(s)+1}^{k(t)+1} (X_{u_i} - X_{u_{i-1}})^2 \bigg| \mathcal{F}_s\right)$$
$$= E(X_t^2|\mathcal{F}_s) - Q_s^\Delta(X) - E\left(\sum_{i=k(s)+1}^{k(t)+1} X_{u_i}^2 - X_{u_{i-1}}^2 \bigg| \mathcal{F}_s\right)$$
$$= X_s^2 - Q_s^\Delta(X)$$

46   Chapter 2   Stochastic Integration

where the second equality follows from (3.4).   □

Looking at (a) and noticing that $Q_t^\Delta(X)$ is increasing except for the last term, the reader can probably leap to the conclusion that we will construct the process A_t desired in (3.1) by taking a sequence of partitions with mesh going to 0 and proving that the limit exists. Lemma (c) is the heart (of darkness) of the proof. To prepare for its proof we note that using (a), taking expected value and using (3.4) gives

(b)
$$E(Q_t^\Delta(X) - Q_s^\Delta(X)|\mathcal{F}_s) = E(X_t^2 - X_s^2|\mathcal{F}_s)$$
$$= E((X_t - X_s)^2|\mathcal{F}_s)$$

To inspire you for the proof of (c), we promise that once (c) is established the rest of the proof of (3.1) is routine.

(c) Lemma. Let X_t be a bounded continuous martingale. Fix $r > 0$ and let Δ_n be a sequence of partitions $0 = t_0^n < t_1^n \ldots < t_{k_n}^n = r$ of $[0,r]$ with mesh $|\Delta_n| = \sup_k |t_k^n - t_{k-1}^n| \to 0$. Then $Q_r^{\Delta_n}(X)$ converges to a limit in L^2.

Proof   If Δ and Δ' are two partitions, we call $\Delta\Delta'$ the partition obtained by taking all the points in Δ and in Δ'. If we apply (a) twice and take differences we see that $Y_t = Q_t^\Delta(X) - Q_t^{\Delta'}(X)$ is a martingale. By definition $Y_t^2 - Q_t^{\Delta\Delta'}(Y)$ is a martingale, so

$$E(Q_r^\Delta(X) - Q_r^{\Delta'}(X))^2 = EY_r^2 = EQ_r^{\Delta\Delta'}(Y)$$

For reasons that will become clear in a moment, we will drop the argument from Q_r^Δ when it is X and we will drop the r when we want to refer to the process $t \to Q_t^\Delta$. Since

(3.5) $(a+b)^2 \leq 2a^2 + 2b^2$ for any real numbers a and b

(since $2a^2 + 2b^2 - (a+b)^2 = (a-b)^2 \geq 0$) we have

$$Q_r^{\Delta\Delta'}(Y) \leq 2Q_r^{\Delta\Delta'}(Q^\Delta) + 2Q_r^{\Delta\Delta'}(Q^{\Delta'})$$

Combining the last two results about Q we see that it is enough to prove that

(d) if $|\Delta| + |\Delta'| \to 0$ then $EQ_r^{\Delta\Delta'}(Q^\Delta) \to 0$.

To do this let $s_k \in \Delta\Delta'$ and $t_j \in \Delta$ so that $t_j \leq s_k < s_{k+1} \leq t_{j+1}$. Recalling the definition of $Q_s^\Delta(X)$ and doing a little arithmetic we have

$$Q_{s_{k+1}}^\Delta - Q_{s_k}^\Delta = (X_{s_{k+1}} - X_{t_j})^2 - (X_{s_k} - X_{t_j})^2$$
$$= (X_{s_{k+1}} - X_{s_k})^2 + 2(X_{s_{k+1}} - X_{s_k})(X_{s_k} - X_{t_j})$$
$$= (X_{s_{k+1}} - X_{s_k})(X_{s_{k+1}} + X_{s_k} - 2X_{t_j})$$

Summing the squares gives
$$Q_r^{\Delta\Delta'}(Q^\Delta) \le Q_r^{\Delta\Delta'}(X)\sup_k(X_{s_{k+1}}+X_{s_k}-2X_{t_{j(k)}})^2$$
where $j(k) = \sup\{j : t_j \le s_k\}$. By the Cauchy-Schwarz inequality
$$E(Q_r^{\Delta\Delta'}(Q^\Delta)) \le \{EQ_r^{\Delta\Delta'}(X)^2\}^{1/2}\left\{E\sup_k(X_{s_{k+1}}+X_{s_k}-2X_{t_{j(k)}})^4\right\}^{1/2}$$

Whenever $|\Delta|+|\Delta'| \to 0$ the second factor goes to 0, since X is bounded and continuous, so it is enough to prove

(e) If $|X_t| \le M$ for all t then $EQ_r^{\Delta\Delta'}(X)^2 \le 12M^4$.

The value 12 is not important but it does make it clear that the bound does not depend on r, Δ, or Δ'.

Proof of (e) If $\Gamma = \Delta\Delta'$ is the partition $0 = s_0 < s_1 \ldots < s_n = r$ then

(e.1)
$$\begin{aligned}Q_r^\Gamma(X)^2 &= \left(\sum_{m=1}^n (X_{s_m}-X_{s_{m-1}})^2\right)^2 \\ &= \sum_{m=1}^n (X_{s_m}-X_{s_{m-1}})^4 \\ &\quad + 2\sum_{m=1}^{n-1}(X_{s_m}-X_{s_{m-1}})^2(Q_r^\Gamma(X)-Q_{s_m}^\Gamma(X))\end{aligned}$$

To bound the first term on the right-hand side we note that if $|X_t| \le M$ for all t then some arithmetic and (3.4) imply

(e.2)
$$\begin{aligned}E\sum_{m=1}^n (X_{s_m}-X_{s_{m-1}})^4 &\le (2M)^2 E\sum_{m=1}^n (X_{s_m}-X_{s_{m-1}})^2 \\ &= 4M^2 E\sum_{m=1}^n X_{s_m}^2 - X_{s_{m-1}}^2 \\ &\le 4M^2 EX_r^2 \le 4M^4\end{aligned}$$

To bound the second term we note that $(X_{s_m}-X_{s_{m-1}})^2 \in \mathcal{F}_{s_m}$ then use (b) and $|X_r - X_{s_m}| \le 2M$ to get

(e.3)
$$\begin{aligned}E((X_{s_m}&-X_{s_{m-1}})^2\{Q_r^\Gamma(X)-Q_{s_m}^\Gamma(X)\}|\mathcal{F}_{s_m}) \\ &= (X_{s_m}-X_{s_{m-1}})^2 E(\{Q_r^\Gamma(X)-Q_{s_m}^\Gamma(X)\}|\mathcal{F}_{s_m}) \\ &= (X_{s_m}-X_{s_{m-1}})^2 E((X_r-X_{s_m})^2|\mathcal{F}_{s_m}) \\ &\le (2M)^2(X_{s_m}-X_{s_{m-1}})^2\end{aligned}$$

48 Chapter 2 Stochastic Integration

Taking expected value in (e.3), summing over m, and using (3.4) as before, we get

(e.4) $$E\sum_{m=1}^{n-1}(X_{s_m} - X_{s_{m-1}})^2 (Q_r^\Gamma(X) - Q_{s_m}^\Gamma(X)) \leq 4M^2 E \sum_{m=1}^{n} X_{s_m}^2 - X_{s_{m-1}}^2$$
$$\leq 4M^2 EX_r^2 \leq 4M^4$$

Plugging this bound and (e.2) into (e.1) proves (e) which completes the proof of (d) and hence of (c). □

It only remains to put the pieces together. Let Δ_n be the partition with points $k2^{-n}r$ with $0 \leq k \leq 2^n$. Since $Q_t^{\Delta_m} - Q_t^{\Delta_n}$ is a martingale (by (a)), using the L^2 maximal inequality (see e.g., (4.3) in Chapter 4 of Durrett (1995)) gives

(f) $$E\left(\sup_{t \leq r} |Q_t^{\Delta_m} - Q_t^{\Delta_n}|^2\right) \leq 4E|Q_r^{\Delta_m} - Q_r^{\Delta_n}|^2$$

From this and (c) it follows that

(g) There is a subsequence so that $Q_t^{\Delta_{n(k)}} \to$ a limit A_t uniformly on $[0, r]$.

Proof Since $Q_r^{\Delta_n}$ converges in L^2, it is a Cauchy sequence in L^2 and we can pick an increasing sequence $n(k)$ so that for $m \geq n(k)$

$$E|Q_r^{\Delta_m} - Q_r^{\Delta_{n(k)}}|^2 \leq 2^{-k}$$

Using (f) and Chebyshev's inequality it follows that

$$P\left(\sup_{t \leq r} |Q_t^{\Delta_{n(k+1)}} - Q_t^{\Delta_{n(k)}}| > 1/k^2\right) \leq k^4 2^{-k}$$

Since the right-hand side is summable, the Borel-Cantelli lemma implies the inequality on the right only fails finitely many times and the desired result follows. □

By taking subsequences of subsequences and using a diagonal argument we can get uniform convergence on $[0, N]$ for all N. Since each Q_t^Δ is continuous, A_t is as well. Because of the last term $(X_t - X_{k(t)})^2$, Q_t^Δ is not increasing. However, if $m \geq n$ then $k \to Q_{k2^{-n}r}^{\Delta_m}$ is increasing, so $k \to A_{k2^{-n}r}$ is increasing. Since n is arbitrary and $t \to A_t$ is continuous it follows that $t \to A_t$ is increasing.

The last detail for the case in which X is a bounded martingale is to check that $M_t^2 - A_t$ is a martingale. This follows from the next result (with $p = 2$).

(3.6) Lemma. Suppose that for each n, Z_t^n is a martingale with respect to $\mathcal{F}_t$, and that for each t, $Z_t^n \to Z_t$ in L^p where $p \geq 1$. Then Z_t is a martingale.

Proof The martingale property implies that if $s < t$ then $E(Z_t^n|\mathcal{F}_s) = Z_s^n$. The right-hand side converges to Z_s in L^p. To check that the left-hand side converges to $E(Z_t|\mathcal{F}_s)$ we note that linearity properties of and Jensen's inequality for conditional expectation (see e.g., (1.1a) and (1.1d) in Chapter 4 of Durrett (1995)) imply

$$E|E(Z_t^n|\mathcal{F}_s) - E(Z_t|\mathcal{F}_s)|^p = E|E(Z_t^n - Z_t|\mathcal{F}_s)|^p$$
$$\leq EE(|Z_t^n - Z_t|^p|\mathcal{F}_s) = E|Z_t^n - Z_t|^p \to 0$$

Taking limits in L^p it follows that $E(Z_t|\mathcal{F}_s) = Z_s$ and the proof is complete. □

Proof of existence in (3.1) when X is a local martingale Our first step in extending the result to local martingales is to prove a result about the quadratic variation of a stopped martingale. Once one remembers the definition of Y_t^T given at the beginning of Section 2.2, the next result should be obvious: the quadratic variation does not increase after the process stops.

(3.7) Lemma. Let X be a bounded martingale, and T be a stopping time. Then $\langle X^T \rangle = \langle X \rangle^T$.

Proof By the optional stopping theorem $(X^T)^2 - \langle X \rangle^T$ is a local martingale so the result follows from uniqueness. □

Let T_n be a sequence of stopping times increasing to ∞ so that

$$Y^n = X^{T_n} \cdot 1_{(T_n > 0)} \quad \text{is a bounded martingale}$$

By the previous result there is a unique continuous predictable increasing process A^n so that $(Y_t^n)^2 - A_t^n$ is a martingale. By (3.7) $A_t^n = A_t^{n+1}$ for $t \leq T_n$ so we can unambiguously define $\langle X \rangle_t = A_t^n$ for $t \leq T_n$. Clearly, $\langle X \rangle_t$ is continuous, predictable, and increasing. The definition implies $X_{T_n \wedge t}^2 \cdot 1_{(T_n > 0)} - \langle X \rangle_{T_n \wedge t}$ is a martingale so $X_t^2 - \langle X \rangle_t$ is a local martingale. □

End of Existence Proof

Finally, we have the following extension of (c) that will be useful later.

(3.8) Theorem. Let X_t be a continuous local martingale. For every t and sequence of subdivisions Δ_n of $[0,t]$ with mesh $|\Delta_n| \to 0$ we have

$$\sup_{s \le t} |Q_s^{\Delta_n}(X) - \langle X \rangle_s| \to 0 \quad \text{in probability}$$

Proof Let $\delta, \epsilon > 0$. We can find a stopping time S so that X_t^S is a bounded martingale and $P(S \le t) \le \delta$. It is clear that $Q^\Delta(X)$ and $Q^\Delta(X^S)$ coincide on $[0, S]$. From the definition and (3.7) it follows that $\langle X \rangle$ and $\langle X^S \rangle$ are equal on $[0, S]$. So we have

$$P\left(\sup_{s \le t} |Q_s^\Delta(X) - \langle X \rangle_s| > \epsilon \right) \le \delta + P\left(\sup_{s \le t} |Q_s^\Delta(X^S) - \langle X^S \rangle_s| > \epsilon \right)$$

Since X^S is a bounded martingale (c) implies that the last term goes to 0 as $|\Delta| \to 0$. $\square$

(3.9) Definition. If X and Y are two continuous local martingales, we let

$$\langle X, Y \rangle_t = \frac{1}{4}(\langle X + Y \rangle_t - \langle X - Y \rangle_t)$$

Remark. If X and Y are random variables with mean zero,

$$\text{cov}(X,Y) = EXY = \frac{1}{4}(E(X+Y)^2 - E(X-Y)^2)$$
$$= \frac{1}{4}(\text{var}(X+Y) - \text{var}(X-Y))$$

so it is natural, I think, to call $\langle X, Y \rangle_t$ the **covariance** of X and Y.

Given the definitions of $\langle X \rangle_t$ and $\langle X, Y \rangle_t$ the reader should not be surprised that if for a given partition $\Delta = \{0 = t_0 < t_1 < t_2 \ldots\}$ with $\lim_n t_n = \infty$, we let $k(t) = \sup\{k : t_k < t\}$ and define

$$Q_t^\Delta(X, Y) = \sum_{k=1}^{k(t)} (X_{t_k} - X_{t_{k-1}})(Y_{t_k} - Y_{t_{k-1}})$$
$$+ (X_t - X_{t_{k(t)}})(Y_t - Y_{t_{k(t)}})$$

then we have

(3.10) Theorem. Let X and Y be continuous local martingales. For every t and sequence of partitions Δ_n of $[0,t]$ with mesh $|\Delta_n| \to 0$ we have

$$\sup_{s \le t} |Q_s^{\Delta_n}(X, Y) - \langle X, Y \rangle_s| \to 0 \quad \text{in probability}$$

Section 2.3 Variance and Covariance Processes

Proof Since

$$Q_s^{\Delta_n}(X,Y) = \frac{1}{4}\left(Q_s^{\Delta_n}(X+Y) - Q_s^{\Delta_n}(X-Y)\right)$$

this follows immediately from the definition of $\langle X, Y \rangle$ and (3.8). □

Our next result is useful for computing $\langle X, Y \rangle_t$.

(3.11) Theorem. Suppose X_t and Y_t are continuous local martingales. $\langle X, Y \rangle_t$ is the unique continuous predictable process A_t that is locally of bounded variation, has $A_0 = 0$, and makes $X_t Y_t - A_t$ a local martingale.

Proof From the definition, it is easy to see that

$$X_t Y_t - \langle X, Y \rangle_t = \frac{1}{4}[(X_t + Y_t)^2 - \langle X+Y \rangle_t - \{(X_t - Y_t)^2 - \langle X-Y \rangle_t\}]$$

is a local martingale. To prove uniqueness, observe that if A_t and A_t' are two processes with the desired properties, then $A_t - A_t' = (X_t Y_t - A_t') - (X_t Y_t - A_t)$ is a continuous local martingale that is locally of bounded variation and hence $\equiv 0$ by (3.3). □

From (3.11) we get several useful properties of the covariance. In what follows, X, Y and Z are continuous local martingales. First since $X_t Y_t = Y_t X_t$,

$$\langle X, Y \rangle_t = \langle Y, X \rangle_t$$

Exercise 3.2. $\langle X + Y, Z \rangle_t = \langle X, Z \rangle_t + \langle Y, Z \rangle_t$.

Exercise 3.3. $\langle X - X_0, Z \rangle_t = \langle X, Z \rangle_t$.

Exercise 3.4. If a, b are real numbers then $\langle aX, bY \rangle_t = ab \langle X, Y \rangle_t$. Taking $a = b$, $X = Y$, and noting $\langle X, X \rangle_t = \langle X \rangle_t$ it follows that $\langle aX \rangle_t = a^2 \langle X \rangle_t$.

Exercise 3.5. $\langle X^T, Y^T \rangle = \langle X, Y \rangle^T$

It is also true that
$$\langle X, Y^T \rangle = \langle X, Y \rangle^T$$
Since we are lazy we will wait until Section 2.5 to prove this. We invite the reader to derive this directly from the definitions. The next two exercises prepare for the third which will be important in what follows:

Exercise 3.6. If X_t is a bounded martingale then $X_t^2 - \langle X \rangle_t$ is a uniformly integrable martingale.

Exercise 3.7. If X is a bounded martingale and S is a stopping time then $Y_t = X_{S+t} - X_S$ is a martingale with respect to $\mathcal{F}_{S+t}$ and $\langle Y \rangle_t = \langle X \rangle_{S+t} - \langle X \rangle_S$.

Exercise 3.8. If $S \leq T$ are stopping times and $\langle X \rangle_S = \langle X \rangle_T$, then X is constant on $[S, T]$.

Exercise 3.9. Conversely, if $S \leq T$ are stopping times and X is constant on $[S, T]$ then $\langle X \rangle_S = \langle X \rangle_T$.

2.4. Integration w.r.t. Bounded Martingales

In this section we will explain how to integrate predictable processes w.r.t. bounded continuous martingales. Once we establish the Kunita-Watanabe inequality in Section 2.5, we will be able to treat a general continuous local martingale in Section 2.6.

As in the case of the Lebesgue integral (i) we will begin with the simplest functions and then take limits to get the general case, and (ii) the sequence of steps used in defining the integral will also be useful in proving results later on.

Step 1: Basic Integrands

We say $H(s, \omega)$ is a basic predictable process if $H(s, \omega) = 1_{(a,b]}(s)C(\omega)$ where $C \in \mathcal{F}_a$. Let $\Pi_0 =$ the set of basic predictable processes. If $H = 1_{(a,b]}C$ and X is continuous, then it is clear that we should define

$$\int H_s dX_s = C(\omega)(X_b(\omega) - X_a(\omega))$$

Here we restrict our attention to integrating over $[0, \infty)$, since once we know how to do that then we can define the integral over $[0, t]$ by

$$(H \cdot X)_t \equiv \int_0^t H_s \, dX_s = \int H_s 1_{[0,t]}(s) dX_s$$

To extend our integral we will need to keep proving versions of the next three results. Under suitable assumptions on H, K, X and Y

(a) $(H \cdot X)_t$ is a continuous local martingale

(b) $((H+K) \cdot X)_t = (H \cdot X)_t + (K \cdot X)_t$, $\quad (H \cdot (X+Y))_t = (H \cdot X)_t + (H \cdot Y)_t$

(c) $\langle H \cdot X, K \cdot Y \rangle_t = \int_0^t H_s K_s \, d\langle X, Y \rangle_s$

In this step we will only prove a version of (a). The other two results will make their first appearance in Step 2. Before plunging into the details recall that at the end of Section 2.2 we defined H_t to be bounded if there was an M so that with probability one $H_t \leq M$ for all $t \geq 0$.

(4.1.a) Theorem. If X is a continuous martingale and $H \in b\Pi_0 = \{H \in \Pi_0 : H \text{ is bounded}\}$, then $(H \cdot X)_t$ is a continuous martingale.

Proof
$$(H \cdot X)_t = \begin{cases} 0 & 0 \leq t \leq a \\ C(X_t - X_a) & a \leq t \leq b \\ C(X_b - X_a) & b \leq t < \infty \end{cases}$$

From this it is clear that $(H \cdot X)_t$ is continuous, $(H \cdot X)_t \in \mathcal{F}_t$ and $E|(H \cdot X)_t| < \infty$. Since $(H \cdot X)_t$ is constant for $t \notin [a, b]$ we can check the martingale property by considering only $a \leq s < t \leq b$. (See Exercise 4.1 below.) In this case,

$$E((H \cdot X)_t | \mathcal{F}_s) - (H \cdot X)_s = E((H \cdot X)_t - (H \cdot X)_s | \mathcal{F}_s)$$
$$= E(C(X_t - X_s) | \mathcal{F}_s)$$
$$= CE(X_t - X_s) | \mathcal{F}_s) = 0$$

where the last two equalities follow from $C \in \mathcal{F}_s$ and $E(X_t | \mathcal{F}_s) = X_s$. □

Exercise 4.1. If Y_t is constant for $t \notin [a, b]$ and $E(Y_t | \mathcal{F}_s) = Y_s$ when $a \leq s < t \leq b$ then Y_t is a martingale.

Step 2: Simple Integrands

We say $H(s, \omega)$ is a **simple predictable process** and write $H \in \Pi_1$ if H can be written as the sum of a finite number of basic predictable processes. It is easy to see that if $H \in \Pi_1$ then H can be written as

$$H(s, \omega) = \sum_{i=1}^{m} 1_{(t_{i-1}, t_i]}(s) C_i(\omega)$$

where $t_0 < t_1 < \ldots < t_m$ and $C_i \in \mathcal{F}_{t_{i-1}}$. In this case we let

$$\int H_s dX_s = \sum_{i=1}^{m} C_i(X_{t_i} - X_{t_{i-1}})$$

The representation in the definition of H is not unique, since one may subdivide the intervals into two or more pieces, but it is easy to see that the right-hand side does not depend on the representation of H chosen.

(4.2.b) Theorem. Suppose X and Y are continuous martingales. If $H, K \in \Pi_1$ then

$$((H + K) \cdot X)_t = (H \cdot X)_t + (K \cdot X)_t$$
$$(H \cdot (X + Y))_t = (H \cdot X)_t + (H \cdot Y)_t$$

Proof Let $H^1 = H$ and $H^2 = K$. By subdividing the intervals in the defintions we can suppose $H_s^j = \sum_{i=1}^m 1_{(t_{i-1}, t_i]}(s) C_i^j$ for $j = 1, 2$. In this case it is easy to see that each side of the first identity is equal to

$$\sum_{i=1}^m (C_i^1 + C_i^2)(X_{t_i} - X_{t_{i-1}})$$

If $H_s = \sum_{i=1}^m 1_{(t_{i-1}, t_i]}(s) C_i$ then each side of the second identity is

$$\sum_{i=1}^m C_i \{(X_{t_i} - X_{t_{i-1}}) + (Y_{t_i} - Y_{t_{i-1}})\} \qquad \square$$

Since the sum of a finite number of continuous martingales is a continuous martingale, it follows from (4.1.a) and (4.2.b) that we have

(4.2.a) Theorem. If X is a continuous martingale and $H \in b\Pi_1 = \{H \in \Pi_1 : H \text{ is bounded}\}$, then $(H \cdot X)_t$ is a continuous martingale.

Turning to our third result

(4.2.c) Theorem. If X and Y are bounded continuous martingales and $H, K \in b\Pi_1$, then

$$\langle H \cdot X, K \cdot Y \rangle_t = \int_0^t H_s K_s d\langle X, Y \rangle_s$$

Consequently, $E((H \cdot X)_t (K \cdot Y)_t) = E \int_0^t H_s K_s d\langle X, Y\rangle_s$ and

$$E(H \cdot X)_t^2 = E \int_0^t H_s^2 d\langle X \rangle_s$$

Remark. Here and in what follows, integrals with respect to $\langle X, Y \rangle_s$ are Lebesgue-Stieltjes integrals. That is, since $s \to \langle X, Y \rangle_s$ is locally of bounded

Section 2.4 Integration w.r.t. Bounded Martingales 55

variation it defines a σ-finite signed measure, so we fix an ω and then integrate with respect to that measure. In the case under consideration, the integrand is piecewise constant so the integral is trivial.

Proof To prove these results it suffices to show that

$$Z_t = (H \cdot X)_t (K \cdot Y)_t - \int_0^t H_s K_s d\langle X, Y \rangle_s$$

is a martingale, for then the first result follows from (3.11), the second by taking expected values, and the third formula follows from the second by taking $H = K$ and $X = Y$. To prove Z_t is a martingale we begin by noting that

$$((H^1 + H^2) \cdot X)_t (K \cdot Y)_t = (H^1 \cdot X)_t (K \cdot Y)_t + (H^2 \cdot X)_t (K \cdot Y)_t$$

$$\int_0^t (H_s^1 + H_s^2) K_s d\langle X, Y \rangle_s = \int_0^t H_s^1 K_s d\langle X, Y \rangle_s + \int_0^t H_s^2 K_s d\langle X, Y \rangle_s$$

The last two observations imply that if the result holds for (H^1, K) and (H^2, K), it holds for $(H^1 + H^2, K)$. Similarly, if the result holds for (H, K^1) and (H, K^2), it holds for $(H, K^1 + K^2)$.

In view of the results in the last paragraph, we can now prove the result by establishing it in the case $H = 1_{(a,b]}C$, $K = 1_{(c,d]}D$, and we can furthermore assume that (i) $b \le c$ or (ii) $a = c$, $b = d$.

Case i. In this case $\int_0^t H_s K_s d\langle X, Y \rangle_s \equiv 0$, so we need to show that $(H \cdot X)_t (K \cdot Y)_t$ is a martingale. To prove this we observe that if $J = C(X_b - X_a)D1_{(c,d]}$ then $(H \cdot X)_t (K \cdot Y)_t = (J \cdot Y)_t$ is a martingale by (4.1.a)

Case ii. In this case

$$Z_s = \begin{cases} 0 & s \le a \\ CD\{(X_s - X_a)(Y_s - Y_a) - (\langle X, Y \rangle_s - \langle X, Y \rangle_a)\} & a \le s \le b \\ CD\{(X_b - X_a)(Y_b - Y_a) - (\langle X, Y \rangle_b - \langle X, Y \rangle_a)\} & s \ge b \end{cases}$$

so it suffices to check the martingale property for $a \le s \le t \le b$. To do this we note

$$Z_t - Z_s = CD\{X_t Y_t - X_s Y_s - X_a(Y_t - Y_s) - Y_a(X_t - X_s) \\ - (\langle X, Y \rangle_t - \langle X, Y \rangle_s)\}$$

Taking expected value and noting $X_a \in \mathcal{F}_s$, $E(Y_t - Y_s | \mathcal{F}_s) = 0$ we have

$$E(Z_t - Z_s | \mathcal{F}_s) = CDE(X_t Y_t - \langle X, Y \rangle_t - \{X_s Y_s - \langle X, Y \rangle_s\} | \mathcal{F}_s) = 0$$

This completes the proof of Case ii and finishes the proof of (4.2.c). □

Step 3: Square Integrable Integrands, Bounded Martingales

Let $\Pi_2(X)$ be the set of all predictable processes H that have

$$\|H\|_X = \left(E \int H_s^2 d\langle X\rangle_s\right)^{1/2} < \infty$$

Exercise 4.2. $\|H\|_X$ is a norm.

Remark. If we define a measure on the predictable σ-field by

$$\mu(A \times (s,t]) = E(\langle X\rangle_t - \langle X\rangle_s; A)$$

when $A \in \mathcal{F}_s$ then $\Pi_2(X) = L^2(\Pi, \mu)$. μ is called the **Doléans measure** after C. Doléans-Dade. For a proof that this is a measure see Chung and Williams (1990), pages 52–53.

Let $\mathcal{M}^2$ be the set of all martingales adapted to $\{\mathcal{F}_t, t \geq 0\}$ that have

$$\|X\|_2 = (\sup_t EX_t^2)^{1/2} < \infty$$

In the proof of (4.6) we will see that $\mathcal{M}^2$ is isomorphic to $L^2(\mathcal{F}_\infty)$.

Exercise 4.3. $X \in \mathcal{M}^2$ if and only if $EX_0^2 < \infty$ and $E\langle X\rangle_\infty < \infty$.

The following is the key to our next extension

(4.4) Isometry Property. If X is a bounded martingale and $H \in b\Pi_1$, then $\|H \cdot X\|_2 = \|H\|_X$.

Remark. We skipped over the number (4.3) since (4.3.a) and (4.3.b) will be proved later in this step.

Proof Recalling the definitions and using the last formula in (4.2.c) gives

$$\|H\|_X^2 = E \int H_s^2 d\langle X\rangle_s = \sup_t E \int_0^t H_s^2 d\langle X\rangle_s$$
$$= \sup_t E(H \cdot X)_t^2 = \|H \cdot X\|_2^2 \qquad \square$$

The first step in extending the integral to $\Pi_2(X)$ is to prove that $b\Pi_1$ is dense in $\Pi_2(X)$. Later we need a slight strengthening of this result so we go ahead and prove the stronger form.

(4.5) Lemma. If for $1 \le i \le k$ we have $X^i \in \mathcal{M}^2$ and $H \in \Pi_2(X^i)$, then there is a sequence $H^n \in b\Pi_1$ with $\|H^n - H\|_{X^i} \to 0$ for $1 \le i \le k$.

Proof Since $X^i \in \mathcal{M}^2$, $E\langle X^i\rangle_t < \infty$ and it follows that $b\Pi_1 \subset \Pi_2(X^i)$. Let $\mathcal{H}_t$ be the collection of predictable G that vanish on (t, ∞) for which the conclusion holds. Clearly, if $r < s \le t$ and $A \in \mathcal{F}_r$ then $G = 1_{(r,s]}1_A \in \mathcal{H}_t$. Suppose now that $0 \le G_n \in \mathcal{H}_t$ and $G_n \uparrow G$ with G bounded. The dominated convergence theorem implies

$$\|G - G_n\|_{X^i}^2 = E\int (G_s - G_s^n)^2 d\langle X^i\rangle_s \to 0$$

as $n \to \infty$, so we can pick n so that the last difference is $< \epsilon^2$ for $1 \le i \le k$. Since $G_n \in \mathcal{H}_t$, we can find a sequence $H^{n,m} \in b\Pi_1$ so that $\|H^{n,m} - G^n\|_{X^i} \to 0$ for $1 \le i \le k$ and then pick m so that $\|H^{n,m} - G^n\|_{X^i} < \epsilon$ for $1 \le i \le k$. The triangle inequality implies that $\|H^{n,m} - G\|_{X^i} < 2\epsilon$ for $1 \le i \le k$. Since ϵ is arbitrary it follows that $G \in \mathcal{H}_t$.

Using the monotone class theorem now ((2.3) in Chapter 1) it follows that $\mathcal{H}_t$ contains all bounded predictable processes that vanish on (t, ∞). Now if $K \in \Pi_2(X^i)$ for $1 \le i \le k$ and we define $K^n = K1_{|K| \le n}1_{[0,n]}$ then the dominated convergence theorem implies that $\|K^n - K\|_{X^i} \to 0$. Since K^n is bounded and vanishes on (n, ∞), another use of the triangle inequality implies K is a limit of $H^n \in b\Pi_1$. $\square$

(4.6) Theorem. $\mathcal{M}^2$ is complete.

Proof Standard martingale convergence theorems imply that if $X \in \mathcal{M}^2$, then as $t \to \infty$, X_t converges almost surely and in L^2 to a limit X_∞ with $EX_\infty^2 = \sup_t EX_t^2$, and the martingale can be recovered from X_∞ by $X_t = E(X_\infty|\mathcal{F}_t)$. Let $\mathcal{F}_\infty = \sigma(\mathcal{F}_t, t \ge 0)$. Since $X_\infty = \lim X_t \in \mathcal{F}_\infty$, the observation above shows that $X \to X_\infty$ maps $\mathcal{M}^2$ one-to-one into $L^2(\mathcal{F}_\infty)$. On the other hand, if $Y \in L^2(\mathcal{F}_\infty)$, then $Y_t = E(Y|\mathcal{F}_t)$ is a martingale with $Y_t \to Y$ as $t \to \infty$, and Jensen's inequality shows that

$$EY_t^2 = E(E(Y|\mathcal{F}_t)^2) \le E(E(Y^2|\mathcal{F}_t)) = EY^2$$

so $Y_t \in \mathcal{M}^2$. Combining this with the previous observation shows that $X \to X_\infty$ is an isometry from $\mathcal{M}^2$ onto $L^2(\mathcal{F}_\infty)$ and proves (4.6). $\square$

With (4.5) and (4.6) established, we can give the

Definition of the integral for $\Pi_2(X)$. To define $H \cdot X$ when X is a bounded continuous martingale and $H \in \Pi_2(X)$, let $H^n \in b\Pi_1$ so that $\|H^n - H\|_X \to 0$.

Since $\|H^n - H^m\|_X \to 0$ as $m, n \to \infty$, $(H^n \cdot X)$ is a Cauchy sequence in $\mathcal{M}^2$ and by (4.6) must converge to a limit in $\mathcal{M}^2$, which we define to be the integral $(H \cdot X)$. To see that the limit is independent of the sequence of approximations chosen suppose $\|H^n - H\|_X \to 0$ and $\|\bar{H}^n - H\|_X \to 0$ with $H^n \cdot X \to Y$ and $\bar{H}^n \cdot X \to \bar{Y}$. Form a third approximating sequence by setting $\tilde{H}^n = H^n$ if n is odd, $\tilde{H}^n = \bar{H}^n$ if n is even. $\|\tilde{H}^n - H\|_X \to 0$ so we have $\tilde{H}^n \cdot X \to Z$. Looking at the even and odd subsequences of $\tilde{H}^n$ it follows that $Y = Z = \bar{Y}$ so the limit is unique. In words, the last argument shows that if the limit exists for ANY sequence of approximations, it must be unique. Note that as a corollary of the construction we get

(4.3.a) Theorem. If X is a bounded continuous martingale and $H \in \Pi_2(X)$ then $H \cdot X \in \mathcal{M}^2$ and is continuous.

Proof The fact that $H \cdot X \in \mathcal{M}^2$ is automatic from the definition. If $H^n \in b\Pi_1$ have $\|H^n - H\|_X \to 0$ then (4.2.a) implies $(H^n \cdot X)_t$ is continuous. Since $\|(H^n \cdot X) - (H \cdot X)\|_2 \to 0$, using Chebyshev and the L^2 maximal inequality we get that

$$P\left(\sup_t |(H^n \cdot X)_t - (H \cdot X)_t| > \epsilon\right) \le \epsilon^{-2} E \sup_t |(H^n \cdot X)_t - (H \cdot X)_t|^2$$
$$\le \epsilon^{-2} \cdot 4\|(H^n \cdot X) - (H \cdot X)\|_2 \to 0$$

Since this holds for any $\epsilon > 0$ we say "$(H^n \cdot X)_t$ converges uniformly to $(H \cdot X)_t$ in probability." By passing to a subsequence we can have

$$\sup_t |(H^n \cdot X)_t - (H \cdot X)_t| \to 0 \quad \text{a.s.}$$

and it follows that $t \to (H \cdot X)_t$ is continuous. $\square$

(4.3.b) Theorem. If X is a bounded continuous martingale, and $H, K \in \Pi_2(X)$ then $H + K \in \Pi_2(X)$ and

$$((H + K) \cdot X)_t = (H \cdot X)_t + (K \cdot X)_t$$

Proof The triangle inequality for the norm $\|\cdot\|_X$ implies that $H + K \in \Pi_2(X)$. Let $H^n, K^n \in b\Pi_1$ with $\|H^n - H\|_X \to 0$ and $\|K^n - K\|_X \to 0$. The triangle inequality implies $\|(H^n + K^n) - (H + K)\|_X \to 0$. (4.2.b) implies

$$((H^n + K^n) \cdot X)_t = (H^n \cdot X)_t + (K^n \cdot X)_t$$

Now let $n \to \infty$ and use the fact that $(G^n \cdot X)_t \to (G \cdot X)_t$ in $\mathcal{M}^2$ when $G = H, K, H + K$. $\square$

The proofs of the remaining two equalities

$$(H \cdot (X+Y))_t = (H \cdot X)_t + (H \cdot Y)_t$$
$$\langle H \cdot X, K \cdot Y \rangle_t = \int_0^t H_s K_s d\langle X, Y \rangle_s$$

will be given in the next section. To develop the results there we will need to extend the isometry property from $b\Pi_1$ to $\Pi_2(X)$. To do this it is useful to recall some of your undergraduate analysis.

Exercise 4.4. If $\|\cdot\|$ is a norm and $\|x_n - x\| \to 0$ then $\|x_n\| \to \|x\|$.

Exercise 4.5. If X is a bounded martingale and $H \in \Pi_2(X)$ then $\|H \cdot X\|_2 = \|H\|_X$.

2.5. The Kunita-Watanabe Inequality

In this section we will prove an inequality due to Kunita and Watanabe and apply this result to extend the formula given in Section 2.4 for the covariance of two stochastic integrals.

(5.1) Theorem. If X and Y are local martingales and H and K are two measurable processes, then almost surely

$$\int_0^\infty |H_s K_s| \, d|\langle X, Y \rangle|_s \leq \left(\int_0^\infty H_s^2 d\langle X \rangle_s \right)^{1/2} \left(\int_0^\infty K_s^2 d\langle Y \rangle_s \right)^{1/2}$$

where $d|\langle X, Y \rangle|_s$ stands for dV_s where V_s is the total variation of $r \to \langle X, Y \rangle_r$ on $[0, s]$.

Remarks. (i) If $X = Y$ and $d\langle X \rangle_s = s$ then $d\langle Y \rangle_s = d|\langle X, Y \rangle|_s = s$ and (5.1) reduces to the Cauchy-Schwarz inequality. (ii) Notice that H and K are not assumed to be predictable. We assume only that $H(s, \omega)$ and $K(s, \omega)$ are measurable with respect to $\mathcal{R} \times \mathcal{F}$ where $\mathcal{R}$ is the Borel subsets of R. The reason we can attain this level of generality is that the notion of martingale does not enter into the proof after the first step.

Proof Step 1: Observe that if $s \leq t$, $\langle X + \lambda Y, X + \lambda Y \rangle_t \geq \langle X + \lambda Y, X + \lambda Y \rangle_s$. If we let $\langle M, N \rangle_s^t = \langle M, N \rangle_t - \langle M, N \rangle_s$, then

$$0 \leq \langle X + \lambda Y, X + \lambda Y \rangle_t - \langle X + \lambda Y, X + \lambda Y \rangle_s$$
$$= \langle X, X \rangle_s^t - 2\lambda \langle X, Y \rangle_s^t + \lambda^2 \langle Y, Y \rangle_s^t$$

60 Chapter 2 Stochastic Integration

for all s, t, and λ. If we fix s and t and throw away a countable number of sets of measure 0 then with probability one the last inequality will hold for all rational λ. Now a quadratic $ax^2 + bx + c$ that is nonnegative at all the rationals and not identically 0 has at most one real root (i.e., $b^2 - 4ac \leq 0$), so we have that

$$(\langle X, Y \rangle_s^t)^2 \leq \langle X, X \rangle_s^t \langle Y, Y \rangle_s^t$$

Step 2: Let $0 = t_0 < t_1 < \cdots < t_n$ be an increasing sequence of times, let $h_i, k_i, 1 \leq i \leq n$, be random variables, and define simple measurable processes

$$H(s, \omega) = \sum_{i=1}^{n} h_i(\omega) 1_{(t_{i-1}, t_i)}(s)$$

$$K(s, \omega) = \sum_{i=1}^{n} k_i(\omega) 1_{(t_{i-1}, t_i)}(s)$$

From the definition of the integral, the result of Step 1, and the Cauchy-Schwarz inequality, it follows that

$$\left| \int_0^\infty H_s K_s d\langle X, Y \rangle_s \right| \leq \sum_{i=1}^{n} |h_i k_i| |\langle X, Y \rangle_{t_{i-1}}^{t_i}|$$

$$\leq \sum_{i=1}^{n} |h_i| \left(\langle X, X \rangle_{t_{i-1}}^{t_i} \langle Y, Y \rangle_{t_{i-1}}^{t_i} \right)^{1/2} |k_i|$$

$$\leq \left(\sum_{i=1}^{n} h_i^2 \langle X, X \rangle_{t_{i-1}}^{t_i} \right)^{1/2} \left(\sum_{i=1}^{n} k_i^2 \langle Y, Y \rangle_{t_{i-1}}^{t_i} \right)^{1/2}$$

proving that for simple measurable processes:

$$(5.2) \quad \left| \int_0^\infty H_s K_s d\langle X, Y \rangle_s \right| \leq \left(\int_0^\infty H_s^2 d\langle X \rangle_s \right)^{1/2} \left(\int_0^\infty K_s^2 d\langle Y \rangle_s \right)^{1/2}$$

which is (5.1) with the absolute values outside the integral.

Step 3: Let M be a large number and let $T = \inf\{t : \langle X \rangle_t \text{ or } \langle Y \rangle_t > M\}$. By the monotone convergence theorem, it suffices to prove (5.1) when $H = K = 0$ for $s \geq T$ and $|H_s|, |K_s| \leq M$ for $s \leq T$. Having restricted our attention to $[0, T]$, $\langle X \rangle$ and $\langle Y \rangle$ are finite measures; so using the bounded convergence theorem, we see that (5.2) holds for bounded measurable processes. To improve (5.2) to (5.1) (and complete the proof), let J_s be a measurable process taking values in $\{-1, 1\}$ such that

$$\int_0^t |d\langle X, Y \rangle_s| = \int_0^t J_s d\langle X, Y \rangle_s$$

To see that this exists, note that J_s is the Radon-Nikodym derivative of $d\langle X, Y\rangle_s$ with respect to $d|\langle X, Y\rangle_s|$. Now apply (5.2) to $H_s = |H_s|$ and $K_s = J_s|K_s|$. □

With (5.1) established, we are now ready to take care of unfinished business from Section 4.

(5.3) Theorem. If $H \in \Pi_2(X) \cap \Pi_2(Y)$ then $H \in \Pi_2(X + Y)$ and

$$(H \cdot (X + Y))_t = (H \cdot X)_t + (H \cdot Y)_t$$

Proof First note that Exercise 3.2 implies

$$\langle X + Y\rangle_t = \langle X\rangle_t + \langle Y\rangle_t + 2\langle X, Y\rangle_t$$

and the Kunita-Watanabe inequality, (5.1), implies

$$|\langle X, Y\rangle|_t \le (\langle X\rangle_t \langle Y\rangle_t)^{1/2} \le (\langle X\rangle_t + \langle Y\rangle_t)/2$$

since $2ab \le a^2 + b^2$, i.e., $0 \le (a - b)^2$. So we have

$$\langle X + Y\rangle_t \le 2(\langle X\rangle_t + \langle Y\rangle_t)$$

(which should remind the reader of $(x + y)^2 \le 2(x^2 + y^2)$) and it follows that $H \in \Pi_2(X + Y)$. To prove the formula now we note that by (4.5) we can find H^n in $b\Pi_1$ so that $\|H^n - H\|_Z \to 0$ for $Z = X, Y, X + Y$. (4.2.b) implies that

$$(H^n \cdot (X + Y))_t = (H^n \cdot X)_t + (H^n \cdot Y)_t$$

Now let $n \to \infty$ and use $(H^n \cdot Z)_t \to (H \cdot Z)_t$ for $Z = X, Y, X + Y$. □

(5.4) Theorem. If X, Y are bounded continuous martingales, $H \in \Pi_2(X)$, $K \in \Pi_2(Y)$ then

$$\langle H \cdot X, K \cdot Y\rangle_t = \int_0^t H_s K_s d\langle X, Y\rangle_s$$

Proof By (3.11), it suffices to show that

(†) $$Z_t = (H \cdot X)_t (K \cdot Y)_t - \int_0^t H_s K_s d\langle X, Y\rangle_s$$

is a martingale. Let H^n and K^n be sequences of elements of $b\Pi_1$ that converge to H and K in $\Pi_2(X)$ and $\Pi_2(Y)$, respectively, and let Z_t^n be the quantity that results when H^n and K^n replace H and K in (†). By (4.2.c), Z_t^n is a martingale. By (3.6), we can complete the proof by showing $Z_s^n \to Z_s$ in L^1. The triangle inequality and (4.3.b) imply that

$$E \sup_t |(H^n \cdot X)_t (K^n \cdot Y)_t - (H \cdot X)_t (K \cdot Y)_t|$$
$$\leq E \sup_t |((H^n - H) \cdot X)_t (K^n \cdot Y)_t|$$
$$+ E \sup_t |(H \cdot X)_t ((K^n - K) \cdot Y)_t|$$

To bound the first term, we use (i) the sup of the product is smaller than the product of the sup's, (ii) the Cauchy-Schwarz inequality, (iii) the L^2 maximal inequality and (iv) the isometry property in Exercise 4.5:

$$\leq E\{\sup_t |((H^n - H) \cdot X)_t| \sup_t |(K^n \cdot Y)_t|\}$$
$$\leq \{E(\sup_t |((H^n - H) \cdot X)_t|^2) E(\sup_t |(K^n \cdot Y)_t|^2)\}^{1/2}$$
$$\leq 4\|(H^n - H) \cdot X\|_2 \|K^n \cdot Y\|_2 = 4\|H^n - H\|_X \|K^n\|_Y \to 0$$

as $n \to \infty$, since $\|H^n - H\|_X \to 0$ and $\|K^n\|_Y \to \|K\|_Y < \infty$. A similar estimate shows

$$E\left(\sup_t |(H \cdot X)_t ((K^n - K) \cdot Y)_t|\right) \to 0$$

so we have shown $(H^n \cdot X)_s (K^n \cdot Y)_s \to (H \cdot X)_s (K \cdot Y)_s$ in L^1.

To estimate the second term in $Z_t^n - Z_t$, we again begin by putting the absolute values inside, replacing the measure by its variation, and then using the triangle inequality

$$\left|\int_0^t H_s^n K_s^n d\langle X, Y\rangle_s - \int_0^t H_s K_s d\langle X, Y\rangle_s\right|$$
$$\leq \int_0^t |H_s^n K_s^n - H_s K_s| d|\langle X, Y\rangle|_s$$
$$\leq \int_0^t |H_s^n - H_s| |K_s^n| d|\langle X, Y\rangle|_s + \int_0^t |H_s| |K_s^n - K_s| d|\langle X, Y\rangle|_s$$

Using the Kunita-Watanabe inequality, (5.1), the first term is

$$\leq \left(\int_0^t |H_s^n - H_s|^2 d\langle X\rangle_s\right)^{1/2} \left(\int_0^t |K_s^n|^2 d\langle Y\rangle_s\right)^{1/2}$$

Taking expected values and using the Cauchy-Schwarz inequality

$$E \int_0^t |H_s^n - H_s| |K_s^n| d|\langle X, Y \rangle|_s$$

$$\leq \left(E \int_0^t |H_s^n - H_s|^2 d\langle X \rangle_s \right)^{1/2} \left(E \int_0^t |K_s^n|^2 d\langle Y \rangle_s \right)^{1/2} \to 0$$

as $n \to \infty$, since $||H^n - H||_X \to 0$ and $||K^n||_Y \to ||K||_Y < \infty$. Combining this with a similar estimate on $\int_0^t |H_s| |K_s^n - K_s| d|\langle X, Y \rangle|_s$ it follows that

$$\int_0^t H_s^n K_s^n d\langle X, Y \rangle_s \to \int_0^t H_s K_s d\langle X, Y \rangle_s$$

in L^1. We have now shown $Z_s^n \to Z_s$ in L^1 and the desired result follows from (3.6). □

Remark. In some treatments (see e.g., Revuz and Yor (1991) p. 130), if $M \in \mathcal{M}^2$ and $H \in \Pi_2(M)$, $H \cdot M$ is DEFINED to be the unique $L \in \mathcal{M}^2$ with $L_0 = 0$ so that
$$\langle L, N \rangle = H \cdot \langle M, N \rangle$$
for all $N \in \mathcal{M}^2$.

Exercise 5.1. Prove the uniqueness in the last claim.

(5.4) allows us to do easily some things that we would have had to work to prove in Section 2.3. Let X and Y be continuous local martingales, and suppose $X_0 = Y_0 = 0$. The last assumption entails no loss of generality. See Exercise 3.3.

Exercise 5.2. $\langle X, Y^T \rangle = \langle X, Y \rangle^T$.

Exercise 5.3. $X_t^T (Y_t - Y_t^T)$ is a local martingale.

2.6. Integration w.r.t. Local Martingales

In this section we will extend our integral so that the integrators can be continuous local martingales and so that the integrands are in

$$\Pi_3(X) = \left\{ H : \int_0^t H_s^2 d\langle X \rangle_s < \infty \text{ a.s. for all } t \geq 0 \right\}$$

64 Chapter 2 Stochastic Integration

We will argue in Section 3.4 that this is the largest possible class of integrands. To extend the integral from bounded martingales to local martingales we begin by proving an obvious fact.

(6.1) Theorem. Suppose X is a bounded continuous martingale, $H, K \in \Pi_2(X)$ and $H_s = K_s$ for $s \leq T$ where T is a stopping time. Then $(H \cdot X)_s = (K \cdot X)_s$ for $s \leq T$.

Proof Write $H_s = H_s^1 + H_s^2$ and $K_s = K_s^1 + K_s^2$ where

$$H_s^1 = K_s^1 = H_s 1_{(s \leq T)} = K_s 1_{(s \leq T)}$$

Clearly, $(H^1 \cdot X)_t = (K^1 \cdot Y)_t$ for all t. Since $H_s^2 = K_s^2 = 0$ for $s \leq T$, (5.4) implies

$$\langle H^2 \cdot X \rangle_s = \langle K^2 \cdot X \rangle_s = 0 \quad \text{for } s \leq T$$

and it follows from Exercise 3.8 that $(H^2 \cdot X)_t = (K^2 \cdot X)_t = 0$ for $t \leq T$. Combining this with the first result and using (4.3.b) gives the desired conclusion. □

Exercise 6.1. Extend the proof of (6.1) to show that if X is a bounded continuous martingale, $H, K \in \Pi_2(X)$ and $H_s = K_s$ for $S \leq s \leq T$ where S, T are stopping times then $(H \cdot X)_s - (H \cdot X)_S = (K \cdot X)_s - (K \cdot X)_S$ for $S \leq s \leq T$.

To extend the integral now, let S_n be a sequence of stopping times with

$$S_n \leq T_n = \inf\left\{t : |X_t| > n \text{ or } \int_0^t H_s^2 d\langle X \rangle_s > n\right\}$$

and $S_n \uparrow \infty$. Let $H_s^n = H_s 1_{(s \leq S_n)}$, and observe that if $m < n$, (6.1) implies that $(H^m \cdot X)_t = (H^n \cdot X)_t$ for $t \leq S_m$, so we can define $H \cdot X$ by setting $(H \cdot X)_t = (H^n \cdot X)_t$ for $t \leq S_n$.

To complete the definition we have to show that if R_n and S_n are two sequences of stopping times $\leq T_n$ with $R_n \uparrow \infty$ and $S_n \uparrow \infty$ then we end up with the same $(H \cdot X)_t$. Let H_t^T be the stopped version of the process defined at the beginning of Section 2.2, let $Q_n = R_n \wedge S_n$ and note that (6.1) implies

$$(H^{R_n} \cdot X)_s = (H^{S_n} \cdot X)_s \quad \text{for } s \leq Q_n$$

Since $Q_n \uparrow \infty$, it follows that $(H \cdot X)_t$ is independent of the sequence of stopping times chosen. The uniqueness result and (6.1) imply

(6.2) Theorem. If X is a continuous local martingale and $H \in \Pi_3(X)$ then

$$H^T \cdot X = (H \cdot X)^T = H \cdot X^T = H^T \cdot X^T$$

In words if we set the integrand $= 0$ after time T or stop the martingale at time T or do both, then this just stops the integral at T.

Our next task is to generalize our abc's to integrands in Π_3.

(6.3) Theorem. If X is a continuous local martingale and $H \in \Pi_3(X)$, then $(H \cdot X)_t$ is a continuous local martingale.

Proof By stopping at $T_n = \inf\{t : \int_0^t H_s^2 d\langle X\rangle_s > n \text{ or } |X_t| > n\}$, it suffices to show that if X is a bounded martingale and $H \in \Pi_2(X)$, then $(H \cdot X)_t$ is a continuous martingale but this follows from (4.3.a). □

(6.4) Theorem. Let X and Y be continuous local martingales. If $H, K \in \Pi_3(X)$ then $H + K \in \Pi_3(X)$ and

$$((H+K) \cdot X)_t = (H \cdot X)_t + (K \cdot X)_t$$

If $H \in \Pi_3(X) \cap \Pi_3(Y)$ then $H \in \Pi_3(X+Y)$ and

$$(H \cdot (X+Y))_t = (H \cdot X)_t + (H \cdot Y)_t$$

Proof To prove the first formula we note that the triangle inequality for the norm $\|\cdot\|_X$ implies $H + K \in \Pi_3(X)$. Stopping at

$$T_n = \inf\left\{t : |X_t|, \int_0^t H_s^2 d\langle X\rangle_s, \text{ or } \int_0^t K_s^2 d\langle Y\rangle_s > n\right\}$$

reduces the result to (4.3.b)

For the second formula we note that the argument in (5.3) shows that $H \in \Pi_3(X+Y)$ and by stopping we can reduce the result to (5.3). □

After seeing the last two proofs the reader can undoubtedly improve (5.4) to

(6.5) Theorem. If X, Y are continuous local martingales, $H \in \Pi_3(X)$ and $K \in \Pi_3(Y)$ then

$$\langle H \cdot X, K \cdot Y\rangle_t = \int_0^t H_s K_s d\langle X, Y\rangle_s$$

Using Exercise 3.2, we can generalize (6.5) to sums of stochastic integrals.

(6.6) Theorem. Let $X = \sum_{i=1}^m H^i \cdot X^i$ and $Y = \sum_{j=1}^n K^j \cdot Y^j$ where the X^i and Y^j are continuous local martingales. If $H^i \in \Pi_3(X^i)$ and $K^j \in \Pi_3(Y^j)$, then

$$\langle X, Y\rangle_t = \sum_{i,j} \int_0^t H_s^i K_s^j d\langle X^i, Y^j\rangle_s$$

66 Chapter 2 Stochastic Integration

We leave it to the reader to make the final extension of the isometry property:

Exercise 6.2. If X is a continuous local martingale and $H \in \Pi_2(X)$ then $H \cdot X \in \mathcal{M}^2$ and $\|H \cdot X\|_2 = \|H\|_X$.

We will conclude this section by computing some stochastic integrals. The first formula should be obvious, but for completeness must be derived from the definitions.

Exercise 6.3. Let X be a continuous local martingale. Let $S \leq T < \infty$ be stopping times, let $C(\omega)$ be bounded with $C(\omega) \in \mathcal{F}_S$, and define $H_s = C$ for $S < s \leq T$ and 0 otherwise. Then $H \in \Pi_3(X)$ and

$$\int H_s \, dX_s = C(X_T - X_S)$$

For the rest of this section, $\Delta_n = \{0 = t_0^n < t_1^n < \ldots < t_{k(n)}^n = t\}$ is a sequence of partitions of $[0, t]$ with mesh $|\Delta_n| = \sup_i |t_i^n - t_{i-1}^n| \to 0$. Our theme here is that if the integrand H_t is continuous then the integral is a limit of Riemann sums, *provided we evaluate the integrand at the left endpoint of the intervals.*

(6.7) Theorem. If X is a continuous local martingale and $t \to H_t$ is continuous then as $n \to \infty$

$$\sum_i H_{t_i^n} \{X(t_{i+1}^n) - X(t_i^n)\} \to \int_0^t H_s \, dX_s \quad \text{in probability}$$

Proof First note that $H \in \Pi_3(X)$ so the integral exists. By stopping at $T = \inf\{s : s, \langle X \rangle_s, \text{ or } |H_s| > M\}$ and making H constant after time T to preserve continuity, we can suppose $\langle X \rangle_t \leq M$ and $|H_s| \leq M$ for all s. Let $H_s^n = H_{t_i^n}$ for $s \in (t_i^n, t_{i+1}^n]$, $H_s^n = H_t$ for $s > t$. Since $s \to H_s$ is continuous on $[0, t]$ we have for each ω that

$$\sup_s |H_s^n - H_s| \to 0$$

The desired result now follows from the following lemma which will be useful later.

(6.8) Lemma. Let X_t be a continuous local martingale with $\langle X \rangle_t \leq M$ for all t. If H^n is a sequence of predictable processes with $|H^n| \leq M$ for all s, ω and with $\sup_s |H_s^n - H_s| \to 0$ in probability then $(H^n \cdot X) \to (H \cdot X)$ in $\mathcal{M}^2$.

Section 2.6 Integration w.r.t. Local Martingales 67

Proof Exercise 6.2 (or 4.4) implies

$$\|(H^n - H) \cdot X\|_2^2 = \|H^n - H\|_X^2 \leq ME\sup_s |H_s^n - H_s|^2 \to 0$$

as $n \to \infty$ by the bounded convergence theorem. □

The rest of the section is devoted to investigating what happens when we evaluate at the right end point or the center of each interval. For simplicity we will only consider the special case in which $H_s = 2X_s$.

Exercise 6.4. If X is a continuous local martingale then

$$\int_0^t 2X_s \, dX_s = X_t^2 - X_0^2 - \langle X \rangle_t$$

Exercise 6.5. Show that if X is a continuous local martingale and we evaluate at the right end point then

$$\sum_i 2X_{t_{i+1}^n}\{X(t_{i+1}^n) - X(t_i^n)\} \to \int_0^t 2X_s \, dX_s + 2\langle X \rangle_t = X_t^2 - X_0^2 + \langle X \rangle_t$$

Comparing the last two exercises you might jump to the conclusion that if we evaluate at the midpoint we get an integral that performs like the one in calculus. However, we will not ask you to prove this in general.

Exercise 6.6. Show that if B_t is a one dimensional Brownian motion starting at 0 then

$$\sum_{k=0}^{2^n-1} 2B((k+1/2)2^{-n}t)\{B((k+1)2^{-n}t) - B(k2^{-n}t)\}$$

converges in probability to $\int_0^t 2B_s \, dB_s + t = B_t^2$.

This approach to integration, called the **Stratonovich integral**, is more convenient than Itô's approach in certain circumstances (e.g., diffusion processes on manifolds) but we will not discuss it further here. Changing topics we have one more consequence of (6.7) that we will need in Chapter 5.

Exercise 6.7. Suppose $h : [0, \infty) \to \mathbf{R}$ is continuous. Show that $\int_0^t h_s \, dB_s$ has a normal distribution with mean 0 and variance $\int_0^t h_s^2 \, ds$.

2.7. Change of Variables, Itô's Formula

This section is devoted to a proof of our first version of Itô's fromula, (7.1). In Section 2.10, we will prove a bigger and better version, (10.2), with a slicker proof, but the mundane proof given here has the advantage of explaining where the term with f'' comes from.

(7.1) Theorem. Suppose X is a continuous local martingale and f has two continuous derivatives. Then with probability 1, for all $t \geq 0$

$$f(X_t) - f(X_0) = \int_0^t f'(X_s)dX_s + \frac{1}{2}\int_0^t f''(X_s)d\langle X \rangle_s$$

Remark. If A_t is a continuous function that is locally of bounded variation, and f has a continuous derivative then (see Exercise 7.2 below)

(7.2) $$f(A_t) - f(A_0) = \int_0^t f'(A_s)dA_s$$

As the reader will see in the proof of (7.1), the second term comes from the fact that local martingale paths have quadratic variation $\langle X \rangle_t$, while the 1/2 in front of it comes from expanding f in a Taylor series.

Proof By stopping at $T_M = \inf\{t : |X_t|$ or $\langle X \rangle_t \geq M\}$, it suffices to prove the result when $|X_t|$ and $\langle X \rangle_t \leq M$. From calculus we know that for any a and b, there is a $c(a,b)$ in between a and b such that

(7.3) $$f(b) - f(a) = (b-a)f'(a) + \frac{1}{2}(b-a)^2 f''(c(a,b))$$

Let t be a fixed positive number. Consider a sequence Δ_n of partitions $0 = t_0^n < t_1^n \ldots < t_{k_n}^n = t$ with mesh $|\Delta_n| \to 0$. From (7.3) it follows that

(7.4)
$$\begin{aligned}
f(X_t) - f(X_0) &= \sum_i f(X_{t_{i+1}^n}) - f(X_{t_i^n}) \\
&= \sum_i f'(X_{t_i^n})(X_{t_{i+1}^n} - X_{t_i^n}) \\
&\quad + \frac{1}{2}\sum_i g_i^n(\omega)(X_{t_{i+1}^n} - X_{t_i^n})^2
\end{aligned}$$

where $g_i^n(\omega) = f''(c(X_{t_i^n}, X_{t_{i+1}^n}))$.

Comparing (7.4) with (7.1), it becomes clear that we want to show

(7.5) $$\sum_i f'(X_{t_i^n})(X_{t_{i+1}^n} - X_{t_i^n}) \to \int_0^t f'(X_s)dX_s$$

(7.6) $$\frac{1}{2}\sum_i g_i(\omega)(X_{t_{i+1}^n} - X_{t_i^n})^2 \to \frac{1}{2}\int_0^t f''(X_s)d\langle X\rangle_s$$

in probability as $n \to \infty$. (7.5) follows from (6.8).

To prove (7.6), we let $G_s^n = g_i^n(\omega) = f''(c(X_{t_i^n}, X_{t_{i+1}^n}))$ when $s \in (t_i^n, t_{i+1}^n]$, $G_s^n = f''(X_t)$ for $s \geq t$ and let

$$A_s^n = \sum_{t_{i+1} \leq s}(X_{t_{i+1}} - X_{t_i})^2$$

so that

$$\sum_i g_i^n(\omega)(X_{t_{i+1}^n} - X_{t_i^n})^2 = \int_0^t G_s^n dA_s^n$$

and what we want to show is

(7.7) $$\int_0^t G_s^\delta dA_s^\delta \to \int_0^t f''(X_s)d\langle X\rangle_s$$

To do this we begin by observing that the uniform continuity of f'' implies that as $n \to \infty$ we have $G_s^n \to f''(X_{s \wedge t})$ uniformly in s, while (3.8) implies that A_s^n converges in probability to $\langle X\rangle_s$. Now by taking subsequences we can suppose that with probability 1, $A_{s \wedge t}^n$ converges weakly to $\langle X\rangle_{s \wedge t}$. In other words, if we fix ω and regard $s \to A_{s \wedge t}^n$ and $s \to \langle X\rangle_{s \wedge t}$ as distribution functions, then the associated measures converge weakly. Having done this, we can fix ω and deduce (7.7) from the following simple result:

(7.8) **Lemma.** If (i) measures μ_n on $[0,t]$ converge weakly to μ_∞, a finite measure, and (ii) g_n is a sequence of functions with $|g_n| \leq M$ that have the property that whenever $s_n \in [0,t] \to s$ we have $g_n(s_n) \to g(s)$, then as $n \to \infty$

$$\int g_n d\mu_n \to \int g\, d\mu_\infty$$

Proof By letting $\mu_n'(A) = \mu_n(A)/\mu_n([0,t])$, we can assume that all the μ_n are probability measures. A standard construction (see (2.1) in Chapter 2 of Durrett (1995)) shows that there is a sequence of random variables X_n with distribution μ_n so that $X_n \to X_\infty$ a.s. as $n \to \infty$ The convergence of g_n to g

implies $g_n(X_n) \to g(X_\infty)$, so the result follows from the bounded convergence theorem. □

(7.8) is the last piece in the proof of (7.1). Tracing back through the proof, we see that (7.8) implies (7.7), which in turn completes the proof of (7.6). So adding (7.5) and using (7.4) gives that for each t,

$$f(X_t) - f(X_0) = \int_0^t f'(X_s)dX_s + \frac{1}{2}\int_0^t f''(X_s)d\langle X\rangle_s \quad \text{a.s.}$$

Since each side of the formula is a continuous function of t, it follows that with probability 1 the equality holds for all $t \geq 0$, the statement made in (7.1). □

(7.9) Remark. In Chapter 3 we will need to use Itô's formula for complex valued f. The extension is trivial. Write $f = u + iv$, apply Itô's formula to u and v, multiply the v formula by i and add the two.

Exercise 7.1. Let A_t be a continuous process with total variation $|A|_t \leq M$ for all t. If K^n is a sequence of measurable processes with $|K^n(s,\omega)| \leq M$ for all s,ω and $\sup_s |K_s^n - K_s| \to 0$ then $\sup_t |(K^n - K) \cdot A)_t| \to 0$.

Exercise 7.2. Use the fact that $f(b) - f(a) = f'(c(a,b))(b-a)$ for some c between a and b, and Exercise 7.1 to prove (7.2).

2.8. Integration w.r.t. Semimartingales

X is said to be a **continuous semimartingale** if X_t can be written as $M_t + A_t$, where M_t is a continuous local martingale and A_t is a continuous adapted process that is locally of bounded variation. A nice feature of continuous semimartingales that is unheard of for their more general counterparts is

(8.1) Theorem. Let X_t be a continuous semimartingale. If the (continuous) processes M_t and A_t are chosen so that $A_0 = 0$, then the decomposition $X_t = M_t + A_t$ is unique.

Proof If $M'_t + A'_t$ is another decomposition, then $A_t - A'_t = M'_t - M_t$ is a continuous local martingale and locally of bounded variation, so by (3.3) $A_t - A'_t$ is constant and hence $\equiv 0$. □

Remark. In what follows we will sometimes write "Let $X_t = M_t + A_t$ be a continuous semimartingale" to specify that the decomposition of X_t consists of the local martingale M_t and the process A_t that is locally of bounded variation.

Section 2.8 Integration w.r.t. Semimartingales

Exercise 8.1. Show that if $X_t = M_t + A_t$ and $X_t' = M_t' + A_t'$ are continuous semimartingales then $X_t + X_t' = (M_t + M_t') + (A_t + A_t')$ is a continuous semimartingale.

In this section we will extend the class of integrators for our stochastic integral from continuous local martingales to continuous semimartingales. There are three reasons for doing this.

(i) If X is a continuous local martingale and f is C^2 then Itô's formula shows us that $f(X_t)$ is always a semimartingale but it is not a local martingale unless $f''(x) = 0$ for all x. In the next section we will prove a generalization of Itô's formula which implies that if X is a continuous semimartingale and f is C^2 then $f(X_t)$ is again a semimartingale.

(ii) It can be argued that any "reasonable integrator" is a semimartingale. To explain this, we begin by defining an **easy integrand** to be a process of the form

$$H = \sum_{i=0}^n H_i 1_{(T_i, T_{i+1}]}$$

where $0 = T_0 \leq T_1 \leq \cdots \leq T_{n+1}$ are stopping times, and the $H_i \in \mathcal{F}_{T_i}$ have $|H_i| < \infty$ a.s. Let $b\Pi_{e,t}$ be the collection of bounded easy predictable processes that vanish on (t, ∞) equipped with the uniform norm

$$\|H\|_u = \sup_{s,\omega} |H_s(\omega)|$$

For easy integrands we define the integral as

$$(H \cdot X) = \sum_{i=1}^n H_i (X_{T_{i+1}} - X_{T_i})$$

Finally, let L^0 be the collection of all random variables topologized by convergence in probability, which comes from the metric $\|X\|_0 = E(|X|/(1+|X|))$. See Exercise 6.4 in Chapter 1 of Durrett (1995).

A result proved independently by Bichteler and Dellacherie states

(8.2) Theorem. If $H \to (H \cdot X)$ is continuous from $b\Pi_{e,t} \to L^0$ for all t then X is a semimartingale.

Since any extension of the integral from $b\Pi_{e,t}$ will involve taking limits, we need our integral to be a continuous function on the simple integrands (in some sense). We have placed a very strong topology on the domain and a weak

72 Chapter 2 Stochastic Integration

topology on the range, so this is a fairly weak requirement. Protter (1990) takes (8.2) as the definition of semimartingale. In his approach one gets some very deep results without much effort but then one must sweat to prove that a semimartingale (defined as a good integrator) is a semimartingale (as defined above).

(iii) Last but not least, it is easy to extend the integral from local martingales to semimartingales if we replace $\Pi_3(X)$ by a slightly smaller class of integrands that does not depend on X.

Getting back to business, let $X_t = M_t + A_t$ be a continuous semimartingale. We say $H \in lb\Pi$ = the set of **locally bounded predictable processes**, if there is a sequence of stopping times $T_n \uparrow \infty$ so that $|H(s,\omega)| \leq n$ for $s \leq T_n$. If $H \in lb\Pi$ we can define

$$(H \cdot A)_t(\omega) = \int_0^t H_s(\omega) dA_s(\omega)$$

as a Lebesgue-Stieltjes integral (which exists for a.e. ω). To integrate with respect to the local martingale M_t we note that

$$\int_0^{T_n} H_s^2 d\langle M \rangle_s \leq n^2 \langle M \rangle_{T_n} < \infty$$

and $T_n \to \infty$. So $lb\Pi \subset \Pi_3(M)$, we can define $(H \cdot M)_t$, and let

$$(H \cdot X)_t = (H \cdot M)_t + (H \cdot A)_t$$

since by the uniqueness of the decomposition this is an unambiguous definition.

From the definition above, it follows immediately that we have

(8.3) Theorem. If X is a continuous semimartingale and $H \in lb\Pi$, then $(H \cdot X)_t$ is a continuous semimartingale.

The second of our abc's is also easy. We name it in honor of the similar relationships between addition and multiplication.

(8.4) Distributive Laws. Suppose X and Y are continuous semimartingales, and $H, K \in \ell b\Pi$. Then

$$((H + K) \cdot X)_t = (H \cdot X)_t + (K \cdot X)_t$$
$$(H \cdot (X + Y))_t = (H \cdot X)_t + (H \cdot Y)_t$$

Section 2.8 Integration w.r.t. Semimartingales   73

Proof This follows easily from the definition of the integral with respect to a semimartingale, the result for local martingales given in (6.4), and the fact that the results are true when X and Y are locally of bounded variation. In proving the second result we also need Exercise 8.1. □

The rest of this section is devoted to defining the covariance for semimartingales and proving the third of our abc's.

(8.5) Definition. If $X = M + A$ and $X' = M' + A'$ are continuous semimartingales we define the covariance $\langle X, X' \rangle_t = \langle M, M' \rangle_t$.

To explain this we recall the approximate quadratic variation $Q_t^{\Delta}(X)$ defined in Section 2.3, and note

(8.6) Theorem. Suppose $X = M + A$ and $X' = M' + A'$ are continuous semimartingales. If Δ_n is a sequence of partitions of $[0, t]$ with mesh $|\Delta_n| \to 0$ then
$$Q_t^{\Delta_n}(X, X') \to \langle M, M' \rangle_t \text{ in probability}$$

Proof Since
$$Q_t^{\Delta_n}(X, X') = Q_t^{\Delta_n}(M, M') + Q_t^{\Delta_n}(A, M') + Q_t^{\Delta_n}(X, A')$$

it suffices to show that if Y_t is a continuous process and V_t is continuous and locally of bounded variation then $Q_t^{\Delta_n}(Y, V) \to 0$ almost surely. To do this we note that
$$Q_t^{\Delta_n}(Y, V) \leq |V|_t \sup_i |Y_{t_{i+1}^n} - Y_{t_i^n}|$$

As $n \to \infty$ the second term $\to 0$ almost surely, while the first one has $|V|_t < \infty$ and the desired result follows. □

(8.7) Theorem. Suppose X^i and Y^j are continuous semimartingales $H^i, K^j \in lb\Pi$, $X = \sum_{i=1}^{m} H^i \cdot X^i$, $Y = \sum_{j=1}^{n} K^j \cdot Y^j$, then

$$\langle X, Y \rangle_t = \sum_{i,j} \int_0^t H_s^i K_s^j \, d\langle X^i, Y^j \rangle_s$$

Proof Let M^i and N^j be the local martingale parts of X^i and Y^j, let $M = \sum_{i=1}^{m} H^i \cdot M^i$, and $N = \sum_{j=1}^{n} K^j \cdot N^j$. It follows from our definition that $\langle X, Y \rangle_t = \langle M, N \rangle_t$, so using (6.6) and then $\langle M^i, N^j \rangle_t = \langle X^i, Y^j \rangle_t$ gives the desired result. □

2.9. Associative Law

Let X_t be a continuous semimartingale. In some computations, it is useful to write the integral relationship

$$Y_t = \int_0^t K_s \, dX_s$$

as the formal equation

$$dY_t = K_t \, dX_t$$

where the dY_t and dX_t are fictitious objects known as "stochastic differentials." A good example is the derivation of the following formula, which (for obvious reasons) we call the **associative law**:

($\star$) $\qquad\qquad\qquad H \cdot (K \cdot X) = (HK) \cdot X$

Proof using stochastic differentials $d(H \cdot Y)_t = H_t dY_t$. Letting $Y_t = (K \cdot X)_t$ and observing $dY_t = K_t dX_t$ gives

$$d(H \cdot (K \cdot X))_t = H_t \, d(K \cdot X)_t = H_t K_t \, dX_t = d((HK) \cdot X)_t \qquad \square$$

The above proof is not rigorous, but the computation is useful because it tells us what the answer should be. Once we know the answer, it is routine to verify by checking that it holds for basic predictable processes and then following the extension process we used for defining the integral to conclude that it holds in general.

(9.1) Lemma. Let X be any process. If $H, K \in \Pi_1$, then ($\star$) holds.

Proof Since the formula is linear in H and in K separately, we can without loss of generality suppose $H = 1_{(a,b]}C$ and $K = 1_{(c,d]}D$ and further that either (i) $b \leq c$ or (ii) $a = c, b = d$. In case (i), both sides of the equation are $\equiv 0$ and hence equal. In case (ii),

$$(K \cdot X)_t = \begin{cases} 0 & 0 \leq t \leq a \\ D(X_t - X_a) & a \leq t \leq b \\ D(X_b - X_a) & b \leq t < \infty \end{cases}$$

so

$$(H \cdot (K \cdot X))_t = \begin{cases} 0 & 0 \leq t \leq a \\ CD(X_t - X_a) & a \leq t \leq b \\ CD(X_b - X_a) & b \leq t < \infty \end{cases}$$

and it follows that $(H \cdot (K \cdot X))_t = ((HK) \cdot X)_t$. $\qquad \square$

Section 2.9 Associative Law

To extend to more general integrands, we will take limits. First, however, we need a technical result.

(9.2) Lemma. Suppose μ is a signed measure with finite total variation, h and k are bounded, and let $\nu([0,t]) = \int_0^t k_s \mu(ds)$. Then

$$\int_0^t h_s \, \nu(ds) = \int_0^t h_s k_s \, \mu(ds)$$

Proof of (9.2) The Radon-Nikodym derivative $d\nu/d\mu = k_s$ so the desired identity is just the well known fact that

$$\int_0^t h_s \, \nu(ds) = \int_0^t h_s \frac{d\nu}{d\mu} \mu(ds)$$

See for example Exercise 8.7 in the Appendix of Durrett (1995). □

To simplify the extension to $\Pi_2(X)$, we will take the limit for K and then for H.

(9.3) Lemma. Let X be a continuous local martingale. If $H \in b\Pi_1$ and $K \in \Pi_2(X)$, then $(\star)$ holds.

Proof Let $K^n \in b\Pi_1$ such that $K^n \to K$ in $\Pi_2(X)$. Since H is bounded, $HK^n \to HK$ in $\Pi_2(X)$, and it follows from Exercise 6.2 that $(HK^n) \cdot X \to (HK) \cdot X$ in $\mathcal{M}^2$. To deal with the left side of $(\star)$, we observe that using (6.4) twice, Exercise 6.2, (6.5) with (9.2), and the fact that H_s is bounded we have

$$\|H \cdot (K \cdot X) - H \cdot (K^n \cdot X)\|_2^2 = \|H \cdot ((K - K^n) \cdot X)\|_2^2$$
$$= E \int_0^\infty H_s^2 \, d\langle (K - K^n) \cdot X \rangle_s$$
$$= E \int_0^\infty H_s^2 (K_s - K_s^n)^2 \, d\langle X \rangle_s$$
$$\leq C \|K - K^n\|_X^2$$

so as $n \to \infty$, $H \cdot (K^n \cdot X) \to H \cdot (K \cdot X)$ in $\mathcal{M}^2$ and the result follows. □

(9.4) Lemma. Let X be a continuous local martingale. If $H \in \Pi_2(K \cdot X)$ and $K \in \Pi_2(X)$, then $(\star)$ holds.

Proof Let $H^n \in b\Pi_1$ such that $H^n \to H$ in $\Pi_2(K \cdot X)$. (6.4) and Exercise 6.2 imply

$$\|H^n \cdot (K \cdot X) - H \cdot (K \cdot X)\|_2 = \|(H^n - H) \cdot (K \cdot X)\|_2 = \|H^n - H\|_{K \cdot X}$$

so $H^n \cdot (K \cdot X) \to H \cdot (K \cdot X)$ in $\mathcal{M}^2$. To deal with the right side of $(\star)$, we observe that using the definition of the norm then (6.5) with (9.2)

$$\|H^n K - HK\|_X^2 = E \int_0^\infty (H_s^n - H_s)^2 K_s^2 d\langle X\rangle_s$$
$$= \|H^n - H\|_{K \cdot X}^2 \to 0$$

so applying Exercise 6.2 we have $H^n K \cdot X \to HK \cdot X$ in $\mathcal{M}^2$. □

(9.5) Lemma. Let X be a continuous local martingale. If $K \in \Pi_3(Y)$ and $H \in \Pi_3(K \cdot X)$ then $(\star)$ holds.

Proof Let $T_n = \inf\{t : \int_0^t K_s^2 d\langle X\rangle_s \text{ or } \int_0^t H_s^2 d\langle K \cdot X\rangle_s > n\}$. Let $\bar{H}_s^{T_n} = H_s 1_{(s \leq T_n)}$ and $\bar{K}_s^{T_n} = K_s 1_{(s \leq K_n)}$. (9.4) implies that

$$\bar{H}^{T_n} \cdot (\bar{K}^{T_n} \cdot X) = (\bar{H}^{T_n} \bar{K}^{T_n}) \cdot X$$

Using (6.2) now it follows that $(H \cdot (K \cdot X))_t = (HK \cdot X)_t$ for $t \leq T_n$, and letting $n \to \infty$ gives the desired result. □

(9.6) Associative Law. Suppose X is a continuous semimartingale. If $H, K \in lb\Pi$ then $H \cdot (K \cdot X) = HK \cdot X$.

Proof In view of (9.5) and the definition of the integral it suffices to prove the result when $X_t = A_t$ is continuous adapted and locally of bounded variation. However, by stopping the first time $|H_s|$, $|K_s|$ or the variation of A on $[0, s]$ exceeds n the statement reduces to something covered by (9.2). □

2.10. Functions of Several Semimartingales

In this section, we will prove a version of Itô's formula for functions of several semimartingales. The key to the proof is the following

(10.1) Integration by Parts. If X and Y are continuous semimartingales then

$$X_t Y_t - X_0 Y_0 = \int_0^t Y_s dX_s + \int_0^t X_s dY_s + \langle X, Y\rangle_t$$

Proof Let $\Delta_n = \{t_i^n\}$ be a sequence of subdivisions of $[0,t]$ with mesh $|\Delta_n|$ going to 0. We can write

$$X_t Y_t - X_0 Y_0 = \sum_i (X_{t_{i+1}^n} - X_{t_i^n})(Y_{t_{i+1}^n} - Y_{t_i^n})$$
$$+ \sum_i (X_{t_{i+1}^n} - X_{t_i^n}) Y_{t_i^n} + \sum_i X_{t_i^n}(Y_{t_{i+1}^n} - Y_{t_i^n})$$

(8.6) implies that

$$\sum_i (X_{t_{i+1}} - X_{t_i})(Y_{t_{i+1}} - Y_{t_i}) \to \langle X, Y \rangle_t$$

in probability, while application of (6.7) to each of the last two sums implies that they converge to $\int_0^t Y_s dX_s$ and to $\int_0^t X_s dY_s$. □

When Y_t is locally of bounded variation then $\langle X, Y \rangle_t \equiv 0$ and this reduces to the ordinary integration by parts formula of Lebesgue-Stieltjes integration.

$$\int_0^t Y_s dX_s = X_t Y_t - X_0 Y_0 - \int_0^t X_s dY_s$$

Note that the right hand side makes sense in a path-by-path sense.

We are now ready to prove the following generalization of (7.1):

(10.2) Itô's formula. Let $f : \mathbf{R}^d \to \mathbf{R}$ and $0 \leq c \leq d$. If $X_t^1, \ldots, X_t^d$ are continuous semimartingales and $X_t^{c+1}, \ldots, X_t^d$ are locally of bounded variation then

$$f(X_t) - f(X_0) = \sum_{i=1}^d \int_0^t D_i f(X_s) dX_s^i + \frac{1}{2} \sum_{1 \leq i,j \leq c} \int_0^t D_{ij} f(X_s) d\langle X^i, X^j \rangle_s$$

provided the derivatives $D_j f$, $1 \leq j \leq d$ and $D_{ij} f$, $1 \leq i, j, \leq c$ exist and are continuous.

Remark. At this point saving a derivative or two may seem like pinching pennies but when we get to Chapter 4 and apply this result to $u(t, B_t)$ we will be very happy to not have to check that the partial derivatives u_{tt} and u_{t,x_i} exist.

Proof By stopping, it suffices to prove the result when $|X_t^i|, \langle X^i \rangle_t \leq M$ for all i, t. Since any function f satisfying the hypotheses can be approximated by

78   Chapter 2   Stochastic Integration

polynomials g_n in such a way that $g_n, D_i g_n, 1 \leq i \leq d$ and $D_{ij} g_n, 1 \leq i, j \leq c$ converge to $f, D_i f$, and $D_{ij} f$ uniformly on $[-M, M]^d$, it suffices to prove the result when f is a polynomial. For then (6.8) and Exercise 7.2 allow us to pass from the result for g_n to that for f.

To prove the result for a polynomial, it suffices by linearity to prove the result when f is a monomial $x^{k_1} x^{k_2} \cdots x^{k_n}$ where $k_1, k_2, \ldots, k_n \in \{1, \ldots, d\}$. (The reader should note that $k_1, \ldots, k_n$ are superscripts, not powers, e.g., our monomial might be $x^1 x^4 x^1 x^1 x^2$.)

If $n = 1$ and $k_1 = k$, then (10.2) says that

$$X_t^k - X_0^k = \int_0^t 1 \, dX_s^k$$

which is trivially true. To prove the result for a general monomial, we use induction. Let $Y_t = \prod_{m=1}^n X_t^{k(m)}$ be a monomial for which (10.2) holds, and let $Z_t = X_t^{k(n+1)}$. Applying (10.1) gives

$$Y_t Z_t - Y_0 Z_0 = \int_0^t Z_s \, dY_s + \int_0^t Y_s \, dZ_s + \langle Y, Z \rangle_t$$

Applying (10.2) to Y gives

(10.3)
$$Y_t - Y_0 = \sum_{i \leq n} \int_0^t \left(\prod_{\substack{m=1 \\ m \neq i}}^n X_s^{k(m)} \right) dX_s^{k(i)}$$
$$+ \frac{1}{2} \sum_{\substack{i,j \leq n \\ i \neq j}} \int_0^t \left(\prod_{\substack{m=1 \\ m \neq i,j}}^n X_s^{k(m)} \right) d\langle X^{k(i)}, X^{k(j)} \rangle_s$$

so using the associative law (9.6),

$$\int_0^t Z_s \, dY_s = \sum_{i \leq n} \int_0^t \left(\prod_{\substack{m=1 \\ m \neq i}}^{n+1} X_s^{k(m)} \right) dX_s^{k(i)}$$
$$+ \frac{1}{2} \sum_{\substack{i,j \leq n \\ i \neq j}} \int_0^t \left(\prod_{\substack{m=1 \\ m \neq i,j}}^{n+1} X_s^{k(m)} \right) d\langle X^{k(i)}, X^{k(j)} \rangle_s$$

By definition,

$$\int_0^t Y_s \, dZ_s = \int_0^t \left(\prod_{m=1}^n X_s^{k(m)} \right) dX_s^{k(n+1)}$$

To evaluate the third term $\langle Y, Z \rangle_t$, we observe that by (10.3) and the formula for the covariance of stochastic integrals, (8.7),

$$\langle Y, Z \rangle_t = \sum_{i \leq n} \int_0^t \left(\prod_{\substack{m=1 \\ m \neq i}}^{n} X_s^{k(m)} \right) d\langle X^{k(i)}, X^{k(n+1)} \rangle_s$$

Adding the last three equalities gives

$$Y_t Z_t - Y_0 Z_0 = \sum_{i \leq n+1} \int_0^t \left(\prod_{\substack{m=1 \\ m \neq i}}^{n+1} X_s^{k(m)} \right) dX_s^{k(i)}$$

$$+ \frac{1}{2} \sum_{\substack{i,j \leq n+1 \\ m \neq i,j}} \int_0^t \left(\prod_{\substack{m=1 \\ m \neq i,j}}^{n+1} X_s^{k(m)} \right) d\langle X^{k(i)}, X^{k(j)} \rangle_s$$

(Notice that for each i in the sum for $\langle Y, Z \rangle_t$, there are two terms $i = i, j = n+1$, and $i = n+1, j = i$ in the last sum.) The proof is complete. □

Chapter Summary

With the completion of the proof of the multivariate form of Itô's formula, we have enough stochastic calculus to carry us through most of the applications in the book, so we pause for a moment to recall what we've just done.

In Section 2.1, we introduced the integrands H_s, the predictable processes and in Section 2.2 the (first) integrators X_t, the continuous local martingales. In Section 2.3 we defined the variance process associated with a local martingale, and more generally the covariance $\langle X, Y \rangle_t$ of two local martingales. Theorem (3.11) tells us that the covariance could have been defined as the unique continuous predictable process A_t that is locally of bounded variation, has $A_0 = 0$, and makes $X_t Y_t - A_t$ is a martingale. The next result gives a pathwise interpretation of $\langle X, Y \rangle_t$:

(3.10) Quadratic Variation. Suppose X and Y are continuous local martingales. If Δ_n is a sequence of partitions of $[0, t]$ with mesh $|\Delta_n| \to 0$ then

$$\sum_{k=1}^{k_n} (X_{t_k} - X_{t_{k-1}}) \cdot (Y_{t_k} - Y_{t_{k-1}}) \to \langle X, Y \rangle_t \text{ in probability}$$

Taking $X = Y$ we see that $\langle X \rangle_t$ is the "quadratic variation" of the path up to time t.

80 Chapter 2 Stochastic Integration

In Section 2.4–2.5, we defined the integral, beginning with bounded continuous martingales and considering a sequence of progressively more general integrands.

$$\Pi_0 \quad 1_{(a,b]}(s)C(\omega) \text{ with } C \in \mathcal{F}_a$$
$$\Pi_1 \quad \text{finite sum of elements of } \Pi_0$$
$$\Pi_2(X) \quad H \text{ with } E\int_0^\infty H_s^2 \, d\langle X\rangle_s < \infty$$
$$\Pi_3(X) \quad H \text{ with } \int_0^t H_s^2 \, d\langle X\rangle_s < \infty \text{ a.s. for each } t$$

A key role in this development was played by the

(5.1) Kunita-Watanabe Inequality. If X and Y are local martingales and H and K are two measurable processes, then almost surely

$$\int_0^\infty |H_s K_s| \, d|\langle X,Y\rangle|_s \leq \left(\int_0^\infty H_s^2 \, d\langle X\rangle_s\right)^{1/2} \left(\int_0^\infty K_s^2 \, d\langle Y\rangle_s\right)^{1/2}$$

where $d|\langle X,Y\rangle|_s$ stands for dV_s where V_s is the total variation of $r \to \langle X,Y\rangle_r$ on $[0,s]$.

In Section 2.6, we generalized the integrators X_s to continuous local martingales and, in Section 2.8, to continuous **semimartingales** which are the sum (in a unique way) of a continuous local martingale M_s and a continuous process A_s that is locally of bounded variation. Further, we defined the covariance of two semimartingales $X_t = M_t + A_t$ and $X_t' = M_t' + A_t'$ to be that of their martingale parts, i.e., $\langle M, M'\rangle_t$, and generalized the quadratic variation interpretation in (3.10), see (8.6).

To be able to integrate with respect to semimartingales we had to restrict the integrands to the class of locally bounded predictable processes, $\ell b\Pi$, but in this generality the integral has many nice properties.

(8.3) Closure under Integration. If X is a continuous semimartingale and $H \in \ell b\Pi$, then $(H \cdot X)_t$ is a continuous semimartingale.

(8.4) Distributive Laws. Let X and Y be continuous semimartingales, and $H, K \in \ell b\Pi$.

$$((H+K) \cdot X)_t = (H \cdot X)_t + (K \cdot X)_t$$
$$(H \cdot (X+Y))_t = (H \cdot X)_t + (H \cdot Y)_t$$

(8.7) **Covariance of Stochastic Integrals.** Suppose X^i and Y^j are continuous semimartingales $H^i, K^j \in lb\Pi$, $X = \sum_{i=1}^{m} H^i \cdot X^i$, $Y = \sum_{j=1}^{n} K^j \cdot Y^j$, then

$$\langle X, Y \rangle_t = \sum_{i,j} \int_0^t H_s^i K_s^j d\langle X^i, Y^j \rangle_s$$

(9.6) **Associative Law.** Let X be a continuous semimartingale. If $H, K \in lb\Pi$ then $H \cdot (K \cdot X) = HK \cdot X$.

As the reader probably remembers, the first three results were only obtained after several improvements

Integrands =	Π_1	Π_2	Π_3	$lb\Pi$
a. Closure	(4.2.a)	(4.3.a)	(6.3)	(8.3)
b. Distributive Laws	(4.2.b)	(4.3.b), (5.3)	(6.4)	(8.4)
c. Covariance Formula	(4.2.c)	(5.4)	(6.5), (6.6)	(8.6), (8.7)

In Section 2.7, we proved a change of variables formula. This led to:

(10.1) **Integration by Parts.** Let X and Y be continuous semimartingales.

$$X_t Y_t - X_0 Y_0 = \int_0^t Y_s dX_s + \int_0^t X_s dY_s + \langle X, Y \rangle_t$$

which in turn led to a bigger and better change of variables formula.

(10.2) **Itô's formula.** Let $f : \mathbf{R}^d \to \mathbf{R}$ and $0 \le c \le d$. If $X_t^1, \ldots, X_t^d$ are continuous semimartingales and $X_t^{c+1}, \ldots, X_t^d$ are locally of bounded variation

$$f(X_t) - f(X_0) = \sum_{i=1}^{d} \int_0^t D_i f(X_s) dX_s^i + \frac{1}{2} \sum_{1 \le i,j \le c} \int_0^t D_{ij} f(X_s) d\langle X^i, X^j \rangle_s$$

provided the derivatives $D_j f$, $1 \le j \le d$ and $D_{ij} f$, $1 \le i,j, \le c$ exist and are continuous.

Coming Attractions

Section 2.11 is devoted to a proof of the Meyer-Tanaka formula, which extends the version of Itô's formula given in (7.1) to f that are a difference of two convex functions. This leads in a nice way to the definition of an occupation time density (or "local time") L_t^a for a continuous semimartingale that is jointly continuous in t and a.

The developments described in the previous paragraph are used only in (9.3) of Chapter 4 and in Example 2.1 of Chapter 5. The material in Section 2.12 concerning Girsanov's formula is not used until Section 5.5. We would like to suggest that the reader proceed now to the beginning of Chapter 3.

2.11. The Meyer-Tanaka Formula, Local Time

In this section we will prove an extension of Itô's formula due to Meyer (1976) that has its roots in work of Tanaka (1963). See (11.4). Loosely speaking it says that Itô's formula is valid if f is a difference of two convex functions. Since such functions are differentiable except at a countable set and have a second derivative that is a nice signed measure, this is a modest generalization. However, this is the best one can do in general: if B_t is a standard Brownian motion and $X_t = f(B_t)$ is a semimartingale then f must be the difference of two convex functions. See Cinlar, Jacod, Protter, and Sharpe (1980). The developments in this section follow those in Sections 4 and 5 of Chapter IV of Protter (1990) but things are simpler here since we have no jumps. Our first step is

(11.1) Theorem. Let $f : \mathbf{R} \to \mathbf{R}$ be convex and let X be a continuous semimartingale. Then $f(X)$ is a semimartingale and

$$f(X_t) - f(X_0) = \int_0^t f'(X_s)\,dX_s + K_t$$

Here $f'(x) = \lim_{h\downarrow 0}(f(x) - f(x-h))/h$ is the left derivative which exists at all x by convexity and K_t is a continuous adapted increasing process.

Proof Let $X_t = M_t + A_t$ be the decomposition of the semimartingale. By stopping we can suppose without loss of generality that $|X_t| \leq N$, $|A|_t \leq N$ and $\langle M \rangle_t \leq N$ for all t. Let g be a C^∞ function with compact support in $[-\infty, 0]$ and having $\int g(s)\,ds = 1$. Let

(a) $$f_n(x) = \int f\left(x + \frac{y}{n}\right) g(y)\,dy$$

Then f_n is convex and C^∞. Using Itô's formula we have

(b) $$f_n(X_t) - f_n(X_0) = \int_0^t f_n'(X_s)\,dX_s + \frac{1}{2}\int_0^t f_n''(X_s)\,d\langle X \rangle_s$$

Our strategy for the proof will be to let $n \to \infty$ in (b) to get the desired formula. Since a convex function is Lipschitz continuous on each bounded

Section 2.11 The Meyer-Tanaka Formula, Local Time   83

interval, it is easy to see that $f_n(x) \to f(x)$ uniformly on compact sets. Since $|X_t| \le N$, it follows that

(c) $$f_n(X_t) \to f(X_t) \quad \text{uniformly in } t$$

To deal with the first term on the right, we differentiate (a) to get

(d) $$f_n'(x) = \int f'\left(x + \frac{y}{n}\right) g(y)\, dy \uparrow f'(x)$$

as $n \to \infty$. Now if $I_t^n = \int_0^t f_n'(X_s)\, dM_s$ and $I_t = \int_0^t f'(X_s)\, dM_s$, then the L^2 maximal inequality for martingales and the isometry property of stochastic integrals (Exercise 6.2) imply

$$E\left(\sup_t (I_t^n - I_t)^2\right) \le 4 \sup_t E(I_t^n - I_t)^2$$
$$= 4E \int_0^\infty (f_n'(X_s) - f'(X_s))^2\, d\langle M\rangle_s \to 0$$

by the bounded convergence theorem. (Recall $\langle M\rangle_\infty \le N$.) By passing to a subsequence we can improve the last conclusion to

(e) $$\sup_t (I_t^{n_k} - I_t)^2 \to 0 \quad \text{almost surely}$$

Let $J_t^n = \int_0^t f_n'(X_s)\, dA_s$ and $J_t = \int_0^t f'(X_s)\, dA_s$. In this case it follows from the bounded convergence theorem that for almost every ω

(f) $$\sup_t |J_t^n - J_t| \le \int_0^\infty |f_n'(X_s) - f'(X_s)|\, dA_s \to 0$$

To take the limit of the third and final term $K_t^n = \frac{1}{2}\int_0^t f_n''(X_s)\, d\langle X\rangle_s$, we note that (b) implies

(g) $$K_t^{n_k} = f_{n_k}(X_t) - f_{n_k}(X_0) - \int_0^t f_{n_k}'(X_s)\, dX_s$$

Combining (c), (e), and (f) we see that almost surely the right hand side of (g) converges to a limit uniformly in t, so the left hand side does also. Since $K_t^{n_k}$ is continuous adapted and increasing the limit is also. □

If we fix X then for a given f we call K the **increasing process associated with** f. The increasing process associated with $|x - a|$ is called the **local**

84   Chapter 2   Stochastic Integration

time at a, and is denoted by L_t^a. To prove the first property that justifies this name, (11.3), we need the following preliminary

(11.2) Lemma. The increasing process associated with $(x-a)^+$ or $(x-a)^-$ is $(1/2)L_t^a$.

Proof Let $f_1(x) = (x-a)^+$, $f_2(x) = (x-a)^-$, and let K_t^i be the increasing process associated with f_i. We begin by observing $f_1+f_2 = |x-a|$, so $K_t^1+K_t^2 = L_t^a$. Our second claim is that since $f_1 - f_2 = x - a$ we have $K_t^1 - K_t^2 = 0$. To see this let $g(x) = x - a$ which has $g'(x) = 1$ and note $X_t - X_0 = \int_0^t 1\,dX_s$ so the associated K_t must be $\equiv 0$. □

(11.3) Theorem. L_t^a only increases when $X_t = a$ or to be precise, if we let ℓ^a be the measure with distribution function $t \to L_t^a$, then ℓ^a is supported by $\{t : X_t = a\}$.

Proof Intuitively, $|X_t - a|$ is a local martingale except when $X_t = a$ so L_t^a is constant when $X_t \neq a$. To prove (11.3), however, it is easier to use the alternative definitions in (11.2). Let $S < T$ be stopping times so that $[S,T] \subset \{t : X_t < a\}$. Applying (11.1) to $f(x) = (x-a)^+$ we have

$$(X_T - a)^+ - (X_S - a)^+ = \int_S^T 1_{\{X_s > a\}}\,dX_s + \frac{1}{2}(L_T^a - L_S^a)$$

The left-hand side and the integral on the right-hand side vanish so $L_T^a = L_S^a$. Since this holds when $S \equiv q$ with q any rational and $T = \inf\{t > S : X_t \geq a - 1/n\}$ for any n, it follows that $\ell^a(\{t : X_t < a\}) = 0$. A similar argument using $(x-a)^-$ instead of $(x-a)^+$ shows $\ell^a(\{t : X_t > a\}) = 0$ and the proof is complete. □

We are now ready for our main result.

(11.4) Meyer-Tanaka formula. Let X be a continuous semimartingale. Let f be the difference of two convex functions, f' be the left derivative of f and suppose $f'' = \mu$ in the sense of distribution (i.e., μ has distribution function f'). Then

$$f(X_t) - f(X_0) = \int_0^t f'(X_s)\,dX_s + \frac{1}{2}\int \mu(da) L_t^a$$

Proof Since f is the difference of two convex functions and the formula is linear in f, we can suppose without loss of generality that f is convex. By

Section 2.11 The Meyer-Tanaka Formula, Local Time

stopping we can suppose $|X_t| \leq N$ and $\langle X \rangle_t \leq N$ for all t. Having done this we can let

$$g(x) = \frac{1}{2} \int_{-N}^{N} \mu(da)|x - a|$$

Since $(f - g)'' = 0$ on $[-N, N]$ it follows that $f(x) - g(x) = a + bx$ for $|x| \leq N$. Since the result is trivial for linear functions, it suffices now to prove the result for g, which is almost trivial. Differentiating the definition of g we have

$$g'(x) = \frac{1}{2} \int_{-N}^{N} \mu(da) \operatorname{sign}(x - a)$$

Starting with the definition of the local time

$$|X_t - a| - |X_0 - a| = \int_0^t \operatorname{sign}(X_s - a) \, dX_s + L_t^a$$

and then integrating $1/2 \int_{-N}^{N} \mu(da)$ gives

$$g(X_t) - g(X_0) = \frac{1}{2} \int_{-N}^{N} \mu(da) \int_0^t \operatorname{sign}(X_s - a) \, dX_s$$
$$+ \frac{1}{2} \int_{-N}^{N} \mu(da) L_t^a$$

(11.5) and (11.6) below will justify interchanging the two integrals in the first term on the right-hand side to give

$$g(X_t) - g(X_0) = \int_0^t g'(X_s) \, dX_s + \frac{1}{2} \int_{-N}^{N} \mu(da) L_t^a$$

Since $L_t^a = 0$ for $|a| > N$ (recall $|X_t| \leq N$ and use (11.3)) this proves the result for g and completes the proof of (11.4). □

For the next two results let S be a set, let $\mathcal{S}$ be a σ-field of subsets of S, and let μ be a finite measure on $(S, \mathcal{S})$. For our purposes it would be enough to take $S = \mathbf{R}$, but it is just as easy to treat a general set. The first step in justifying the interchange of the order of integration is to deal with a measurability issue: if we have a family $H_t^a(\omega)$ of integrands for $a \in S$ then we can define the integrals $\int_0^t H_s^a \, dX_s$ to be a measurable function of (a, t, ω).

(11.5) **Lemma.** Let X be a continuous semimartingale with $X_0 = 0$ and let $H_t^a(\omega) = H(a, t, \omega)$ be bounded and $\mathcal{S} \times \Pi$ measurable. Then there is a

$Z(a,t,\omega) \in \mathcal{S} \times \Pi$ such that for μ-almost every a $Z(a,t,\omega)$ is a continuous version of $\int_0^t H_s^a \, dX_s$.

Proof Let $X_t = M_t + A_t$ be the decomposition of the semimartingale. By stopping we can suppose that $|A|_t \leq N$, $|M_t| \leq N$ and $\langle X \rangle_t = \langle M \rangle_t \leq N$ for all t.

Let $\mathcal{H}$ be the collection of bounded $H \in \mathcal{S} \times \Pi$ for which the conclusion holds. We will check the assumptions of the Monotone Class Theorem, (2.3) in Chapter 1. Clearly $\mathcal{H}$ is a vector space. To check (i) of the MCT suppose $H(a,t,\omega) = f(a)K(t,\omega)$ where $K(t,\omega) \in b\Pi$ and $f(a) \in b\mathcal{S}$. In this case

$$\int_0^t H_s^a \, dX_s = f(a) \int_0^t K_s \, dX_s$$

so $fK \in \mathcal{H}$. Taking f and K to be indicator functions of sets in $\mathcal{S}$ and Π respectively, we have checked (i) of the MCT with $\mathcal{A} = \{A \times B : A \in \mathcal{S}, B \in \Pi\}$.

To check (ii) of the MCT let $0 \leq H^n \in \mathcal{H}$ with $H^n \uparrow H$ and H bounded. The L^2 maximal inequality for martingales and the isometry property of the stochastic integral imply

$$E\left(\sup_t |(H^{n,a} \cdot M)_t - (H^a \cdot M)_t|^2\right) \leq 4 \sup_t E|(H^{n,a} \cdot M)_t - (H^a \cdot M)_t|^2$$

$$= 4E \int (H_s^{n,a} - H_s^a)^2 \, d\langle M \rangle_s \to 0$$

By passing to a subsequence we have for μ almost every a

$$\sup_t |(H^{n_k,a} \cdot M)_t - (H^a \cdot M)_t| \to 0$$

To deal with the bounded variation part we note that

$$E \sup_t |(H^{n,a} \cdot A)_t - (H^a \cdot A)_t| \leq E \int_0^\infty |H_t^{n,a} - H_t^a| \, d|A|_t \to 0$$

as $n \to \infty$ by the bounded convergence theorem. Combining the last two results completes the proof. $\square$

(11.6) Fubini's Theorem. Let X be a continuous semimartingale, let $H_t^a \in b(\mathcal{S} \times \Pi)$, and let $Z_t^a \in \mathcal{S} \times \Pi$ be a continuous version of $\int_0^t H_s^a \, dX_s$ for μ-almost every a. Then $Y_t = \int_S Z_t^a \, \mu(da)$ is a version of $H \cdot X$ where $H_t = \int_S H_t^a \, \mu(da)$. Less formally,

$$\int_S \int_0^t H_s^a \, dX_s \, \mu(da) = \int_0^t \int_S H_s^a \, \mu(da) \, dX_s$$

Section 2.11 The Meyer-Tanaka Formula, Local Time

Proof Let $X_t = M_t + A_t$ be the decomposition of the semimartingale. By stopping we can suppose that $|A|_t \leq N$, $|M_t| \leq N$ and $\langle X \rangle_t = \langle M \rangle_t \leq N$ for all t. As noted in the proof of (11.5), the result in the special case $X = A$ is just Fubini's theorem from measure theory, so we will suppose for the rest of the proof that $X = M$.

Let $\mathcal{H}$ be the collection of bounded $H \in \mathcal{S} \times \Pi$ for which the conclusion holds. Again, we will check the assumptions of the Monotone Class Theorem, (2.3) in Chapter 1. Clearly $\mathcal{H}$ is a vector space. To check (i) of the MCT, suppose $H(a,t,\omega) = f(a)K(t,\omega)$ where $K(t,\omega) \in b\Pi$ and $f(a) \in b\mathcal{S}$. In this case $Z_t^a = f(a)(K \cdot X)_t$ so

$$\int_S Z_t^a \, \mu(da) = (K \cdot X)_t \int_S f(a) \, \mu(da)$$

$$= (\{\int_S f(a)\,\mu(da)\, K\} \cdot X)_t = (H \cdot X)_t$$

so $fK \in \mathcal{H}$. Taking f and K to be indicator functions of sets in $\mathcal{S}$ and Π respectively, we have checked (i) of the MCT with $\mathcal{A} = \{A \times B : A \in \mathcal{S}, B \in \Pi\}$.

To check (ii) of the MCT let $0 \leq H^n \in \mathcal{H}$ with $H^n \uparrow H$ and H bounded. Letting $\|\mu\|$ be the total mass of μ applying Jensen's inequality to the probability measure $\mu(\cdot)/\|\mu\|$, then multiplying each side by $\|\mu\|$ we have

$$\frac{1}{\|\mu\|} E\left(\int_S \sup_t |Z_t^{n,a} - Z_t^a| \,\mu(da)\right)^2$$

$$\leq E \int_S \sup_t |Z_t^{n,a} - Z_t^a|^2 \,\mu(da)$$

Using Fubini's theorem, the L^2 maximal inequality, and the isometry property the right-hand side

$$= \int_S E \sup_t |Z_t^{n,a} - Z_t^a|^2 \,\mu(da)$$

$$\leq 4 \int_S \sup_t E|Z_t^{n,a} - Z_t^a|^2 \,\mu(da)$$

$$= 4 \int_S E\left(\int_0^\infty (H_s^{n,a} - H_s^a)^2 \, d\langle X \rangle_s \right) \mu(da) \to 0$$

by using the bounded convergence theorem three times. Putting the absolute values inside the integral we have

$$E\left(\sup_t \left|\int_S Z_t^{n,a}\mu(da) - \int_S Z_t^a\mu(da)\right|\right)^2$$

$$\leq E\left(\int_S \sup_t |Z_t^{n,a} - Z_t^a|\mu(da)\right)^2 \to 0$$

Taking $H_t^n = \int_S H_t^{n,a} \mu(da)$ and passing to a subsequence n_k we have that with probability one $(H^{n_k} \cdot X)_t = \int_S Z_t^{n_k,a} \mu(da)$ converges uniformly to $\int_S Z_t^a \mu(da)$. The last detail is to check that $H^n \cdot X$ converges to $H \cdot X$ in $\mathcal{M}^2$, for then it follows that $\int_S Z_t^a \mu(da) = (H \cdot X)_t$. In view of the isometry property, we can complete the last detail by showing that $\|H^n - H\|_X \to 0$ but this is routine.

$$\frac{1}{\|\mu\|} E \int_0^\infty \left(\int_S H_s^{n,a} \mu(da) - \int_S H_s^a \mu(da) \right)^2 d\langle X \rangle_s$$

$$\leq E \int_0^\infty \int_S (H_s^{n,a} - H_s^a)^2 \mu(da) \, d\langle X \rangle_s \to 0$$

by using the bounded convergence theorem three times. □

We can now complete the identification of L_t^a as the local time at a.

(11.7) Theorem. Let X be a continuous semimartingale with local time L_t^a. If g is a bounded Borel measurable function then

$$\int_{-\infty}^\infty L_t^a g(a) \, da = \int_0^t g(X_s) d\langle X \rangle_s$$

Proof Suppose first that g is continuous and let $f \in C^2$ with $f'' = g$. In this case comparing (11.4) with Itô's formula proves the identity in question. Since the identity holds for any continuous function, using the Monotone Class Theorem proves the result for a bounded measurable g. □

When X is a Brownian motion the identity in (11.7) becomes

$$\int_{-\infty}^\infty L_t^a g(a) \, da = \int_0^t g(X_s) \, ds$$

From this, we see that L_t^a can be thought of as the time spent at a up to time t, or to be precise it is a density function for the occupation time measure $\nu_t(A) = \int_0^t 1_A(X_s) \, ds$.

There are many ways of defining local time for Brownian motion. The one we have given above is famous for making it easy to prove that there is a version in which L_t^a is a jointly continuous function of a and t. It is somewhat remarkable that this property is true for a continuous local martingale and it is not any more difficult to prove the result in this generality.

(11.8) Theorem. Let X be a continuous local martingale. There is a version of the L_t^a in which $(a,t) \to L_t^a$ is continuous.

Section 2.11 The Meyer-Tanaka Formula, Local Time

Example 11.1. (11.8) is not true for $X_t = |B_t|$, which is a semimartingale by (11.1). Introducing subscripts to indicate the process we are referring to, we clearly have $L_X^a(t) = L_B^a(t) + L_B^{-a}(t)$ for $a > 0$ and $L_X^a(t) = 0$ for $a < 0$. Since $L_B^0(t) \not\equiv 0$ it follows that $a \to L_X^a(t)$ is discontinuous at $a = 0$.

One can complain that this example is trivial since 0 is an end point of the set of possible values. A related example with 0 in the middle is provided by **skew Brownian motion**. Start with $|B_t|$ and flip coins with probability $p > 1/2$ of $+$ and $1 - p$ of $-$ to assign signs to excursions of B_t, i.e., maximal time intervals (a, b) on which $|B_t| > 0$. If we call the resulting process Y_t then $L_Y^a(t) \to 2pL_B^0(t)$ as $a \downarrow 0$ and $L_Y^a(t) \to 2(1 - p)L_B^0(t)$ as $a \uparrow 0$. For details see Walsh (1978).

Proof As usual, by stopping we can suppose that $|X_t| \le N$ and $\langle X \rangle_t \le N$ for all t. By (11.2)

$$\frac{1}{2}L_t^a = (X_t - a)^+ - (X_0 - a)^+ - \int_0^t 1_{\{X_s > a\}}\, dX_s$$

It is clear that $(a, t) \to (X_t - a)^+ - (X_0 - a)^+$ is continuous, so we only have to investigate the joint continuity of the stochastic integral

$$I_t^a = \int_0^t 1_{\{X_s > a\}}\, dX_s$$

Fix a time $T < \infty$ and regard $a \to I^a$ as a mapping from $\mathbf{R}$ to $C([0, T], \mathbf{R})$, the real valued functions continuous on $[0, T]$, which we equip with the sup norm $\|f\| = \sup_{0 \le t \le T} |f(s)|$. In view of Kolmogorov's continuity theorem ((1.6) in Chapter 1) it suffices to show that for some $\alpha, \beta > 0$ we have

$$E\|I_\cdot^a - I_\cdot^b\|^\beta \le C|a - b|^{1+\alpha}$$

Suppose without loss of generality that $a < b$. One of the Burkholder Davis Gundy inequalities ((5.1) in Chapter 3) implies that

$$E\|I_\cdot^a - I_\cdot^b\|^4 \le CE\left(\int_0^T 1_{\{a < X_s \le b\}}\, d\langle X \rangle_s\right)^2$$

To bound the right-hand side we note that using (11.7) and then the Cauchy-Schwarz inequality and Fubini's theorem:

$$= E\left(\int_a^b L_T^x\, dx\right)^2 \le (b - a)E\int_a^b (L_T^x)^2\, dx$$

$$= (b - a)\int_a^b E(L_T^x)^2\, dx \le (b - a)^2 \sup_{x \in (a, b)} E(L_T^x)^2$$

To bound the sup, we recall the definition of L_T^x.

$$L_T^x = |X_T - x| - |X_0 - x| - \int_0^T \text{sgn}(X_s - x)\,dX_s$$

which with the trivial inequalities $(a+b)^2 \le 2a^2 + 2b^2$ and $|y-x|-|z-x| \le |y-z|$ implies that

$$E(L_T^x)^2 \le 2E(X_T - X_0)^2 + 2E\left(\int_0^T \text{sign}(X_s - x)\,dX_s\right)^2$$

$$\le 2(2N)^2 + 2E(\langle X \rangle_T) \le 8(N^2 + N)$$

by the isometry property and the bounds we have imposed on $|X_t|$ and $\langle X \rangle_t$ by stopping. This gives us what we need to check the inequality in Kolmogorov's continuity theorem and completes the proof. □

Remark. The proof above almost works for semimartingales $X_t = M_t + A_t$. If we define $I_t^a = \int_0^t 1_{\{X_s > a\}}\,dM_s$ and $J_t^a = \int_0^t 1_{\{X_s > a\}}\,dA_s$, then the argument above shows that $(a,t) \to I_t^a$ is continuous, so we will get the desired conclusion if we adopt assumptions that imply $(a,t) \to J_t^a$ is continuous.

2.12. Girsanov's Formula

In this section, we will show that the collection of semimartingales and the definition of the stochastic integral are not affected by a locally equivalent change of measure. For concreteness, we will work on the canonical probability space $(C, \mathcal{C})$, with $\mathcal{F}_t$ the filtration generated by the coordinate maps $X_t(\omega) = \omega_t$. Two measures Q and P defined on a filtration $\mathcal{F}_t$ are said to be **locally equivalent** if for each t their restrictions to $\mathcal{F}_t$, Q_t and P_t are equivalent, i.e., mutually absolutely continuous. In this case we let $\alpha_t = dQ_t/dP_t$. The reasons for our interest in this quantity will become clear as the story unfolds.

(12.1) Lemma. Y_t is a (local) martingale/Q if and only if $\alpha_t Y_t$ is a (local) martingale/P.

Proof The parentheses are meant to indicate that the statement is true if the two locals are removed. Let Y_t be a martingale/Q, $s < t$ and $A \in \mathcal{F}_s$. Now if $Z \in \mathcal{F}_t$, then (i) $\int Z\,dP = \int Z\,dP_t$, and (ii) $\int Z\alpha_t\,dP_t = \int Z\,dQ_t$. So using (i), (ii), the fact that Y is a martingale/Q, (ii), and (i) we have

$$\int_A \alpha_t Y_t\,dP = \int_A \alpha_t Y_t\,dP_t = \int_A Y_t\,dQ_t$$

$$= \int_A Y_s\,dQ_s = \int_A \alpha_s Y_s\,dP_s = \int_A \alpha_s Y_s\,dP$$

Section 2.12 Girsanov's Formula 91

This shows $\alpha_t Y_t$ is a martingale/P. If Y is a local martingale/Q then there is a sequence of stopping times $T_n \uparrow \infty$ so that $Y_{t \wedge T_n}$ is a martingale/Q and hence $\alpha_t Y_{t \wedge T_n}$ is a martingale/P. The optional stopping theorem implies that $\alpha_{t \wedge T_n} Y_{t \wedge T_n}$ is a martingale/P and it follows that $\alpha_t Y_t$ is a local martingale/P.

To prove the converse, observe that (a) interchanging the roles of P and Q and applying the last result shows that if $\beta_t = dP_t/dQ_t$, and Z_t is a (local) martingale/P, then $\beta_t Z_t$ is a (local) martingale/Q and (b) $\beta_t = \alpha_t^{-1}$, so letting $Z_t = \alpha_t Y_t$ we have the desired result. □

Since 1 is a martingale/Q we have

(12.2) **Corollary.** $\alpha_t = dQ_t/dP_t$ is a martingale/P.

There is a converse of (12.2) which will be useful in constructing examples.

(12.3) **Lemma.** Given α_t a nonnegative martingale/P there is a unique locally equivalent probability measure Q so that $dQ_t/dP_t = \alpha_t$.

Proof The last equation in the lemma defines the restriction of Q to $\mathcal{F}_t$ for any t. To see that this defines a unique measure on $\mathcal{C}$, let $t_1 < t_2 \ldots < t_n$, A_i be Borel subsets of $\mathbf{R}^d$, and let $B = \{\omega : \omega(t_i) \in A_i \text{ for } 1 \leq i \leq n\}$. Define the finite dimensional distributions of a measure Q by setting

$$Q(B) = \int_B \alpha_t \, dP$$

whenever $t \geq t_n$. The martingale property of α_t implies that the finite dimensional distributions are consistent so we have defined a unique measure on $(C, \mathcal{C})$. □

We are ready to prove the main result of the section.

(12.4) **Girsanov's formula.** If X is a local martingale/P and we let $A_t = \int_0^t \alpha_s^{-1} d\langle \alpha, X \rangle_s$, then $X_t - A_t$ is a local martingale/Q.

Proof Although the formula for A looks a little strange, it is easy to see that it must be the right answer. If we suppose that A is locally of b.v. and has $A_0 = 0$, then integrating by parts, i.e., using (10.1), and noting $\langle \alpha, A \rangle_t = 0$ since A has bounded variation gives

(12.5) $\alpha_t(X_t - A_t) - \alpha_0 X_0 = \int_0^t (X_s - A_s) d\alpha_s + \int_0^t \alpha_s dX_s - \int_0^t \alpha_s dA_s + \langle \alpha, X \rangle_t$

92   Chapter 2   Stochastic Integration

At this point, we need the assumption made in Section 2.2 that our filtration only admits continuous martingales, so there is a continuous version of α to which we can apply our integration-by-parts formula.

Since α_s and X_s are local martingales/P, the first two terms on the right in (12.5) are local martingales/P. In view of (12.1), if we want $X_t - A_t$ to be a local martingale/Q, we need to choose A so that the sum of the third and fourth terms $\equiv 0$, that is,

$$\int_0^t \alpha_s \, dA_s = \langle \alpha, X \rangle_t$$

From the last equation and the associative law (9.6), it is clear that it is necessary and sufficient that

$$A_t = \int_0^t \alpha_s^{-1} d\langle \alpha, X \rangle_s$$

The last detail remaining is to prove that the integral that defines A_t exists. Let $T_n = \inf\{t : \alpha_t \leq n^{-1}\}$. If $t \leq T_n$, then

$$\int_0^t \frac{d|\langle \alpha, X \rangle|_s}{\alpha_s} \leq n \int_0^t d|\langle \alpha, X \rangle|_s < \infty$$

by the Kunita-Watanabe inequality. So if $T = \lim_{n\to\infty} T_n$ then A_t is well defined for $t \leq T$. The optional stopping theorem implies that $E\alpha_{t \wedge T_n} = E\alpha_t$, so noting $\alpha_{t \wedge T_n} = \alpha_t$ on $T_n > t$ we have

$$E(\alpha_t; T_n \leq t) = E(\alpha_{T_n}; T_n \leq t)$$

Since $\alpha_t \geq 0$ is continuous and $T_n \leq T$, it follows that

$$E(\alpha_t; T \leq t) \leq E(\alpha_t; T_n \leq t) = E(\alpha_{T_n}; T_n \leq t) \leq 1/n$$

Letting $n \to \infty$, we see that $\alpha_t = 0$ a.s. on $\{T \leq t\}$, so

$$Q_t(T \leq t) = E(\alpha_t; T \leq t) = 0$$

But P_t is equivalent to Q_t, so $0 = P_t(T \leq t) = P(T \leq t)$. Since t is arbitrary $P(T < \infty) = 0$, and the proof is complete. □

(12.4) shows that the collection of semimartingales is not affected by change of measure. Our next goal is to show that if X is a semimartingale/P, Q is locally equivalent to P, and $H \in \ell b\Pi$, then the integral $(H \cdot X)_t$ is the same under P and Q. The first step is

(12.6) Theorem. The quadratic variation $\langle X \rangle_t$ and hence the covariance $\langle X, Y \rangle_t$ is the same under P and Q.

Proof The second conclusion follows from the first and (3.9). To prove the first we recall (8.6) shows that if $\Delta_n = \{0 = t_0^n < t_1^n \ldots < t_{k_n}^n = t\}$ is a sequence of partitions of $[0, t]$ with mesh $|\Delta_n| \to 0$ then

$$\sum_i (X_{t_{i+1}^n} - X_{t_i^n})^2 \to \langle X \rangle_t$$

in probability for any semimartingale/P. Since convergence in probability is not affected by locally equivalent change of measure, the desired conclusion follows. □

(12.7) Theorem. If $H \in \ell b\Pi$ then $(H \cdot X)_t$ is the same under P and Q.

Proof The value is clearly the same for simple integrands. Let M and N be the local martingale parts and A and B be the locally bounded variation parts of X under P and Q respectively. Note that (12.6) implies $\langle M \rangle_t = \langle N \rangle_t$. Let

$$T_n = \inf\{t : \langle M \rangle_t, |A|_t \text{ or } |B|_t \geq n\}$$

If $H \in \ell b\Pi$ and $H_t = 0$ for $t \geq T_n$ then by (4.5) we can find simple H^m so that

$$\|H^m - H\|_M, \|H^m - H\|_N \to 0 \quad \text{and} \quad \int |H_s^m - H_s|\, d(|A| + |B|)_t \to 0$$

These conditions allow us to pass to the limit in the equality

$$(H^m \cdot (M + A))_t = (H^m \cdot (N + B))_t$$

(the left-hand side being computed under P and the right under Q) to conclude that for any $H \in \ell b\Pi$ the integrals under P and Q agree up to time T_n. Since n is arbitrary and $T_n \to \infty$ the proof is complete. □

3 Brownian Motion, II

In this chapter we will use Itô's formula to deepen our understanding of Brownian motion or, more generally, continuous local martingales.

3.1. Recurrence and Transience

If B_t is a d-dimensional Brownian motion and B_t^i is the ith component then B_t^i is a martingale with $\langle B^i \rangle_t = t$. If $i \neq j$ Exercise 2.2 in Chapter 1 tells us that $B_t^i B_t^j$ is a martingale, so (3.11) in Chapter 2 implies $\langle B^i, B^j \rangle_t = 0$. Using this information in Itô's formula we see that if $f : \mathbf{R}^d \to \mathbf{R}$ is C^2 then

$$f(B_t) - f(B_0) = \sum_i \int_0^t D_i f(B_s) dB_s^i + \frac{1}{2} \sum_i \int_0^t D_{ii} f(B_s) ds$$

Writing $\nabla f = (D_1 f, \ldots, D_d f)$ for the gradient of f, and $\Delta f = \sum_{i=1}^d D_{ii} f$ for the Laplacian of f, we can write the last equation more neatly as

$$(1.1) \qquad f(B_t) - f(B_0) = \int_0^t \nabla f(B_s) \cdot dB_s + \frac{1}{2} \int_0^t \Delta f(B_s) ds$$

Here, the dot in the first term stands for the inner product of two vectors and the precise meaning of that is given in the previous equation.

Functions with $\Delta f = 0$ are called **harmonic**. (1.1) shows that if we compose a harmonic function with Brownian motion the result is a local martingale. The next result (and judicious choices of harmonic functions) is the key to deriving properties of Brownian motion from Itô's formula.

(1.2) Theorem. Let G be a bounded open set and $\tau = \inf\{t : B_t \notin G\}$. If $f \in C^2$ and $\Delta f = 0$ in G, and f is continuous on the closure of G, $\bar{G}$, then for $x \in G$ we have $f(x) = E_x f(B_\tau)$.

Proof Our first step is to prove that $P_x(\tau < \infty) = 1$. Let $K = \sup\{|x - y| : x, y \in G\}$ be the diameter of G. If $x \in G$ and $|B_1 - x| > K$ then $B_1 \notin \bar{G}$ and $\tau < 1$. Thus

$$P_x(\tau < 1) \geq P_x(|B_1 - x| > K) = P_0(|B_1| > K) = \epsilon_K > 0$$

This shows $\sup_x P_x(\tau \geq k) \leq (1-\epsilon_K)^k$ holds when $k = 1$. To prove the last result by induction on k we observe that if $p_t(x,y)$ is the transition probability for Brownian motion, the Markov property implies

$$P_x(\tau \geq k) \leq \int_G p_1(x,y) P_y(\tau \geq k-1)\,dy$$
$$\leq (1-\epsilon_K)^{k-1} P_x(B_1 \in G) \leq (1-\epsilon_K)^k$$

where the last two inequalities follow from the induction assumption and the fact that $|B_1 - x| > K$ implies $B_1 \notin G$. The last result implies that $P_x(\tau < \infty) = 1$ for all $x \in G$ and moreover that

(1.3) $$\sup_x E_x \tau^p < \infty \qquad \text{for all } 0 < p < \infty$$

To get the last conclusion recall that (see e.g., (5.7) in Chapter 1 of Durrett (1995))

$$E_x \tau^p = \int_0^\infty p t^{p-1} P_x(\tau > t)\,dt$$

When $\Delta f = 0$ in G, (1.1) implies that $f(B_t)$ is a local martingale on $[0, \tau)$. We have assumed that f is continuous on $\bar{G}$ and G is bounded, so f is bounded and if we apply the time change γ defined in Section 2.2, $X_t = f(B_{\gamma(t)})$ is a bounded martingale (with respect to $\mathcal{G}_t = \mathcal{F}_{\gamma(t)}$). Being a bounded martingale X_t converges almost surely to a limit X_∞ which has $X_t = E_x(X_\infty|\mathcal{G}_t)$ and hence $E_x X_t = E_x X_\infty$. Since $\tau < \infty$ and f is continuous on $\bar{G}$, $X_\infty = f(B_\tau)$. Taking $t = 0$ it follows that $f(x) = E_x X_0 = E_x X_\infty = E_x f(B_\tau)$. □

In the rest of this section we will use (1.2) to prove some results concerning the range of Brownian motion $\{B_t : t \geq 0\}$. We start with the one-dimensional case.

(1.4) Theorem. Let $a < x < b$ and $T = \inf\{t : B_t \notin (a,b)\}$.

$$P_x(B_T = a) = \frac{b-x}{b-a} \qquad P_x(B_T = b) = \frac{x-a}{b-a}$$

Proof $f(x) = (b-x)/(b-a)$ has $f'' = 0$ in (a,b), is continuous on $[a,b]$, and has $f(a) = 1$, $f(b) = 0$, so (1.2) implies $f(x) = E_x f(B_T) = P_x(B_T = a)$. □

Exercise 1.1 Deduce the last result by noting that B_t is a martingale and using the optional stopping theorem at time T.

Let $T_x = \inf\{t : B_t = x\}$. From (1.4), it follows immediately that

(1.5) Theorem. For all x and y, $P_x(T_y < \infty) = 1$.

Proof Since $P_x(T_y < \infty) = P_{x-y}(T_0 < \infty)$, it suffices to prove the result when $y = 0$. A little reflection (pun intended) shows we can also suppose $x > 0$. Now using (1.4) $P_x(T_0 < T_{Mx}) = (M-1)/M$, and the right-hand side approaches 1 as $M \to \infty$. □

It is trivial to improve (1.5) to conclude that

(1.6) Theorem. For any $s < \infty$, $P_x(B_t = y \text{ for some } t \geq s) = 1$.

Proof By the Markov property,

$$P_x(B_t = y \text{ for some } t \geq s) = E_x(P_{B(s)}(T_y < \infty)) = 1 \qquad \square$$

The conclusion of (1.6) implies (argue by contradiction) that for any y with probability 1 there is a sequence of times $t_n \uparrow \infty$ (which will depend on the outcome ω) so that $B_{t_n} = y$, a conclusion we will hereafter abbreviate as "$B_t = y$ infinitely often" or "$B_t = y$ i.o." In the terminology of the theory of Markov processes, what we have shown is that one-dimensional Brownian motion is recurrent.

Exercise 1.2 Use (1.6) to conclude

$$\limsup_{t \to \infty} B_t = \infty \qquad \liminf_{t \to \infty} B_t = -\infty$$

In order to study Brownian motion in $d \geq 2$, we need to find some appropriate harmonic functions. In view of the spherical symmetry of Brownian motion, an obvious way to do this is to let $\varphi(x) = f(|x|^2)$ and try to pick $f : \mathbf{R} \to \mathbf{R}$ so that $\Delta \varphi = 0$. We use $|x|^2 = x_1^2 + \cdots x_d^2$ rather than $|x|$ since it is easier to differentiate:

$$D_i f(|x|^2) = f'(|x|^2) 2x_i$$
$$D_{ii} f(|x|^2) = f''(|x|^2) 4x_i^2 + 2f'(|x|^2)$$

Therefore, for $\Delta \varphi = 0$ we need

$$0 = \sum_i \{f''(|x|^2) 4x_i^2 + 2f'(|x|^2)\}$$
$$= 4|x|^2 f''(|x|^2) + 2d\, f'(|x|^2)$$

Letting $y = |x|^2$, we can write the above as $4yf''(y) + 2df'(y) = 0$ or, if $y > 0$,

$$f''(y) = \frac{-d}{2y} f'(y)$$

Taking $f'(y) = Cy^{-d/2}$ guarantees $\Delta \varphi = 0$ for $x \neq 0$, so by choosing C appropriately we can let

$$\varphi(x) = \begin{cases} \log |x| & d = 2 \\ |x|^{2-d} & d \geq 3 \end{cases}$$

We are now ready to use (1.2) in $d \geq 2$. Let $S_r = \inf\{t : |B_t| = r\}$ and $r < R$. Since φ has $\Delta \varphi = 0$ in $G = \{x : r < |x| < R\}$, and is continuous on $\bar{G}$, (1.2) implies

$$\varphi(x) = E_x \varphi(B_\tau) = \varphi(r) P_x(S_r < S_R) + \varphi(R)(1 - P_x(S_r < S_R))$$

where $\varphi(r)$ is short for the value of $\varphi(x)$ on $\{x : |x| = r\}$. Solving now gives

(1.7) $$P_x(S_r < S_R) = \frac{\varphi(R) - \varphi(x)}{\varphi(R) - \varphi(r)}$$

In $d = 2$, the last formula says

(1.8) $$P_x(S_r < S_R) = \frac{\log R - \log |x|}{\log R - \log r}$$

If we fix r and let $R \to \infty$ in (1.8), the right-hand side goes to 1. So

$$P_x(S_r < \infty) = 1 \quad \text{for any } x \text{ and any } r > 0$$

and repeating the proof of (1.6) shows that

(1.9) Theorem. Two-dimensional Brownian motion is **recurrent** in the sense that if G is any open set, then $P_x(B_t \in G \text{ i.o.}) \equiv 1$.

If we fix R, let $r \to 0$ in (1.8), and let $S_0 = \inf\{t > 0 : B_t = 0\}$, then for $x \neq 0$

$$P_x(S_0 < S_R) \leq \lim_{r \to 0} P_x(S_r < S_R) = 0$$

Since this holds for all R and since the continuity of Brownian paths implies $S_R \uparrow \infty$ as $R \uparrow \infty$, we have $P_x(S_0 < \infty) = 0$ for all $x \neq 0$. To extend the last result to $x = 0$ we note that the Markov property implies

$$P_0(B_t = 0 \text{ for some } t \geq \epsilon) = E_0[P_{B_\epsilon}(T_0 < \infty)] = 0$$

for all $\epsilon > 0$, so $P_0(B_t = 0$ for some $t > 0) = 0$, and thanks to our definition of $S_0 = \inf\{t > 0 : B_t = 0\}$, we have

(1.10) $$P_x(S_0 < \infty) = 0 \quad \text{for all } x$$

Thus, in $d \geq 2$ Brownian motion will not hit 0 at a positive time even if it starts there.

Exercise 1.3 Use the continuity of the Brownian path and $P_x(S_0 = \infty) = 1$ to conclude that if $x \neq 0$ then $P_x(S_r \uparrow \infty$ as $r \downarrow 0) = 1$.

For $d \geq 3$, formula (1.7) says

(1.11) $$P_x(S_r < S_R) = \frac{R^{2-d} - |x|^{2-d}}{R^{2-d} - r^{2-d}}$$

There is no point in fixing R and letting $r \to 0$, here. The fact that two dimensional Brownian motion does not hit points implies that three dimensional Brownian motion does not hit points and indeed will not hit the line $\{x : x_1 = x_2 = 0\}$. If we fix r and let $R \to \infty$ in (1.11) we get

(1.12) $$P_x(S_r < \infty) = (r/|x|)^{d-2} < 1 \quad \text{if } |x| > r$$

From the last result it follows easily that for $d \geq 3$, Brownian motion is **transient**, i.e. it does not return infinitely often to any bounded set.

(1.13) Theorem. As $t \to \infty$, $|B_t| \to \infty$ a.s.

Proof Let $A_n = \{|B_t| > n^{1/2}$ for all $t \geq S_n\}$ and note that $S_n < \infty$ by (1.3). The strong Markov property implies

$$P_x(A_n^c) = E_x(P_{B(S_n)}(S_{n^{1/2}} < \infty)) = (n^{1/2}/n)^{d-2} \to 0$$

as $n \to \infty$. Now $\limsup A_n = \cap_{N=1}^\infty \cup_{n=N}^\infty A_n$ has

$$P(\limsup A_n) \geq \limsup P(A_n) = 1$$

So infinitely often the Brownian path never returns to $\{x : |x| \leq n^{1/2}\}$ after time S_n and this implies the desired result. □

Dvoretsky and Erdös (1951) have proved the following result about how fast Brownian motion goes to ∞ in $d \geq 3$.

(1.14) **Theorem.** Suppose $g(t)$ is positive and decreasing. Then

$$P_0(|B_t| \le g(t)\sqrt{t} \text{ i.o. as } t \uparrow \infty) = 1 \text{ or } 0$$

according as $\int^\infty g(t)^{d-2}/t\, dt = \infty$ or $< \infty$.

Here the absence of the lower limit implies that we are only concerned with the behavior of the integral "near ∞." A little calculus shows that

$$\int^\infty t^{-1} \log^{-\alpha} t\, dt = \infty \text{ or } < \infty$$

according as $\alpha \le 1$ or $\alpha > 1$, so B_t goes to ∞ faster than $\sqrt{t}/(\log t)^{\alpha/d-2}$ for any $\alpha > 1$. Note that in view of the Brownian scaling relationship $B_t =_d t^{1/2} B_1$ we could not sensibly expect escape at a faster rate than $\sqrt{t}$. The last result shows that the escape rate is not much slower.

Review. At this point, we have derived the basic facts about the recurrence and transience of Brownian motion. What we have found is that

(i) $P_x(|B_t| < 1 \text{ for some } t \ge 0) \equiv 1$ if and only if $d \le 2$

(ii) $P_x(B_t = 0 \text{ for some } t > 0) \equiv 0$ in $d \ge 2$.

The reader should observe that these facts can be traced to properties of what we have called φ, the (unique up to linear transformations) spherically symmetric function that has $\Delta\varphi(x) = 0$ for all $x \ne 0$, that is :

$$\varphi(x) = \begin{cases} |x| & d = 1 \\ \log|x| & d = 2 \\ |x|^{2-d} & d \ge 3 \end{cases}$$

and the features relevant for (i) and (ii) above are

(i) $\varphi(x) \to \infty$ as $|x| \to \infty$ if and only if $d \le 2$

(ii) $|\varphi(x)| \to \infty$ as $x \to 0$ in $d \ge 2$.

3.2. Occupation Times

Let $D = B(0,r) = \{y : |y| < r\}$ the ball of radius r centered at 0. In Section 3.1, we learned that B_t will return to D i.o. in $d \le 2$ but not in $d \ge 3$. In this section we will investigate the occupation time $\int_0^\infty 1_D(B_t)\,dt$ and show that for any x

(2.1) $\quad P_x\left(\int_0^\infty 1_D(B_t)\,dt = \infty\right) = 1 \quad$ in $d \le 2$

(2.2) $\quad E_x \int_0^\infty 1_D(B_t)\,dt < \infty \quad$ in $d \ge 3$

Proof of (2.1) Let $T_0 = 0$ and $G = B(0, 2r)$. For $k \ge 1$, let

$$S_k = \inf\{t > T_{k-1} : B_t \in D\}$$
$$T_k = \inf\{t > S_k : B_t \in G\}$$

Writing τ for T_1 and using the strong Markov property, we get for $k \ge 1$

$$P_x\left(\int_{S_k}^{T_k} 1_D(B_t)\,dt \ge s \middle| \mathcal{F}_{S_k}\right) = P_{B(S_k)}\left(\int_0^\tau 1_D(B_t)\,dt \ge s\right) = H(s)$$

From this and (4.5) in Chapter 1 it follows that

$$\int_{S_k}^{T_k} 1_D(B_t)\,dt \quad \text{are i.i.d.}$$

Since these random variables have positive mean it follows from the strong law of large numbers that

$$\int_0^\infty 1_D(B_t)\,dt \ge \lim_{n \to \infty} \sum_{k=1}^n \int_{S_k}^{T_k} 1_D(B_t)\,dt = \infty \quad \text{a.s.}$$

proving the desired result. $\square$

Proof of (2.2) If f is a nonnegative function, then Fubini's theorem implies

$$E_x \int_0^\infty f(B_t)\,dt = \int_0^\infty E_x f(B_t)\,dt = \int_0^\infty \int p_t(x,y) f(y)\,dy\,dt$$
$$= \int \int_0^\infty p_t(x,y)\,dt\, f(y)\,dy$$

where $p_t(x,y) = (2\pi t)^{-d/2} e^{-|x-y|^2/2t}$ is the transition density for Brownian motion. As $t \to \infty$, $p_t(x,y) \sim (2\pi t)^{-d/2}$, so if $d \le 2$ then $\int p_t(x,y)\,dt = \infty$.

When $d \geq 3$, changing variables $t = |x-y|^2/2s$ gives

(2.3)
$$\int_0^\infty p_t(x,y)\, dt = \int_0^\infty \frac{1}{(2\pi t)^{d/2}} e^{-|y-x|^2/2t}\, dt$$
$$= \int_\infty^0 \left(\frac{s}{\pi |x-y|^2}\right)^{d/2} e^{-s}\left(-\frac{|x-y|^2}{2s^2}\right)\, ds$$
$$= \frac{|x-y|^{2-d}}{2\pi^{d/2}} \int_0^\infty s^{(d/2)-2} e^{-s}\, ds$$
$$= \frac{\Gamma(\frac{d}{2}-1)}{2\pi^{d/2}} |x-y|^{2-d}$$

where $\Gamma(\alpha) = \int_0^\infty s^{\alpha-1} e^{-s}\, ds$ is the usual gamma function. If we define

$$G(x,y) = \int_0^\infty p_t(x,y)\, dt$$

then in $d \geq 3$, $G(x,y) < \infty$ for $x \neq y$, and

(2.4)
$$E_x \int_0^\infty f(B_t)\, dt = \int G(x,y) f(y)\, dy$$

To complete the proof of (2.2) now we observe that taking $f = 1_D$ with $D = B(0,r)$ and changing to polar coordinates

$$\int_D G(0,y)\, dy = \int_0^r s^{d-1} C_d s^{2-d}\, ds = \frac{C_d}{2} r^2 < \infty$$

To extend the last conclusion to $x \neq 0$ observe that applying the strong Markov property at the exit time τ from $B(0, |x|)$ for a Brownian motion starting at 0 and using the rotational symmetry we have

$$E_x \int_0^\infty 1_D(B_s)\, ds = E_0 \int_\tau^\infty 1_D(B_s)\, ds \leq E_0 \int_0^\infty 1_D(B_s)\, ds \qquad \square$$

We call $G(x,y)$ the **potential kernel**, because $G(\cdot, y)$ is the electrostatic potential of a unit charge at y. See Chapter 3 of Port and Stone (1978) for more on this. In $d \leq 2$, $\int_0^\infty p_t(x,y)\, dt \equiv \infty$ so we have to take another approach to define a useful G:

$$G(x,y) = \int_0^\infty (p_t(x,y) - a_t)\, dt$$

where the a_t are constants we will choose to make the integral converge (at least when $x \neq y$). To see why this modified definition might be useful note that if $\int f(y)\,dy = 0$ then (assuming we can use Fubini's theorem)

$$\int G(x,y)f(y)\,dy = \int_0^\infty E_x f(B_t)\,dt$$

When $d = 1$, we let $a_t = p_t(0,0)$. With this choice,

$$G(x,y) = \frac{1}{\sqrt{2\pi}} \int_0^\infty (e^{-(y-x)^2/2t} - 1) t^{-1/2}\,dt$$

and the integral converges, since the integrand is ≤ 0 and $\sim -(y-x)^2/2t^{3/2}$ as $t \to \infty$. Changing variables $t = (y-x)^2/2u$ gives

(2.5)
$$\begin{aligned}
G(x,y) &= \frac{1}{\sqrt{2\pi}} \int_\infty^0 (e^{-u} - 1) \left(\frac{2u}{(y-x)^2}\right)^{1/2} \frac{-(y-x)^2}{2u^2}\,du \\
&= -\frac{|y-x|}{2\sqrt{\pi}} \int_0^\infty \left(\int_0^u e^{-s}\,ds\right) u^{-3/2}\,du \\
&= -\frac{|y-x|}{\sqrt{\pi}} \int_0^\infty ds\,e^{-s} \int_s^\infty \frac{u^{-3/2}}{2}\,du \\
&= -\frac{|y-x|}{\sqrt{\pi}} \int_0^\infty ds\,e^{-s} s^{-1/2} = -|y-x|
\end{aligned}$$

since

$$\int_0^\infty ds\,e^{-s} s^{-1/2} = \int_0^\infty dr\,e^{-r^2/2} \sqrt{2} = \frac{1}{2}\sqrt{2} \cdot \sqrt{2\pi} = \sqrt{\pi}$$

The computation is almost the same for $d = 2$. The only thing that changes is the choice of a_t. If we try $a_t = p_t(0,0)$ again, then for $x \neq y$ the integrand $\sim -t^{-1}$ as $t \to 0$ and the integral diverges, so we let $a_t = p_t(0,e_1)$ where $e_1 = (1,0)$. With this choice of a_t, we get

(2.6)
$$\begin{aligned}
G(x,y) &= \frac{1}{2\pi} \int_0^\infty (e^{-|x-y|^2/2t} - e^{-1/2t}) t^{-1}\,dt \\
&= \frac{1}{2\pi} \int_0^\infty \left(\int_{|x-y|^2/2t}^{1/2t} e^{-s}\,ds\right) t^{-1}\,dt \\
&= \frac{1}{2\pi} \int_0^\infty ds\,e^{-s} \int_{|x-y|^2/2s}^{1/2s} t^{-1}\,dt \\
&= \frac{1}{2\pi} \left(\int_0^\infty ds\,e^{-s}\right)(-\log(|x-y|^2)) = \frac{-1}{\pi}\log(|x-y|)
\end{aligned}$$

To sum up, the potential kernels are given by

(2.7) $$G(x,y) = \begin{cases} (\Gamma(d/2-1)/2\pi^{d/2}) \cdot |x-y|^{2-d} & d \geq 3 \\ (-1/\pi) \cdot \log(|x-y|) & d = 2 \\ -1 \cdot |x-y| & d = 1 \end{cases}$$

The reader should note that in each case $G(x,y) = C\varphi(|x-y|)$ where φ is the harmonic function we used in Section 3.1. This is, of course, no accident. $x \to G(x,0)$ is obviously spherically symmetric and, as we will see, satisfies $\Delta G(x,0) = 0$ for $x \neq 0$, so the results above imply $G(x,0) = A + B\varphi(|x|)$.

The formulas above correspond to $A = 0$, which is nice and simple. But what about the weird looking B's? What is special about them? The answer is simple: They are chosen to make $\frac{1}{2}\Delta G(x,0) = -\delta_0$ (a point mass at 0) in the distributional sense. It is easy to see that this happens in $d = 1$. In that case $\varphi(x) = |x|$ then

$$\varphi'(x) = \begin{cases} 1 & x > 0 \\ 1 & x < 0 \end{cases}$$

so if $B = -1$ then $B\varphi''(x) = -2\delta_0$. More sophisticated readers can check that this is also true in $d \geq 2$. (See F. John (1982), pages 96-97.)

To explain why we want to define the potential kernels, we return to Brownian motion in the half space, first considered in Section 1.4. Let $H = \{y : y_d > 0\}$, $\tau = \inf\{t : B_t \notin H\}$, and let

$$\bar{y} = (y_1, \ldots, y_{d-1}, -y_d)$$

be the reflection of y through the plane $\{y \in \mathbf{R}^d : y_d = 0\}$

(2.8) **Theorem.** If $x \in H$, $f \geq 0$ has compact support, and $\{x : f(x) > 0\} \subset H$ then

$$E_x\left(\int_0^\tau f(B_t)dt\right) = \int G(x,y)f(y)dy - \int G(x,\bar{y})f(y)dy$$

Proof Using Fubini's theorem which is justified since $f \geq 0$, then using (4.9) from Chapter 1 and the fact that $\{x : f(x) > 0\} \subset H$ we have

$$E_x \int_0^\tau f(B_t)\,dt = \int_0^\infty E_x(f(B_t); \tau > t)\,dt$$
$$= \int_0^\infty \int_H (p_t(x,y) - p_t(x,\bar{y}))f(y)\,dy\,dt$$
$$= \int_0^\infty \int_H (p_t(x,y) - a_t)f(y)\,dy\,dt$$
$$- \int_0^\infty \int_H (p_t(x,\bar{y}) - a_t)f(y)\,dy\,dt$$
$$= \int G(x,y)f(y)\,dy - \int G(x,\bar{y})f(y)\,dy$$

The compact support of f and the formulas for G imply that $\int |G(x,y)f(y)|dy$ and $\int |G(x,\bar{y})f(y)|dy$ are finite, so the last two equalities are valid. □

The proof given above simplifies considerably in the case $d \geq 3$; however, part of the point of the proof above is that, with the definition we have chosen for G in the recurrent case, the formulas and proofs can be the same for all d.

Let $G_H(x,y) = G(x,y) - G(x,\bar{y})$. We think of $G_H(x,y)$ as the "expected occupation time (density) at y for a Brownian motion starting at x and killed when it leaves H." The rationale for this interpretation is that (for suitable f)

$$E_x\left(\int_0^\tau f(B_t)dt\right) = \int G_H(x,y)f(y)\,dy$$

With this interpretation for G_H introduced, we invite the reader to pause for a minute and imagine what $y \to G_H(x,y)$ looks like in one dimension. If you don't already know the answer, you will probably not guess the behavior as $y \to \infty$. So much for small talk. The computation is easier than guessing the answer:

$$G(x,y) = -|x-y|$$

so $G_H(x,y) = -|x-y| + |x+y|$. Separating things into cases, we see that

$$G_H(x,y) = \begin{cases} -(x-y) + (x+y) = 2y & \text{when } 0 < y < x \\ -(y-x) + (x+y) = 2x & \text{when } x < y \end{cases}$$

so we can write

(2.9) $$G_H(x,y) = 2(x \wedge y) \quad \text{for all } x, y > 0$$

It is somewhat surprising that $y \to G_H(x,y)$ is constant $= 2x$ for $y \geq x$, that is, all points $y > x$ have the same expected occupation time!

3.3. Exit Times

In this section we investigate the moments of the exit times $\tau = \inf\{t : B_t \notin G\}$ for various open sets. We begin with $G = \{x : |x| < r\}$ in which case $\tau = S_r$ in the notation of Section 3.1.

(3.1) Theorem. If $|x| \leq r$ then $E_x S_r = (r^2 - |x|^2)/d$

Proof The key is the observation that

$$|B_t|^2 - dt = \sum_{i=1}^d \{(B_t^i)^2 - t\}$$

being the sum of d martingales is a martingale. Using the optional stopping theorem at the bounded stopping time $S_r \wedge t$ we have

$$|x|^2 = E_x\left\{|B_{S_r \wedge t}|^2 - (S_r \wedge t)d\right\}$$

(1.3) tells us that $E_x S_r < \infty$, and we have $|B_{S_r \wedge t}|^2 \leq r^2$, so letting $t \to \infty$ and using the dominated convergence theorem gives $|x|^2 = E_x\left(r^2 - S_r d\right)$ which implies the desired result. □

Exercise 3.1. Let $a, b > 0$ and $T = \inf\{t : B_t \notin (-a, b)\}$. Show $E_0 T = ab$.

To get more formulas like (3.1) we need more martingales. Applying Itô's formula, (10.2) in Chapter 2, with $X_t^1 = X_t$, a continuous local martingale, and $X_t^2 = \langle X \rangle_t$ we obtain

(3.2)
$$\begin{aligned}
f(X_t, \langle X \rangle_t) - f(X_0, 0) &= \int_0^t D_1 f(X_s, \langle X \rangle_s) dX_s \\
&+ \int_0^t D_2 f(X_s, \langle X \rangle_s) d\langle X \rangle_s \\
&+ \frac{1}{2} \int_0^t D_{11} f(X_s, \langle X \rangle_s) d\langle X \rangle_s
\end{aligned}$$

From (3.2) we see that if $(\frac{1}{2}D_{11} + D_2)f = 0$, then $f(X_t, \langle X \rangle_t)$ is a local martingale. Examples of such functions are

$$f(x, y) = x, \quad x^2 - y, \quad x^3 - 3xy, \quad x^4 - 6x^2 y + 3y^2 \ldots$$

or to expose the pattern

$$f_n(x, y) = \sum_{0 \leq m \leq [n/2]} c_{n,m} x^{n-2m} y^m$$

where $[n/2]$ denotes the largest integer $\leq n/2$, $c_{n,0} = 1$ and for $0 \leq m < [n/2]$ we pick

$$\frac{1}{2} c_{n,m}(n - 2m)(n - 2m - 1) = -(m + 1)c_{n,m+1}$$

so that $D_{xx}/2$ of the mth term is cancelled by D_y of the $(m+1)$th. (D_{xx} of the $[n/2]$th term is 0.)

The first two of our functions give us nothing new (X_t and $X_t^2 - \langle X \rangle_t$ are local martingales), but after that we get some new local martingales:

$$X_t^3 - 3X_t \langle X \rangle_t, \quad X_t^4 - 6X_t^2 \langle X \rangle_t + 3\langle X \rangle_t^2, \quad \ldots$$

These local martingales are useful for computing expectations for one dimensional Brownian motion.

(3.3) Theorem. Let $\tau_a = \inf\{t : |B_t| \geq a\}$. Then
(i) $E_0\tau_a = a^2$, (ii) $E_0\tau_a^2 = 5a^4/3$.

The dependence of the moments on a is easy to explain: the Brownian scaling relationship $B_{ct} =_d c^{1/2}B_t$ implies that $\tau_a/a^2 =_d \tau_1$.

Proof (i) follows from (3.1), so we will only prove (ii). To do this let $X_t = B_t^4 - 6B_t^2 t + 3t^2$ and $T_n \leq n$ be stopping times so that $T_n \uparrow \infty$ and $X_{t \wedge T_n}$ is a martingale. Since $T_n \leq n$ the optional stopping theorem implies

$$0 = E_0\{B_{\tau_a \wedge T_n}^4 - 6B_{\tau_a \wedge T_n}^2 (\tau_a \wedge T_n) + 3(\tau_a \wedge T_n)^2\}$$

Now $|B_{\tau_a \wedge T_n}| \leq a$, so using (1.3) and the dominated convergence theorem we can let $n \to \infty$ to conclude $0 = a^4 - 6a^2 E_0\tau_a + 3E_0\tau_a^2$. Using (i) and rearranging gives (ii). □

Exercise 3.2 Find a, b, c so that $B_t^6 - aB_t^4 t + bB_t^2 t^2 - ct^3$ is a (local) martingale and use this to compute $E_0\tau_a^3$.

Our next result is a special case of the Burkholder Davis Gundy inequalities, (5.1), but is needed to prove (4.4) which is a key step in our proof of (5.1).

(3.4) Theorem. If X_t is a continuous local martingale with $X_0 = 0$ then

$$E\left(\sup_t X_t^4\right) \leq 381 E\langle X\rangle_\infty^2$$

Proof First suppose that $|X_t|$ and $\langle X\rangle_t$ are $\leq M$ for all t. In this case (2.5) in Chapter 2 implies $X_t^4 - 6X_t^2 \langle X\rangle_t + 3\langle X\rangle_t^2$ is a martingale, so its expectation is 0. Rearranging and using the Cauchy-Schwarz inequality

$$EX_t^4 + 3E\langle X\rangle_t^2 = 6E(X_t^2 \langle X\rangle_t) \leq 6(EX_t^4)^{1/2}(E\langle X\rangle_t^2)^{1/2}$$

Using the L^4 maximal inequality ((4.3) in Chapter 4 of Durrett (1995)) and the fact that $(4/3)^4 \leq 3.1605 < 19/6$ we have

$$E\sup_{s \leq t} X_s^4 \leq (4/3)^4 EX_t^4 \leq 19\left(E\sup_{s \leq t} X_s^4\right)^{1/2} (E\langle X\rangle_t^2)^{1/2}$$

Since $|X_s| \le M$ for all s we can divide each side by $E(\sup_{s\le t} X_s^4)^{1/2}$ then square to get

$$E\left(\sup_{s\le t} X_s^4\right) \le 381 (E\langle X\rangle_t^2)^{1/2}$$

The last inequality holds for a general martingale if we replace t by $T_n = \inf\{t : t, |X_t|, \text{ or } \langle X\rangle_t \ge n\}$. Using that conclusion, letting $n \to \infty$ and using the monotone convergence theorem, we have the desired result. □

If we notice that $f(x, y) = \exp(x - y/2)$ satisfies $(\frac{1}{2}D_{11} + D_2)f = 0$, then we get another useful result.

(3.5) The Exponential Local Martingale. If X is a continuous local martingale, then $\mathcal{E}(X)_t = \exp(X_t - \frac{1}{2}\langle X\rangle_t)$ is a local martingale.

If we let $Y_t = \exp(X_t - \frac{1}{2}\langle X\rangle_t)$, then (3.2) says that

(3.6) $$Y_t - Y_0 = \int_0^t Y_s\, dX_s$$

or, in stochastic differential notation, that $dY_t = Y_t dX_s$. This property gives Y_t the right to be called the martingale exponential of Y_t. As in the case of the ordinary differential equation

$$f'(t) = f(t)a(t) \qquad f(0) = 1$$

which for a given continuous function $a(t)$ has unique solution $f(t) = \exp(A_t)$, where $A_t = \int_0^t a(s)\, ds$. It is possible to prove (under suitable assumptions) that Z is the only solution of (3.6). See Doléans-Dade (1970) for details.

The exponential local martingale will play an important role in Section 5.3. Then (and now) it will be useful to know when the exponential local martingale is a martingale. The next result is not very sophisticated (see (1.14) and (1.15) in Chapter VIII of Revuz and Yor (1991) for better results) but is enough for our purposes.

(3.7) Theorem. Suppose X_t is a continuous local martingale with $\langle X\rangle_t \le Mt$ and $X_0 = 0$. Then $Y_t = \exp(X_t - \frac{1}{2}\langle X\rangle_t)$ is a martingale.

Proof Let $Z_t = \exp(2X_t - \frac{1}{2}\langle 2X\rangle_t)$, which is a local martingale by (3.5). Now,

$$Y_t^2 = \exp(2X_t - \langle X\rangle_t) = Z_t \exp(\langle X\rangle_t)$$

So if T_n is a sequence of times that reduces Z_t, the L^2 maximal inequality applied to $Y_{t \wedge T_n}$ gives

$$E\left(\sup_{s \leq t} Y_{s \wedge T_n}^2\right) \leq 4 E Y_{t \wedge T_n}^2 \leq 4 e^{Mt} E(Z_{t \wedge T_n}) = 4 e^{Mt}$$

Letting $n \uparrow \infty$ and using the monotone convergence theorem we have

$$4 e^{Mt} \geq E\left(\sup_{s \leq t} Y_s^2\right) \geq \left(E \sup_{s \leq t} |Y_s|\right)^2$$

by Jensen's inequality. Using (2.5) in Chapter 2 now, we see that Y_t is a martingale. □

Remark. It follows from the last proof that if X_t is a continuous local martingale with $X_0 = 0$ and $\langle X \rangle_t \leq M$ for all t then $Y_t = \exp(X_t - \frac{1}{2}\langle X \rangle_t)$ is a martingale in $\mathcal{M}^2$.

Letting $\theta \in \mathbf{R}$ and setting $X_t = \theta B_t$ in (3.6), where B_t is a one dimensional Brownian motion, gives us a family of martingales $\exp(\theta B_t - \theta^2 t/2)$. These martingales are useful for computing the distribution of hitting times associated with Brownian motion.

(3.8) Theorem. Let $T_a = \inf\{t : B_t = a\}$. Then for $a > 0$ and $\lambda \geq 0$

$$E_0 \exp(-\lambda T_a) = e^{-a\sqrt{2\lambda}}$$

Remark. If you are good at inverting Laplace transforms, you can use (3.8) to prove (4.1) in Chapter 1:

$$P_0(T_a \leq t) = \int_0^t (2\pi s^3)^{1/2} a e^{-a^2/2s} \, ds$$

Proof $P_0(T_a < \infty) = 1$ by (1.5). Let $X_t = \exp(\theta B_t - \theta^2 t/2)$ and $S_n \leq n$ be stopping times so that $S_n \uparrow \infty$ and $X_{t \wedge S_n}$ is a martingale. Since $S_n \leq n$, the optional stopping theorem implies

$$1 = E_0 \exp(\theta B_{T_a \wedge S_n} - \theta^2 (T_a \wedge S_n)/2)$$

If $\theta \geq 0$ the right-hand side is $\leq \exp(\theta a)$ so letting $n \to \infty$ and using the bounded convergence theorem we have $1 = E_0 \exp(\theta a - \theta^2 T_a/2)$. Taking $\theta = \sqrt{2\lambda}$ now gives the desired result. □

(3.9) Theorem. Let $\tau_a = \inf\{t : |B_t| \geq a\}$. Then for $a > 0$ and $\lambda \geq 0$,

$$E_0 \exp(-\lambda \tau_a) = 2e^{-a\sqrt{2\lambda}}/(1 + e^{-2a\sqrt{2\lambda}})$$

Proof Let $\psi_a(\lambda) = E_0 \exp(-\lambda T_a)$. Applying the strong Markov property at time τ_a (and dropping the subscript a to make the formula easier to typeset) gives

$$E_0 \exp(-\lambda T_a) = E_0(\exp(-\lambda \tau); B_\tau = a)$$
$$+ E_0(\exp(-\lambda \tau)\psi_{2a}(\lambda); B_\tau = -a)$$

Symmetry dictates that $(\tau, B_\tau) =_d (\tau, -B_\tau)$. Since $B_\tau \in \{-a, a\}$ it follows that τ and B_τ are independent, and we have

$$\psi_a(\lambda) = \frac{1}{2}(1 + \psi_{2a}(\lambda))E_0 \exp(-\lambda \tau)$$

Using the expression for $\psi_a(\lambda)$ given in (3.8) now gives the desired result. □

Another consequence of (3.8) is a bit of calculus that will come in handy in Section 7.2.

(3.10) Theorem. If $\theta > 0$ then

$$\int_0^\infty \frac{1}{\sqrt{2\pi t}} e^{-z^2/2t} e^{-\theta t} \, dt = \frac{1}{\sqrt{2\theta}} \exp(-|z|\sqrt{2\theta})$$

Proof Changing variables $t = 1/2s$, $dt = -ds/2s^2$ the integral above

$$= \int_0^\infty \frac{\sqrt{2s}}{\sqrt{2\pi}} e^{-z^2 s} e^{-\theta/2s} \frac{ds}{2s^2}$$
$$= \frac{1}{\sqrt{2\theta}} \int_0^\infty \frac{1}{\sqrt{2\pi s^3}} \sqrt{\theta} e^{-\theta/2s} e^{-z^2 s} \, ds$$

Using (3.8) now with $a = \sqrt{\theta}$ and $\lambda = z^2$ and consulting the remark after (3.8) for the density function of T_a we see that the last expression is equal to the right-hand side of (3.10). □

The exponential martingale can also be used to study a one dimensional Brownian motion B_t plus drift. Let $Z_t = \sigma B_t + \mu t$ where $\sigma > 0$ and μ is real. $X_t = Z_t - \mu t$ is a martingale with $\langle X \rangle_t = \sigma^2 t$ so (3.7) implies that

$$\exp(\theta(Z_t - \mu t) - \theta^2 \sigma^2 t/2)$$

is a martingale. If $\theta = -2\mu/\sigma^2$ then $-\theta\mu - \theta^2\sigma^2/2 = 0$ and $\exp(-(2\mu/\sigma^2)Z_t)$ is a local martingale. Repeating the proof of (3.8) one gets

Exercise 3.3. Let $T_{-a} = \inf\{t : Z_t = -a\}$. If $a, \mu > 0$ then

$$P_0(T_{-a} < \infty) = \exp(-2a\mu/\sigma^2)$$

3.4. Change of Time, Lévy's Theorem

In this section we will prove Lévy's characterization of Brownian motion (4.1) and use it to show that every continuous local martingale is a time change of Brownian motion.

(4.1) Theorem. If X_t is a continuous local martingale with $X_0 = 0$ and $\langle X \rangle_t \equiv t$, then X_t is a one dimensional Brownian motion.

Proof By (4.5) in Chapter 1 it suffices to show

(4.2) Lemma. For any s and t, $X_{s+t} - X_s$ is independent of $\mathcal{F}_s$ and has a normal distribution with mean 0 and variance t.

Proof of (4.2) Applying the complex version of Itô's formula, (7.9) in Chapter 2, to $X'_r = X_{s+r} - X_s$ and $f(x) = e^{i\theta x}$, we get

$$e^{i\theta X'_t} - 1 = i\theta \int_0^t e^{i\theta X'_u} dX'_u - \frac{\theta^2}{2} \int_0^t e^{i\theta X'_u} du$$

Let $\mathcal{F}'_r = \mathcal{F}_{s+r}$ and let $A \in \mathcal{F}_s = \mathcal{F}'_0$. The first term on the right, which we will call Y_t, is a local martingale with respect to $\mathcal{F}'_t$. To get rid of that term let $T_n \uparrow \infty$ be a sequence of stopping times that reduces Y_t, replace t by $t \wedge T_n$, and integrate over A. The definition of conditional expectation implies

$$E(Y_{t \wedge T_n}; A) = E\left(E\left(Y_{t \wedge T_n} | \mathcal{F}'_0\right); A\right) = E(Y_0; A) = 0$$

since $Y_0 = 0$. So we have

$$E(e^{i\theta X'_{t \wedge T_n}}; A) - P(A) = 0 - \frac{\theta^2}{2} E\left(\int_0^{t \wedge T_n} e^{i\theta X'_u} du; A\right)$$

Since $|e^{i\theta x}| = 1$, letting $n \to \infty$ and using the bounded convergence theorem gives

$$E(e^{i\theta X'_t}; A) - P(A) = -\frac{\theta^2}{2} E\left(\int_0^t e^{i\theta X'_u} du; A\right)$$

$$= -\frac{\theta^2}{2} \int E\left(e^{i\theta X'_u}; A\right) du$$

by Fubini's theorem. (The integrand is bounded and the two measures are finite.) Writing $j(t) = E(e^{i\theta X'_t}; A)$, the last equality says

$$j(t) - P(A) = -\frac{\theta^2}{2} \int_0^t j(u)\, du$$

Since we know that $|j(s)| \leq 1$, it follows that $|j(t) - j(u)| \leq |t - u|\theta^2/2$, so j is continuous and we can differentiate the last equation to conclude j is differentiable with

$$j'(t) = \frac{-\theta^2}{2} j(t)$$

Together with $j(0) = P(A)$, this shows that $j(t) = P(A)e^{-\theta^2 t/2}$, or

$$E(e^{i\theta X'_t}; A) = \int_A e^{-\theta^2 t/2}\, dP$$

Since this holds for all $A \in \mathcal{F}'_0$ it follows that

(4.3) $$E(e^{i\theta X'_t}|\mathcal{F}'_0) = e^{-\theta^2 t/2}$$

or in words, the conditional characteristic function of X'_t is that of the normal distribution with mean 0 and variance t.

To get from this to (4.2) we first take expected values of both sides to conclude that X'_t has a normal distribution with mean 0 and variance t. The fact that the conditional characteristic function is a constant suggests that X'_t is independent of $\mathcal{F}'_0$. To turn this intuition into a proof let g be a C^1 function with compact support, and let

$$\varphi(\theta) = \int e^{i\theta x} g(x)\, dx$$

be its Fourier transform. We have assumed more than enough to conclude that φ is integrable and hence

$$g(x) = \frac{1}{2\pi} \int e^{i\theta x} \varphi(-\theta)\, dx$$

Multiplying each side of (4.3) by $\varphi(-\theta)$ and integrating we see that $E(g(X'_t)|\mathcal{F}'_0)$ is a constant and hence

$$E(g(X'_t)|\mathcal{F}'_0) = Eg(X'_t)$$

A monotone class argument now shows that the last conclusion is true for any bounded measurable g. Taking $g = 1_B$ and integrating the last equality over $A \in \mathcal{F}'_0$ we have

$$P(A)P(X'_t \in B) = \int_A E(1_B(X'_t)|\mathcal{F}'_0)\,dP = P(X'_t \in B)$$

by the definition of conditional expectation, and we have proved the desired independence. □

Exercise 4.1. Suppose X^i_t, $1 \le i \le d$ are continuous local martingales with $X_0 = 0$ and

$$\langle X^i, X^j \rangle_t = \begin{cases} t & \text{if } i = j \\ 0 & \text{otherwise} \end{cases}$$

then $X_t = (X^1_t, \ldots, X^d_t)$ is a d-dimensional Brownian motion.

An immediate consequence of (4.1) is:

(4.4) Theorem. Every continuous local martingale with $X_0 = 0$ and having $\langle X \rangle_\infty \equiv \infty$ is a time change of Brownian motion. To be precise if we let $\gamma(u) = \inf\{t : \langle X \rangle_t > u\}$ then $B_u = X_{\gamma(u)}$ is a Brownian motion and $X_t = B_{\langle X \rangle_t}$.

Proof Since $\gamma(\langle X \rangle_t) = t$ the second equality is an immediate consequence of the first. To prove that we note Exercise 3.8 of Chapter 2 implies that $u \to B_u$ is continuous, so it suffices to show that B_u and $B_u^2 - u$ are local martingales.

(4.5) Lemma. $B_u, \mathcal{F}_{\gamma(u)}, u \ge 0$ is a local martingale.

Proof of (4.5) Let $T_n = \inf\{t : |X_t| > n\}$. The optional stopping theorem implies that if $u < v$ then

$$E(X_{\gamma(v) \wedge T_n}|\mathcal{F}_{\gamma(u)}) = X_{\gamma(u) \wedge T_n}$$

where we have used Exercise 2.1 in Chapter 2 to replace $\mathcal{F}_{\gamma(u) \wedge T_n}$ by $\mathcal{F}_{\gamma(u)}$. To let $n \to \infty$ we observe that the L^2 maximal inequality, the fact that $X^2_{\gamma(v) \wedge T_n} - \langle X \rangle_{\gamma(v) \wedge T_n}$ is a martingale, and the definition of $\gamma(v)$ imply

$$E \sup_n X^2_{\gamma(v) \wedge T_n} \le 4 \sup_n E X^2_{\gamma(v) \wedge T_n}$$
$$= 4 \sup_n E \langle X \rangle_{\gamma(v) \wedge T_n} \le 4v$$

The last result and the dominated convergence theorem imply that as $n \to \infty$, $X_{\gamma(t) \wedge T_n} \to X_{\gamma(t)}$ in L^2 for $t = u, v$. Since conditional expectation is a contraction in L^2 it follows that $E(X_{\gamma(v) \wedge T_n} | \mathcal{F}_{\gamma(u)}) \to E(X_{\gamma(v)} | \mathcal{F}_{\gamma(u)})$ in L^2 and the proof is complete. □

To complete the proof of (4.4) now it remains to show

(4.6) Lemma. $B_u^2 - u$, $\mathcal{F}_{\gamma(u)}$, $u \geq 0$ is a local martingale.

Proof of (4.6) As in the proof of (4.5), the optional stopping theorem implies that if $u < v$ then

$$E(X_{\gamma(v) \wedge T_n}^2 - \langle X \rangle_{\gamma(v) \wedge T_n} | \mathcal{F}_{\gamma(u)}) = X_{\gamma(u) \wedge T_n}^2 - \langle X \rangle_{\gamma(u) \wedge T_n}$$

To let $n \to \infty$ we observe that using $(a+b)^2 \leq 2a^2 + 2b^2$, (3.4), and the definition of $\gamma(v)$ then

$$E \sup_n \left(X_{\gamma(v) \wedge T_n}^2 - \langle X \rangle_{\gamma(v) \wedge T_n} \right)^2 \leq 2E \sup_n X_{\gamma(v) \wedge T_n}^4 + 2E \langle X \rangle_{\gamma(v)}^2$$

$$\leq CE \langle X \rangle_{\gamma(v)}^2 \leq Cv^2$$

The proof can now be completed as in (4.5) by using the dominated convergence theorem, and the fact that conditional expectation is a contraction in L^2. □

Our next goal is to extend (4.4) to X_t with $P(\langle X \rangle_\infty < \infty) > 0$. In this case $X_{\gamma(u)}$ is a Brownian motion run for an amount of time $\langle X \rangle_\infty$. The first step in making this precise is to prove

(4.7) Lemma. $\lim_{t \uparrow \infty} X_t$ exists almost surely on $\{\langle X \rangle_\infty < \infty\}$.

Proof Let $T_n = \inf\{t : \langle X \rangle_t \geq n\}$. (3.7) in Chapter 2 implies that

$$\langle X^{T_n} \rangle_t = \langle X \rangle_{t \wedge T_n} \leq n$$

Using this with Exercise 4.3 in Chapter 2 we get $X_{t \wedge T_n} \in \mathcal{M}^2$ so $\lim_{t \to \infty} X_{t \wedge T_n}$ exists almost surely and in L^2. The last statement shows $\lim_{t \to \infty} X_t$ exists almost surely on $\{T_n = \infty\} \supset \{\langle X \rangle_\infty < n\}$. Letting $n \to \infty$ now gives (4.7). □

To prove the promised extension of (4.4) now, let $\gamma(u) = \inf\{t : \langle X \rangle_t > u\}$ when $u < \langle X \rangle_\infty$, let $X_\infty = \lim_{t \to \infty} X_t$ on $\{\langle X \rangle_\infty < \infty\}$, let B_t be a Brownian motion which is independent of $\{X_t, t \geq 0\}$, and let

$$Y_u = \begin{cases} X_{\gamma(u)} & u < \langle X \rangle_\infty \\ X_\infty + B(u - \langle X \rangle_\infty) & u \geq \langle X \rangle_\infty \end{cases}$$

(4.8) **Theorem.** Y is a Brownian motion.

Proof By (4.1) it suffices to show that Y_u and $Y_u^2 - u$ are local martingales with respect to the filtration $\sigma(Y_t : t \leq u)$. This holds on $[0, \langle X \rangle_\infty]$ for reasons indicated in the proof of (4.4). It holds on $[\langle X \rangle_\infty, \infty)$ because B is a Brownian motion independent of X. □

The reason for our interest in (4.8) is that it leads to a converse of (4.7).

(4.9) **Theorem.** The following sets are equal almost surely:

$$C = \{\lim_{t \to \infty} X_t \text{ exists }\} \qquad B = \{\sup_t |X_t| < \infty\}$$
$$A = \{\langle X \rangle_\infty < \infty\} \qquad B_+ = \{\sup_t X_t < \infty\}$$

Proof Clearly, $C \subset B \subset B_+$. In Section 3.1 we showed that Brownian motion has $\limsup_{t \to \infty} B_t = \infty$. This and (4.8) implies that $A^c \subset B_+^c$, or $B_+ \subset A$. Finally (4.7) shows that $A \subset C$. □

The result in (4.8) can be used to justify the assertion we made at the beginning of Sections 2.6 that $\Pi_3(X)$ is the largest possible class of integrands. Suppose $H \in \Pi$ and let $T = \sup\{t : \int_0^t H_s^2 d\langle X \rangle_s < \infty\}$. $(H \cdot X)_t$ can be defined for $t < T$ and has

$$\langle H \cdot X \rangle_t = \int_0^t H_s^2 d\langle X \rangle_s$$

so on $\{\langle H \cdot X \rangle_T = \infty\} \supset \{T < \infty\}$ we have

$$\limsup_{t \uparrow T} (H \cdot X)_t = \infty \qquad \liminf_{t \uparrow T} (H \cdot X)_t = -\infty$$

and there is no reasonable way to continue to define $(H \cdot X)_t$ for $t \geq T$.

Convergence is not the only property of local martingales that can be studied using (4.9). Almost any almost-sure property concerning the Brownian path can be translated into a corresponding result for local martingales. This immediately gives us a number of theorems about the behavior of paths of local martingales. We will state only:

(4.10) **Law of the Iterated Logarithm.** Let $L(t) = \sqrt{2t \log \log t}$ for $t \geq e$. Then on $\{\langle X \rangle_\infty = \infty\}$,

$$\limsup_{t \to \infty} X_t / L(\langle X \rangle_t) = 1 \quad \text{a.s.}$$

Proof This follows from (4.9) and the result for Brownian motion proved in Section 7.9 of Durrett (1995). □

Finally we have a distributional result that can be derived by time change.

Exercise 4.2. Suppose $h : [0, \infty) \to \mathbf{R}$ is measurable and locally bounded. Use (4.4) to generalize Exercise 6.7 in Chapter 2 and conclude that

$$X_t = \int_0^t h_s \, dB_s \quad \text{is normal with mean 0 and variance} \quad \int_0^t h_s^2 \, ds$$

3.5. Burkholder Davis Gundy Inequalities

Let X_t be a local martingale with $X_0 = 0$ and let $X_t^* = \sup_{s \leq t} |X_s|$. This section is devoted to a proof of the following inequalities.

(5.1) Theorem. For any $0 < p < \infty$ there are constants $0 < c, C < \infty$ so that

$$cE\langle X \rangle_t^{p/2} \leq E(X_t^*)^p \leq CE\langle X \rangle_t^{p/2}$$

Remark. This result should be contrasted with the L^p maximal inequality for martingales that only holds for $1 < p < \infty$

$$E(X_t^*)^p \leq \left(\frac{p}{p-1}\right)^p E|X_t|^p$$

Lévy's Theorem, (4.4), tells us that any continuous local martingale is a time change of Brownian motion B_t, so it suffices to let $B_t^* = \sup_{s \leq t} |B_s|$ and show that

(5.2) Theorem. For any $0 < p < \infty$ there are constants $0 < c, C < \infty$ so that for any stopping time τ

$$cE\tau^{p/2} \leq E(B_\tau^*)^p \leq CE\tau^{p/2}$$

Proof of (5.2) The key is the following pair of odd looking inequalities.

(5.3) Lemma. Let $\beta > 1$ and $\delta > 0$. Then for any $\lambda > 0$

(a) $P(B_\tau^* > \beta\lambda, \tau^{1/2} \leq \delta\lambda) \leq \frac{\delta^2}{(\beta-1)^2} P(B_\tau^* > \lambda)$

(b) $P(\tau^{1/2} > \beta\lambda, B_\tau^* \le \delta\lambda) \le \frac{\delta^2}{\beta^2-1} P(\tau^{1/2} > \lambda)$

Remark. The inequalities above are called "good λ" inequalities, although the reason for the name is obscured by our formulation (which is from Burkholder (1973)). The name "good λ" comes from the fact that early versions of this and similar inequalities (see Theorems 3.1 and 4.1 in Burkholder, Gundy, and Silverstein (1971)) were formulated as $P(f > \lambda) \le C_{\beta,K} P(g > \lambda)$ for all λ that satisfy $P(g > \lambda) \le KP(g > \beta\lambda)$. Here $\beta, K > 1$.

Proof It is enough to prove the result for bounded τ for if the result holds for $\tau \wedge n$ for all n, it also holds for τ. Let

$$S_1 = \inf\{t : |B(t \wedge \tau)| > \lambda\}$$
$$S_2 = \inf\{t : |B(t \wedge \tau)| > \beta\lambda\}$$
$$T = \inf\{t : (t \wedge \tau)^{1/2} > \delta\lambda\}$$

Since $B_\tau^* > \beta\lambda$ implies $S_1 < S_2 < \tau$ and $\tau^{1/2} \le \delta\lambda$ implies $T = \infty$ we have

$P(B_\tau^* > \beta\lambda, \tau^{1/2} \le \delta\lambda)$
$\quad \le P(|B(\tau \wedge S_2 \wedge T) - B(\tau \wedge S_1 \wedge T)| \ge (\beta-1)\lambda)$
$\quad \le (\beta-1)^{-2}\lambda^{-2} E\{(B(\tau \wedge S_2 \wedge T) - B(\tau \wedge S_1 \wedge T))^2\}$

where the second inequality is due to Chebyshev. Now if $R_1 \le R_2$ are bounded stopping times then

$$E\{B(R_1)B(R_2)\} = E\{B(R_1)E(B(R_2)|\mathcal{F}_{R_1})\} = E\,B(R_1)^2$$

So we have

$$E(B(R_2) - B(R_1))^2 = EB(R_2)^2 - 2EB(R_1)B(R_2) + EB(R_1)^2$$
$$= EB(R_2)^2 - EB(R_1)^2 = E(R_2 - R_1)$$

since $B_t^2 - t$ is a martingale. Resuming our first computation we find

$$= (\beta-1)^{-2}\lambda^{-2} E\{(\tau \wedge S_2 \wedge T) - (\tau \wedge S_1 \wedge T)\}$$
$$\le (\beta-1)^{-2}\lambda^{-2}(\delta\lambda)^2 P(S_1 < \infty)$$
$$= (\beta-1)^{-2}\delta^2 P(B_\tau^* > \lambda)$$

since $T \wedge \tau \le (\delta\lambda)^2$ proving (a).

118   Chapter 3   Brownian Motion, II

To prove (b) we interchange the roles of $B(t \wedge \tau)$ and $(t \wedge \tau)^{1/2}$ in the first set of definitions and let

$$S_1 = \inf\{t : (\tau \wedge t)^{1/2} > \lambda\}$$
$$S_2 = \inf\{t : (\tau \wedge t)^{1/2} > \beta\lambda\}$$
$$T = \inf\{t : |B(\tau \wedge t)| > \delta\lambda\}$$

Reversing the roles of B_s and $s^{1/2}$ in the proof, it is easy to check that

$$P(\tau^{1/2} > \beta\lambda, B_\tau^* \leq \delta\lambda) \leq P((\tau \wedge S_2 \wedge T) - (\tau \wedge S_1 \wedge T) \geq (\beta^2 - 1)\lambda^2)$$
$$\leq (\beta^2 - 1)^{-1}\lambda^{-2} E\{(\tau \wedge S_2 \wedge T) - (\tau \wedge S_1 \wedge T)\}$$

and using the stopping time result mentioned in the proof of (a) it follows that the above is

$$= (\beta^2 - 1)^{-1}\lambda^{-2} E\{B(\tau \wedge S_2 \wedge T)^2 - B(\tau \wedge S_1 \wedge T)^2\}$$
$$\leq (\beta^2 - 1)^{-1}\lambda^{-2}(\delta\lambda)^2 P(S_1 < \infty)$$
$$\leq (\beta^2 - 1)^{-1}\delta^2 P(\tau^{1/2} > \lambda)$$

since $|B(\tau \wedge T)| \leq (\delta\lambda)^2$ proving (b). $\square$

It will take one more lemma to extract (5.2) from (5.3). First, we need a definition. A function φ is said to be **moderately increasing** if φ is a nondecreasing function with $\varphi(0) = 0$ and if there is a constant K so that $\varphi(2\lambda) \leq K\varphi(\lambda)$. It is easy to see that $\varphi(x) = x^p$ is moderately increasing ($K = 2^p$) but $\varphi(x) = e^{ax} - 1$ is not for any $a > 0$. To complete the proof of (5.2) now it suffices to show (take $\beta = 2$ in (5.3) and note $1/3 \leq 1$)

(5.4) Lemma. If $X, Y \geq 0$ satisfy $P(X > 2\lambda, Y \leq \delta\lambda) \leq \delta^2 P(X > \lambda)$ for all $\delta \geq 0$ and φ is a moderately increasing function, then there is a constant C that only depends on the growth rate K so that

$$E\varphi(X) \leq CE\varphi(Y)$$

Proof It is enough to prove the result for bounded φ for if the result holds for $\varphi \wedge n$ for all $n \geq 1$, it also holds for φ. Now φ is the distribution function of a measure on $[0, \infty)$ that has

$$\varphi(h) = \int_0^h d\varphi(\lambda) = \int_0^\infty 1_{(h > \lambda)} d\varphi(\lambda)$$

Replacing h by a nonnegative random variable Z, taking expectations and using Fubini's theorem gives

$$(5.5) \qquad E\varphi(Z) = \int_0^\infty P(Z > \lambda)\, d\varphi(\lambda)$$

From our assumption it follows that

$$P(X > 2\lambda) = P(X > 2\lambda, Y \leq \delta\lambda) + P(X > 2\lambda, Y > \delta\lambda)$$
$$\leq \delta^2 P(X > \lambda) + P(Y > \delta\lambda)$$

Integrating $d\varphi(\lambda)$ and using (5.5) with $Z = X/2, X, Y/\delta$

$$E\varphi(X/2) \leq \delta^2 E\varphi(X) + E\varphi(Y/\delta)$$

Pick δ so that $K\delta^2 < 1$ and then pick $N \geq 0$ so that $2^N > \delta^{-1}$. From the growth condition and the monotonicity of φ, it follows that

$$E\varphi(Y/\delta) \leq K^N E\varphi(Y)$$

Combining this with the previous inequality and using the growth condition again gives

$$E\varphi(X) \leq KE\varphi(X/2) \leq K\delta^2 E\varphi(X) + K^{N+1} E\varphi(Y)$$

Solving for $E\varphi(X)$ now gives

$$E\varphi(X) \leq \frac{K^{N+1}}{1 - K\delta^2} E\varphi(Y) \qquad \square$$

381 Revisited. To compare with (3.4), suppose $p = 4$. In this case $K = 2^4 = 16$. Taking $\delta = 1/8$ to make $1 - K\delta^2 = 3/4$ and $N = 3$, we get a constant which is

$$16^4 \cdot 4/3 = 87{,}381.333\ldots$$

Of course we really don't care about the value, just that positive finite constants $0 < c, C < \infty$ exist in (5.1).

3.6. Martingales Adapted to Brownian Filtrations

Let $\{\mathcal{B}_t, t \geq 0\}$ be the filtration generated by a d-dimensional Brownian motion B_t with $B_0 = 0$, defined on some probability space $(\Omega, \mathcal{F}, P)$. In this section we will show (i) all local martingales adapted to $\{\mathcal{B}_t, t \geq 0\}$ are continuous and

(ii) every random variable $X \in L^2(\Omega, \mathcal{B}_\infty, P)$ can be written as a stochastic integral.

(6.1) Theorem. All local martingales adapted to $\{\mathcal{B}_t, t \geq 0\}$ are continuous.

Proof Let X_t be a local martingale adapted to $\{\mathcal{B}_t, t \geq 0\}$ and let $T_n \leq n$ be a sequence of stopping times that reduces X. It suffices to show that for each n, $X(t \wedge T_n)$ is continuous, or in other words, it is enough to show that the result holds for martingales of the form $Y_t = E(Y|\mathcal{B}_t)$ where $Y \in \mathcal{B}_n$. We build up to this level of generality in three steps.

Step 1. Let $Y = f(B_n)$ where f is a bounded continuous function. If $t \geq n$, then $Y_t = f(B_n)$, so $t \to Y_t$ is trivially continuous for $t > n$. If $t < n$, the Markov property implies $Y_t = E(Y|\mathcal{B}_t) = h(n-t, B_t)$ where

$$h(s,x) = \int \frac{1}{(2\pi s)^{d/2}} e^{-|y-x|^2/2s} f(y) \, dy$$

It is easy to see that $h(s,x)$ is a continuous function on $(0,\infty) \times R$, so Y_t is continuous for $t < n$. To check continuity at $t = n$, observe that changing variables $y = x + z\sqrt{s}$, $dy = s^{d/2} dz$ gives

$$h(s,x) = \int \frac{1}{(2\pi)^{d/2}} e^{-|z|^2/2} f(x + z\sqrt{s}) \, dz$$

so the dominated convergence theorem implies that as $t \uparrow n$, $h(n-t, B_t) \to f(B_n)$.

Step 2. Let $Y = f_1(B_{t_1}) f_2(B_{t_2})$ where $t_1 < t_2 \leq n$ and f_1, f_2 are bounded and continuous. If $t \geq t_1$, then

$$Y_t = f_1(B_{t_1}) E(f_2(B_{t_2})|\mathcal{B}_t)$$

so the argument from step 1 implies that Y_t is continuous on $[t_1, \infty)$. On the other hand, if $t \leq t_1$, then $Y_t = E(Y_{t_1}|\mathcal{B}_t)$ and

$$Y_{t_1} = f_1(B_{t_1}) E(f_2(B_{t_2})|\mathcal{B}_{t_1}) = f_1(B_{t_1}) g(B_{t_1})$$

where

$$g(x) = \int \frac{1}{(2\pi(t_2 - t_1))^{d/2}} e^{-|y-x|^2/2(t_2-t_1)} f_2(y) \, dy$$

is a bounded continuous function, so

$$Y_t = E(f_1(B_{t_1}) g(B_{t_1})|\mathcal{B}_t) \quad \text{for } t \leq t_1$$

Section 3.6 Martingales Adapted to Brownian Filtrations

and it follows from step 1 that Y_t is continuous on $[0, t_1]$. Repeating the argument above and using induction, it follows that the result holds if $Y = f_1(B_{t_1}) \cdots f_k(B_{t_k})$ where $t_1 < t_2 < \cdots < t_k \leq n$ and $f_1, \ldots, f_k$ are bounded continuous functions.

Step 3. Let $Y \in \mathcal{B}_n$ with $E|Y| < \infty$. It follows from a standard application of the monotone class theorem that for any $\epsilon > 0$, there is a random variable X^ϵ of the form considered in step 2 that has $E|X^\epsilon - Y| < \epsilon$. Now

$$|E(X^\epsilon|\mathcal{B}_t) - E(Y|\mathcal{B}_t)| \leq E(|X^\epsilon - Y| | \mathcal{B}_t)$$

and if we let $Z_t = E(|X^\epsilon - Y| | \mathcal{B}_t)$, it follows from Doob's inequality (see e.g., (4.2) in Chapter 4 of Durrett (1995)) that

$$\lambda P \left(\sup_{t \leq n} Z_t > \lambda \right) \leq EZ_n = E|X^\epsilon - Y| < \epsilon$$

Now $X^\epsilon(t) = E(X^\epsilon|\mathcal{B}_t)$ is continuous, so letting $\epsilon \to 0$ we see that for a.e. ω, $Y_t(\omega)$ is a uniform limit of continuous functions, so Y_t is continuous. □

Remark. I would like to thank Michael Sharpe for telling me about the proof given above.

We now turn to our second goal. Let $B_t = (B_t^1, \ldots, B_t^d)$ with $B_0 = 0$ and let $\{\mathcal{B}_t^i, t \geq 0\}$ be the filtrations generated by the coordinates.

(6.2) Theorem. For any $X \in L^2(\Omega, \mathcal{B}_\infty, P)$ there are unique $H^i \in \Pi_2(B^i)$ with

$$X = EX + \sum_{i=1}^d \int_0^\infty H_s^i \, dB_s^i$$

Proof We follow Section V.3 of Revuz and Yor (1991). The uniqueness is immediate since the isometry property, Exercise 6.2 of Chapter 2, implies there is only one way to write the zero random variable, i.e., using $H^i(s, \omega) \equiv 0$. To prove the existence of H_s, the first step is to reduce to one dimension by proving:

(6.3) Lemma. Let $X \in L^2(\Omega, \mathcal{B}_\infty, P)$ with $EX = 0$, and let $X_i = E(X|\mathcal{B}_\infty^i)$. Then $X = X_1 + \cdots + X_d$.

Proof Geometrically, see (1.4) in Chapter 4 of Durrett (1995), X_i is the projection of X onto L_i^2 the subspace of mean zero random variables measurable with respect to $\mathcal{B}_\infty^i$. If $Y \in L_i^2$ and $Z \in L_j^2$ then Y and Z are independent so $EYZ = EY \cdot EZ = 0$. The last result shows that the spaces L_i^2 and L_j^2 are orthogonal. Since $L_1^2, \ldots, L_d^2$ together span all the elements in L^2 with mean 0, (6.7) follows. □

Proof in one dimension Let $\mathcal{I}$ be the set of integrands which can be written as $\sum_{j=1}^n \lambda_j 1_{(s_{j-1}, s_j]}$ where the λ_j and s_j are (nonrandom!) real numbers and $0 = s_0 < s_1 < \ldots < s_n$. For any integrand $H \in \mathcal{I} \subset \Pi_2(B)$, we can define the stochastic integral $Y_t = \int_0^t H_s \, dB_s$, and use the isometry property, Exercise 6.2 of Chapter 2, to conclude $Y_t \in \mathcal{M}^2$. Using (3.7) and the remark after its proof, we can further define the martingale exponential of the integral, $Z_t = \mathcal{E}(Y)_t$ and conclude that $Z_t \in \mathcal{M}^2$. Itô's formula, see (3.6), implies that $Z_t - Z_0 = \int_0^t Z_s \, dY_s$. Recalling $Y_t = (H \cdot B)_t$ and using the associative law, (9.6) in Chapter 2, we have

$$Z_t - Z_0 = \int_0^t Z_s H_s \, dB_s$$

$Y_0 = 0$ and $\langle Y \rangle_0 = 0$, so $Z_0 = 1$. Since $Z_t \in \mathcal{M}^2$ we have $EZ_t = 1 = Z_0$ and it follows that

$$Z_t = EZ_t + \int_0^\infty Z_s H_s 1_{[0,t]}(s) \, dB_s$$

i.e., the desired representation holds for each of the random variables in the set $\mathcal{J} = \{\mathcal{E}(H \cdot B)_t : H \in \mathcal{I}, t \geq 0\}$.

To complete the proof of (6.2) now, it suffices to show

(6.4) Lemma. If $W \in L^2$ has $E(ZW) = 0$ for all $Z \in \mathcal{J}$ then $W \equiv 0$.

(6.5) Lemma. Let $\mathcal{G}$ be the collection of L^2 random variables X that can be written as $EX + \int_0^\infty H_s \, dB_s$. Then $\mathcal{G}$ is a closed subspace of L^2.

Proof of (6.4) We will show that the measure $W \, dP$ (i.e., the measure μ with $d\mu/dP = W$) is the zero measure. To do this, it suffices to show that $W \, dP$ is the zero measure on the σ-field $\sigma(B_{t_1}, \ldots, B_{t_n})$ for any finite sequence $0 = t_0 < t_1 < t_2 < \ldots < t_n$. Let λ_j, $1 \leq j \leq n$ be real numbers and z be a complex number. The function

$$\varphi(z) = E\left[W \exp\left(z \sum_{j=1}^n \lambda_j (B_{t_j} - B_{t_{j-1}})\right)\right]$$

is easily seen to be analytic in $\mathbf{C}$, i.e., $\varphi(z)$ is represented by an absolutely convergent power series. By the assumed orthogonality, we have $\varphi(x) = 0$ for all real x, so complex variable theory tells us that φ must vanish identically. In particular $\varphi(i) = 0$, when $i = \sqrt{-1}$. The last equality implies that the image of $W\,dP$ under the map

$$\omega \to (B_{t_1}(\omega) - B_{t_0}(\omega), \ldots, B_{t_n}(\omega) - B_{t_{n-1}}(\omega))$$

is the zero measure since its Fourier transform vanishes. This shows that $W\,dP$ vanishes on $\sigma(B_{t_1} - B_{t_0}, \ldots, B_{t_n} - B_{t_{n-1}}) = \sigma(B_{t_1}, \ldots, B_{t_n})$ and the proof of (6.4) is complete. □

Proof of (6.5) It is clear that $\mathcal{G}$ is a subspace. Let $X_n \in \mathcal{G}$ with

$$(6.6) \qquad X_n - EX_n = \int_0^\infty H_s^n \, dB_s$$

and suppose that $X_n \to X$ in L^2. Using an elementary fact about the variance, then (6.6), and the isometry property of the stochastic integral, Exercise 6.2 in Chapter 2, we have

$$E(X_n - X_m)^2 - (EX_n - EX_m)^2 = E\{(X_n - X_m) - (EX_n - EX_m)\}^2$$
$$= E \int_0^\infty (H_s^n - H_s^m)^2 \, ds$$

Since $X_n \to X$ in L^2 implies $E(X_n - X_m)^2 \to 0$ and $EX_n \to EX$, it follows that $\|H^m - H^n\|_B \to 0$.

The completeness of $\Pi_2(B)$, see the remark at the beginning of Step 3 in Section 2.4, implies that there is a predictable H so that $\|H^n - H\|_B \to 0$. Another use of the isometry property implies that $H^m \cdot B$ converges to $H \cdot B$ in $\mathcal{M}^2$. Taking limits in (6.6) gives the representation for X. This completes the proof of (6.5) and thus of (6.2). □

4 Partial Differential Equations

A. Parabolic Equations

In the first third of this chapter, we will show how Brownian motion can be used to construct (classical) solutions of the following equations:

$$u_t = \frac{1}{2}\Delta u$$

$$u_t = \frac{1}{2}\Delta u + g$$

$$u_t = \frac{1}{2}\Delta u + cu$$

in $(0,\infty) \times \mathbf{R}^d$ subject to the boundary condition: u is continuous at each point of $\{0\} \times \mathbf{R}^d$ and $u(0,x) = f(x)$ for $x \in \mathbf{R}^d$. Here,

$$\Delta u = \frac{\partial^2 u_1}{\partial x_1^2} + \cdots + \frac{\partial^2 u_d}{\partial x_d^2}$$

and by a classical solution, we mean one that has enough derivatives for the equation to make sense. That is, $u \in C^{1,2}$, the functions that have one continuous derivative with respect to t and two with respect to each of the x_i. The continuity in u in the boundary condition is needed to establish a connection between the equation which holds in $(0,\infty) \times \mathbf{R}^d$ and $u(0,x) = f(x)$ which holds on $\{0\} \times \mathbf{R}^d$. Note that the boundary condition cannot possibly hold unless $f : \mathbf{R}^d \to \mathbf{R}$ is continuous.

We will see that the solutions to these equations are (under suitable assumptions) given by

$$E_x f(B_t)$$

$$E_x \left(f(B_t) + \int_0^t g(t-s, B_s)\, ds \right)$$

$$E_x \left(f(B_t) \exp\left(\int_0^t c(t-s, B_s)\, ds \right) \right)$$

In words, the solutions may be described as follows:

(i) To solve the heat equation, run a Brownian motion and let $u(t,x) = E_x f(B_t)$.

(ii) To introduce a term $g(x)$ add the integral of g along the path.

(iii) To introduce cu, multiply $f(B_t)$ by $m_t = \exp(\int_0^t c(t-s, B_s) ds)$ before taking expected values. Here, we think of the Brownian particle as having mass 1 at time 0 and changing mass according to $m'_s = c(t-s, B_s) m_s$, and when we take expected values, we take the particle's mass into account. An alternative interpretation when $c \leq 0$ is that $\exp(\int_0^t c(t-s, B_s)\,ds)$ is the probability the particle survives until time t, or $-c(r,x)$ gives the killing rate when the particle is at x at time r.

In the first three sections of this chapter, we will say more about why the expressions we have written above solve the indicated equations. In order to bring out the similarities and differences between these equations and their elliptic counterparts discussed in Sections 4.4 to 4.6, we have adopted a rather robotic style. Formulas (m.2) through (m.6) and their proofs have been developed in parallel in the first six sections, and at the end of most sections we discuss what happens when something becomes unbounded.

4.1. The Heat Equation

In this section, we will consider the following equation:

(1.1a) $u \in C^{1,2}$ and $u_t = \frac{1}{2}\Delta u$ in $(0, \infty) \times \mathbf{R}^d$.

(1.1b) u is continuous at each point of $\{0\} \times \mathbf{R}^d$ and $u(0, x) = f(x)$.

This equation derives its name, the heat equation, from the fact that if the units of measurement are chosen suitably then the solution $u(t,x)$ gives the temperature at the point $x \in R^d$ at time t when the temperature profile at time 0 is given by $f(x)$.

The first step in solving (1.1), as it will be six times below, is to find a local martingale.

(1.2) Theorem. If u satisfies (1.1a), then $M_s = u(t-s, B_s)$ is a local martingale on $[0, t)$.

Proof Applying Itô's formula, (10.2) in Chapter 2, to $u(x_0, \ldots, x_d)$ with

$X_s^0 = t - s$ and $X_s^i = B_s^i$ for $1 \le i \le d$ gives

$$u(t - s, B_s) - u(t, B_0) = \int_0^s -u_t(t - r, B_r)\, dr$$
$$+ \int_0^s \nabla u(t - r, B_r) \cdot dB_r$$
$$+ \frac{1}{2} \int_0^s \Delta u(t - r, B_r)\, dr$$

To check this note that $dX_r^0 = -dr$ and X_r^0 has bounded variation, while the X_r^i with $1 \le i \le d$ are independent Brownian motions, so

$$\langle X^i, X^j \rangle_r = \begin{cases} r & 1 \le i = j \le d \\ 0 & \text{otherwise} \end{cases}$$

(1.2) follows easily from the Itô's formula equation since $-u_t + \frac{1}{2}\Delta u = 0$ and the second term on the right-hand side is a local martingale. □

Our next step is to prove a uniqueness theorem.

(1.3) Theorem. If there is a solution of (1.1) that is bounded then it must be

$$v(t, x) \equiv E_x f(B_t)$$

Here $\equiv$ means that the last equation defines v. We will always use u for a generic solution of the equation and v for our special solution.

Proof If we now assume that u is bounded, then $M_s, 0 \le s < t$, is a bounded martingale. The martingale convergence theorem implies that

$$M_t \equiv \lim_{s \uparrow t} M_s \quad \text{exists a.s.}$$

If u satisfies (1.1b), this limit must be $f(B_t)$. Since M_s is uniformly integrable, it follows that

$$u(t, x) = E_x M_0 = E_x M_t = v(t, x) \qquad \square$$

Now that (1.3) has told us what the solution must be, the next logical step is to find conditions under which v is a solution. It is (and always will be) easy to show that if v is smooth enough then it is a classical solution.

(1.4) Theorem. Suppose f is bounded. If $v \in C^{1,2}$ then it satisfies (1.1a).

Proof The Markov property implies, see Exercise 2.1 in Chapter 1, that

$$E_x(f(B_t)|\mathcal{F}_s) = E_{B_s}(f(B_{t-s})) = v(t-s, B_s)$$

The left-hand side is a martingale, so the right-hand side is also. If $v \in C^{1,2}$, then repeating the calculation in the proof of (1.2) shows that

$$v(t-s, B_s) - v(t, B_0) = \int_0^s (-v_t + \frac{1}{2}\Delta v)(t-r, B_r)\,dr$$
$$+ \text{a local martingale}$$

The left-hand side is a local martingale, so the integral on the right-hand side is also. However, the integral is continuous and locally of bounded variation, so by (3.3) in Chapter 2 it must be $\equiv 0$ almost surely. Since v_t and Δv are continuous, it follows that $-v_t + \frac{1}{2}\Delta v \equiv 0$. For if it were $\neq 0$ at some point (t, x), then it would be $\neq 0$ on an open neighborhood of that point, and, hence, with positive probability the integral would not be $\equiv 0$, a contradiction. □

It is easy to give conditions that imply that v satisfies (1.1b). In order to keep the exposition simple, we first consider the situation when f is bounded.

(1.5) Theorem. If f is bounded and continuous, then v satisfies (1.1b).

Proof $(B_t - B_0) \stackrel{d}{=} t^{1/2} N$, where N has a normal distribution with mean 0 and variance 1, so if $t_n \to 0$ and $x_n \to x$, the bounded convergence theorem implies that

$$v(t_n, x_n) = Ef(x_n + t_n^{1/2} N) \to f(x) \qquad \square$$

The final step in showing that v is a solution is to find conditions that guarantee that it is smooth. In this case, the computations are not very difficult.

(1.6) Theorem. If f is bounded, then $v \in C^{1,2}$ and hence satisfies (1.1a).

Proof By definition,

$$v(t, x) = E_x f(B_t) = \int p_t(x, y) f(y)\,dy$$

where $p_t(x,y) = (2\pi t)^{-d/2} e^{-|x-y|^2/2t}$. Writing $D_i = \partial/\partial x_i$ and $D_t = \partial/\partial t$, a little calculus gives

$$D_i p_t(x,y) = \frac{-(x_i - y_i)}{t} p_t(x,y)$$

$$D_{ii} p_t(x,y) = \frac{(x_i - y_i)^2 - t}{t^2} p_t(x,y)$$

$$D_{ij} p_t(x,y) = \frac{(x_i - y_i)(x_j - y_j)}{t^2} p_t(x,y) \qquad i \neq j$$

$$D_t p_t(x,y) = \left(\frac{-d/2}{t} + \frac{|x-y|^2}{2t^2} \right) p_t(x,y)$$

If f is bounded, then it is easy to see that for $\alpha = i, ij$, or t

$$\int |D_\alpha p_t(x,y) f(y)| \, dy < \infty$$

and is continuous in $\mathbf{R}^d$, so (1.6) follows from the next result on differentiating under the integral sign. This result is lengthy to state but short to prove since we assume everything we need for the proof to work. Nonetheless we will see that this result is useful.

(1.7) Lemma. Let $(S, \mathcal{S}, m)$ be a σ-finite measure space, and $g : S \to \mathbf{R}$ be measurable. Suppose that for $x \in G$ an open subset of $\mathbf{R}^d$ and some $h_0 > 0$ we have:

(a) $u(x) = \int_S K(x,y) g(y) m(dy)$

where K and $\partial K / \partial x_i : G \times S \to \mathbf{R}$ are measurable functions with

(b) $K(x^* + he_i, y) - K(x^*, y) = \int_0^h \frac{\partial K}{\partial x_i}(x^* + \theta e_i, y) \, d\theta$ for $|h| \leq h_0$ and $y \in S$

(c) $u_i(x) = \int_S \frac{\partial K}{\partial x_i}(x,y) g(y) \, m(dy)$ is continuous at x^*

and (d) $\int_S \int_{-h_0}^{h_0} \left| \frac{\partial K}{\partial x_i}(x^* + \theta e_i, y) g(y) \right| d\theta \, m(dy) < \infty$.

Then $\partial u / \partial x_i$ exists at x^* and equals $u_i(x^*)$.

Proof Using the definition of u in (a), then (b) and Fubini's theorem, which is justified for $|h| \leq h_0$ by (d), we have

$$u(x^* + he_i) - u(x^*) = \int_S (K(x^* + he_i, y) - K(x^*, y)) g(y) \, m(dy)$$

$$= \int_0^h \int_S \frac{\partial K}{\partial x_i}(x^* + \theta e_i, y) g(y) \, m(dy) \, d\theta$$

Dividing by h and letting $h \to 0$ the desired result follows from (c). □

Later we will need a result about differentiating sums. Taking $S = \mathbf{Z}$ with $\mathcal{S} = $ all subsets of S, and μ is counting measure in (1.7), then setting $g \equiv 1$ and $f_n(x) = K(x,n)$ gives the following. Note that in (a) and (c) of (1.7) it is implicit that the integrals exist, so here in (a) and (c) we assume that the sums converge absolutely.

(1.8) Lemma. Suppose that for $x \in G$ an open subset of $\mathbf{R}^d$ and some $h_0 > 0$ we have:

(a) $u(x) = \sum_n f_n(x)$

where f_n and $\partial f_n/\partial x_i : G \to \mathbf{R}$, $n \in \mathbf{Z}$ are measurable functions with

(b) $f_n(x^* + he_i) - f_n(x^*) = \int_0^h \frac{\partial f_n}{\partial x_i}(x^* + \theta e_i)\,d\theta$ for $|h| \le h_0$ and $n \in \mathbf{Z}$

(c) $u_i(x) = \sum_n \frac{\partial f_n}{\partial x_i}(x)$ is continuous at x^*

and (d) $\sum_n \int_{-h_0}^{h_0} \left|\frac{\partial f_n}{\partial x_i}(x^* + \theta e_i)\right| d\theta < \infty$.

Then $\partial u/\partial x_i$ exists at x^* and equals $u_i(x^*)$.

Unbounded f. For some applications, the assumption that f is bounded is too restrictive. To see what type of unbounded f we can allow, we observe that, at the bare minimum, we need $E_x|f(B_t)| < \infty$ for all t. Since

$$E_x|f(B_t)| = \int \frac{1}{(2\pi t)^{d/2}} e^{-|x-y|^2/2t} |f(y)|\,dy$$

a condition that guarantees this for locally bounded f is

(*) $\qquad |x|^{-2} \log^+ |f(x)| \to 0 \quad \text{as } x \to \infty$

Replacing the bounded convergence theorem in (1.5) and (1.6) by the dominated convergence theorem, it is not hard to show:

(1.9) Theorem. If f is continuous and satisfies (*), then v satisfies (1.1).

4.2. The Inhomogeneous Equation

In this section, we will consider what happens when we add a function $g(t,x)$ to the equation we considered in the last section. That is, we will study

(2.1a) $u \in C^{1,2}$ and $u_t = \frac{1}{2}\Delta u + g$ in $(0,\infty) \times \mathbf{R}^d$

Section 4.2 The Inhomogeneous Equation 131

(2.1b) u is continuous at each point of $\{0\} \times \mathbf{R}^d$ and $u(0,x) = f(x)$.

We observed in Section 4.1 that (2.1b) cannot hold unless f is continuous. Here $g = u_t - \frac{1}{2}\Delta u$ so the equation in (2.1a) cannot hold with $u \in C^{1,2}$ unless $g(t,x)$ is continuous.

Our first step in treating the new equation is to observe that if u_1 is a solution of the equation with $f = f_0$ and $g = 0$ which we studied in the last section, and u_2 is a solution of the equation with $f = 0$ and $g = g_0$ then $u_1 + u_2$ is a solution of the equation with $f = f_0$ and $g = g_0$ so we can restrict our attention to the case $f \equiv 0$.

Having made this simplification, we will now study the equation above by following the procedure used in the last section. The first step is to find an associated local martingale.

(2.2) Theorem. If u satisfies (2.1a), then

$$M_s = u(t-s, B_s) + \int_0^s g(t-r, B_r)\, dr$$

is a local martingale on $[0,t)$.

Proof Applying Itô's formula as in the proof of (1.2) gives

$$u(t-s, B_s) - u(t, B_0) = \int_0^s \left(-u_t + \frac{1}{2}\Delta u\right)(t-r, B_r)\, dr$$
$$+ \int_0^s \nabla u(t-r, B_r) \cdot dB_r$$

which proves (2.2), since $-u_t + \frac{1}{2}\Delta u = -g$ and the second term on the right-hand side is a local martingale. □

Again the next step is a uniqueness result.

(2.3) Theorem. Suppose g is bounded. If there is a solution of (2.1) that is bounded on $[0,T] \times \mathbf{R}^d$ for any $T < \infty$, it must be

$$v(t,x) \equiv E_x\left(\int_0^t g(t-s, B_s)\, ds\right)$$

Proof Under the assumptions on g and u, $M_s, 0 \leq s < t$, defined in (2.2) is a bounded martingale and $u(0,x) \equiv 0$ so

$$M_t \equiv \lim_{s \uparrow t} M_s = \int_0^t g(t-s, B_s)\, ds$$

132 Chapter 4 Partial Differential Equations

Since M_s is uniformly integrable,
$$u(t,x) = E_x M_0 = E_x M_t = v(t,x) \qquad \square$$

Again, it is easy to show that if v is smooth enough it is a solution.

(2.4) Theorem. Suppose g is bounded and continuous. If $v \in C^{1,2}$, then it satisfies (2.1a) in $(0,\infty) \times \mathbf{R}^d$.

Proof Using the Markov property, see Exercise 2.6 in Chapter 1, gives
$$E_x \left(\int_0^t g(t-r, B_r)\, dr \,\Big|\, \mathcal{F}_s \right)$$
$$= \int_0^s g(t-r, B_r)\, dr + E_{B_s} \left(\int_0^{t-s} g(t-s-u, B_u)\, du \right)$$
$$= \int_0^s g(t-r, B_r)\, dr + v(t-s, B_s)$$

The left-hand side is a martingale, so the right-hand side is also. If $v \in C^{1,2}$, then repeating the calculation in the proof of (2.2) shows that
$$v(t-s, B_s) - v(t, B_0) + \int_0^s g(t-r, B_r)\, dr$$
$$= \int_0^s (-v_t + \tfrac{1}{2}\Delta v + g)(t-r, B_r)\, dr$$
$$\quad + \text{ a local martingale}$$

The left-hand side is a local martingale, so the integral on the right-hand side is also. Since the integral is continuous and locally of bounded variation, (3.3) in Chapter 2 implies it must be $\equiv 0$ almost surely. Again this implies that $(-v_t + \tfrac{1}{2}\Delta v + g) \equiv 0$, for our assumptions imply that this quantity is continuous and if it were $\ne 0$ at some point (t,x) we would have a contradiction. $\square$

The next step is to give a condition that guarantees that v satisfies (2.1b). As in the last section, we will begin by considering what happens when everything is bounded.

(2.5) Theorem. If g is bounded, then v satisfies (2.1b).

Proof If $|g| \le M$, then as $t \to 0$
$$|v(t,x)| \le E_x \int_0^t |g(t-s, B_s)|\, ds \le Mt \to 0 \qquad \square$$

Section 4.2 The Inhomogeneous Equation   133

The final step in showing that v is a solution is to check that $v \in C^{1,2}$. Since the calculations necessary to establish these properties are quite tedious, we will content ourselves to state what the results are and do enough of the proofs to indicate why they are true. If you get dazed and confused and bail out before the end, the

Take home message is: it is not enough to assume g is continuous to have $v \in C^{1,2}$. We must assume g to be **Hölder continuous locally in** t. That is, for any $N < \infty$ there are constants $C, \alpha \in (0, \infty)$, which may depend on N, such that $|g(t, x) - g(t, y)| \leq C|x - y|^\alpha$ whenever $t \leq N$.

The reason for this assumption can be found in the proof of (2.6c) and again in (2.6d).

The first step in showing $v \in C^{1,2}$ is to assume g is bounded and use Fubini's theorem to conclude

$$v(t, x) = \int_0^t ds \int p_s(x, y) g(t - s, y)\, dy$$

where $p_s(x, y) = (2\pi s)^{-d/2} e^{-|y-x|^2/2s}$. The expression we have just written for v above is what Friedman (1964) would call a volume potential and would write as

$$V(x, t) = \int_{T_0}^t \int_D Z(x, t; \xi, \tau) g(\xi, \tau) d\xi\, d\tau$$

To translate between notations, set $T_0 = 0, D = \mathbf{R}^d$,

$$Z(x, t; \xi, \tau) = p_{t-\tau}(x, \xi)$$

and change variables $s = t - \tau, y = \xi$. Because of their importance for the parametrix method, the differentiability properties of volume potentials are well known. The results we will state are just Theorems 2 to 5 in Chapter 1 of Friedman (1964), so the reader who is interested in knowing the whole story can find the missing details there.

(2.6a) Theorem. If g is a bounded measurable function, then $v(t, x)$ is continuous on $(0, \infty) \times \mathbf{R}^d$.

Proof Use the bounded convergence theorem. □

(2.6b) Theorem. There is a constant C so that if $|g| \leq M$ is measurable, then the partial derivatives $D_i v = \partial v / \partial x_i$ have $|D_i v| \leq CMt^{1/2}$, are continuous, and are given by

$$D_i v = \int_0^t \int D_i p_s(x, y) g(t - s, y)\, dy\, ds$$

Proof Using the formula for $D_i p_s$ from (1.6), the right-hand side is

$$-\int_0^t \int (2\pi s)^{-d/2} \frac{(x_i - y_i)}{s} e^{-|x-y|^2/2s} g(t-s,y) \, dy \, ds$$

$$= -\int_0^t \frac{ds}{s} E_x[(x_i - B_s^i)g(t-s, B_s)]$$

Although the last formula looks suspicious because we are integrating s^{-1} near 0, everything is really all right. If $|g| \leq M$, then

$$E_x|(x_i - B_s^i)g(t-s, B_s)| \leq M E_x|x_i - B_s^i| = CMs^{1/2}$$

so we have

$$\int_0^t \frac{ds}{s} E_x|(x_i - B_s^i)g(t-s, B_s)| \leq 2CMt^{1/2} < \infty$$

Using our result on differentiating under the integral sign, (1.7), it follows that the partial derivatives $D_i v$ exist, are continuous, and have the indicated form.
□

Things get even worse when we take second derivatives.

(2.6c) Theorem. Suppose that g is bounded and Hölder continuous locally in t. Then the partial derivatives $D_{ij} v = \partial^2 v / \partial x_i \partial x_j$ are continuous, and

$$D_{ij} v = \int_0^t \int D_{ij} p_s(x, y) g(t-s, y) \, dy \, ds$$

Proof Suppose for simplicity that $i = j$. Consulting (1.6) for the formula for $D_{ii} p_s$, the right-hand side is

$$\int_0^t \int (2\pi s)^{-d/2} \left(\frac{(x_i - y_i)^2 - s}{s^2} \right) e^{-|x-y|^2/2s} g(t-s,y) \, dy \, ds$$

$$= \int_0^t E_x \left\{ \left(\frac{(x_i - B_s^i)^2 - s}{s^2} \right) g(t-s, B_s) \right\} ds$$

This time, however, $E_x|(x_i - B_s^i)^2 - s| = s E_0|(B_1^i)^2 - 1|$ so

$$\int_0^t ds \, E_x \left| \frac{(x_i - B_s^i)^2 - s}{s^2} \right| = \infty$$

We can overcome this problem if g is Hölder continuous at x, because the fact that $E_x(x_i - B_s^i)^2 = s$ allows us to write

$$E_x\left[\left(\frac{(x_i - B_s^i)^2 - s}{s^2}\right)g(t-s, B_s)\right]$$

$$= E_x\left[\left(\frac{(x_i - B_s^i)^2 - s}{s^2}\right)\{g(t-s, B_s) - g(t-s, x)\}\right]$$

Using the Hölder continuity of g one can show the above is $\leq Cs^{-1+\alpha/2}$, so its integral from $s = 0$ to t converges absolutely, and with a little work (2.6c) follows. (See Friedman (1964), pages 10-12, for more details.) □

The last detail now is:

(2.6d) Theorem. Let g be as in (2.6c). Then $\partial v/\partial t$ exists, and

$$\frac{\partial v}{\partial t}(t, x) = g(t, x) + \int_0^t dr \int \frac{\partial}{\partial t} p_{t-r}(x, y) g(r, y)\, dy$$

Proof To take the derivative w.r.t. t, we rewrite v as

$$v(t, x) = \int_0^t \int p_{t-r}(x, y) g(r, y)\, dy\, dr$$

Differentiating the right-hand side w.r.t. t gives two terms. Differentiating the upper limit of the integral gives $g(t, x)$. Differentiating the integrand and using a formula from (1.6) gives

$$\int_0^t \int \frac{\partial}{\partial t} p_{t-r}(x, y) g(r, y)\, dy\, dr$$

$$= \int_0^t \int \frac{-d/2}{(2\pi)^{d/2}(t-r)^{(d+2)/2}} \exp\left(\frac{-|x-y|^2}{2(t-r)}\right) g(r, y)\, dy\, dr$$

$$+ \int_0^t \int (2\pi(t-r))^{-d/2} \frac{|x-y|^2}{2(t-r)^2} \exp\left(\frac{-|x-y|^2}{2(t-r)}\right) g(r, y)\, dy\, dr$$

$$= \frac{-d}{2} \int_0^t \frac{dr}{t-r} E_x g(r, B_{t-r}) + \int_0^t \frac{dr}{2(t-r)^2} E_x(|x - B_{t-r}|^2 g(r, B_{t-r}))$$

In the second integral, we can use the fact that $E_x(|x - B_{t-r}|^2) = C(t-r)$ to cancel one of the $t-r$'s and make the second expression like the first, but even if we do this,

$$\int_0^t \frac{dr}{t-r} = \infty$$

This is the difficulty that we experienced in the proof of (2.6c), and the remedy is the same: We can save the day if g is Hölder continuous locally in t. For further details, see pages 12-13 of Friedman (1964). □

Unbounded g. To see what type of unbounded g's can be allowed, we will restrict our attention to the temporally homogeneous case $g(t,x) = h(x)$. At the bare minimum, we need

$$E_x \int_0^t |h(B_s)| ds < \infty$$

and if we want (2.1b) to hold, we need to know that if $t_n \downarrow 0$ and $x_n \to x$, then

$$E_{x_n} \int_0^{t_n} h(B_s) \, ds \to 0$$

If we put absolute values inside the integral and strengthen the last result to uniform convergence for $x \in \mathbf{R}^d$, then we get a definition that is essentially due to Kato (1973). A function h is said to be in K_d if

$$\lim_{t \downarrow 0} \sup_x E_x \left(\int_0^t |h(B_s)| ds \right) = 0$$

By Fubini's theorem, we can write the above as

(*)
$$\lim_{t \downarrow 0} \sup_x \int k_t(x,y) |h(y)| dy = 0$$

where

$$k_t(x,y) = \int_0^t (2\pi s)^{-d/2} e^{-|x-y|^2/2s} ds$$

By considering the asymptotic behavior of $k_t(x,y)$ as $t \to 0$ and $|x-y|^2/t \to c$, we can cast this condition in a more analytical form as

(**)
$$\lim_{\alpha \downarrow 0} \sup_x \int_{|x-y| \le \alpha} \varphi(|x-y|) |h(y)| dy = 0$$

where

$$\varphi(r) = \begin{cases} r^{-(d-2)} & d \ge 3 \\ -\log r & d = 2 \\ r & d = 1 \end{cases}$$

The equivalence of (*) and (**) is Theorem 4.5 of Aizenman and Simon (1982) or Theorem 3.6 of Chung and Zhao (1995). Section 3.1 of the latter source can be consulted for a wealth of information about these spaces. We will content ourselves here with an

Example 2.1. Let $h(z) = k(|z|)$ where $k(r) = r^{-p}$ for $r \le 1$ and 0 for $r > 1$. We will now show that

(2.7) Theorem. $h \in K_d$ if $p < 2$.

The special case $p = 1$ is the Coulomb potential which is important in physics.

Proof If $\alpha < 1$ then changing to polar coordinates and noticing the integral is largest when $x = 0$ we have

$$\sup_x \int_{|x-y| \le \alpha} \varphi(|x-y|)|h(y)|\, dy \le C_d \int_0^\alpha \varphi(r)k(r)r^{d-1}\, dr$$

If we replace $k(r)$ by r^{-p} and $\varphi(r)$ by $r^{-(d-2)}$, which holds in $d \ne 2$, then the above

$$= C_d \int_0^\alpha r^{1-p}\, dr \to 0$$

when $p < 2$. In $d = 2$ when $p < 2$ we get

$$= C_d \int_0^\alpha r^{1-p} \log r\, dr \to 0 \qquad \square$$

Remark. We will have more to say about these spaces at the end of the next section. For the developments there, we will also need the space K_d^{loc}, which is defined in the obvious way: $f \in K_d^{\text{loc}}$ if for every $R < \infty$, the spatially truncated function $f 1_{(|x|<R)} \in K_d$.

4.3. The Feynman-Kac Formula

In this section, we will consider what happens when we add cu to the right-hand side of the heat equation. That is, we will study

(3.1a) $u \in C^{1,2}$ and $u_t = \frac{1}{2}\Delta u + cu$ in $(0, \infty) \times \mathbf{R}^d$.

(3.1b) u is continuous at each point of $\{0\} \times \mathbf{R}^d$ and $u(0, x) = f(x)$.

If $c(t, x) \le 0$, then this equation describes heat flow with cooling. That is, $u(t, x)$ gives the temperature at the point $x \in \mathbf{R}^d$ at time t, when the heat at x at time t dissipates at the rate $-c(t, x)$.

The first step, as usual, is to find a local martingale.

(3.2) Theorem. Let $c_s^t = \int_0^s c(t-r, B_r)\, dr$. If u satisfies (3.1a), then

$$M_s = u(t-s, B_s) \exp(c_s^t)$$

is a local martingale on $[0, t)$.

Proof Applying Itô's formula with $X_s^0 = t-s$, $X_s^i = B_s^i$ for $1 \le i \le d$, and $X_s^{d+1} = c_s^t$ gives that

$$u(t-s, B_s) \exp(c_s^t) - u(t, B_0)$$
$$= \int_0^s -u_t(t-r, B_r) \exp(c_r^t)\, dr + \int_0^s \exp(c_r^t) \nabla u(t-r, B_r) \cdot dB_r$$
$$+ \int_0^s u(t-r, B_r) \exp(c_r^t)\, dc_r^t + \frac{1}{2} \int_0^s \Delta u(t-r, B_r) \exp(c_r^t)\, dr$$

since we have

$$\langle X^i, X^j \rangle_t = \begin{cases} t & \text{if } 1 \le i = j \le d \\ 0 & \text{otherwise} \end{cases}$$

Using $dc_r^t = c(t-r, B_r)\, dr$ and rearranging, the right-hand side is

$$= \int_0^s \left(-u_t + cu + \frac{1}{2}\Delta u \right) (t-r, B_r) \exp(c_r^t)\, dr$$
$$+ \int_0^s \exp(c_r^t) \nabla u(t-r, B_r) \cdot dB_r$$

which proves (3.2), since $-u_t + cu + \frac{1}{2}\Delta u = 0$ and the second term is a local martingale. □

The next step, again, is a uniqueness result.

(3.3) Theorem. Suppose that c is bounded. If there is a solution of (3.1) that is bounded on $[0, T] \times \mathbf{R}^d$ for any $T < \infty$, it must be

$$v(t, x) \equiv E_x\{f(B_t) \exp(c_t^t)\}$$

Proof Under our assumptions on c and u, M_s, $0 \le s < t$, is a bounded martingale and $M_t \equiv \lim_{s \to t} M_s = f(B_t) \exp(c_t^t)$. Since M_s is uniformly integrable it follows that

$$u(t, x) = E_x M_0 = E_x M_t = v(t, x) \qquad \square$$

Section 4.3 The Feynman-Kac Formula

As before, it is easy to show that if v is smooth enough it is a solution.

(3.4) Theorem. Suppose f is bounded and that c is bounded and continuous. If $v \in C^{1,2}$, then it satisfies (3.1a).

Proof The Markov property implies, see Exercise 2.7 in Chapter 1 and take $h(r,x) = c(t-r,x)$, that

$$E_x(f(B_t)\exp(c_t^t)|\mathcal{F}_s) = \exp(c_s^t)E_{B_s}(f(B_{t-s})\exp(c_{t-s}^{t-s}))$$
$$= \exp(c_s^t)v(t-s, B_s)$$

The left-hand side is a martingale, so the right-hand side is also. If $v \in C^{1,2}$, then repeating the calculation in the proof of (3.2) shows that

$$v(t-s, B_s)\exp(c_s^t) - v(t, B_0)$$
$$= \int_0^s (-v_t + cv + \frac{1}{2}\Delta v)(t-r, B_r)\exp(c_r^t)\,dr$$
$$+ \text{ a local martingale}$$

The left-hand side is a local martingale, so the integral on the right-hand side is also. Since the integral is continuous and locally of bounded variation, (3.3) in Chapter 2 implies that it must be $\equiv 0$ almost surely. Again this implies that $(-v_t + cv + \frac{1}{2}\Delta v)(t-r, B_r)\exp(c_r^t) \equiv 0$ for our assumptions imply this quantity is continuous and if it were $\neq 0$ at some point we would have a contradiction. □

The next step is to give a condition that guarantees that v satisfies (3.1b). As before, we begin by considering what happens when everything is bounded.

(3.5) Theorem. If c is bounded and f is bounded and continuous, then v satisfies (3.1b).

Proof If $|c| \leq M$, then $e^{-Mt} \leq \exp(c_t^t) \leq e^{Mt}$, so $\exp(c_t^t) \to 1$ as $t \to 0$. Letting $\|f\|_\infty = \sup_x |f(x)|$ this result implies

$$|E_x \exp(c_t^t)f(B_t) - E_x f(B_t)| \leq \|f\|_\infty E_x|\exp(c_t^t) - 1| \to 0$$

(1.5) implies that $(t, x) \to E_x f(B_t)$ is continuous at each point of $\{0\} \times \mathbf{R}^d$ and the desired result follows. □

This brings us to the problem of determining when v is smooth enough to be a solution.

(3.6) Theorem. Suppose that f is bounded and Hölder continuous. If c is bounded and Hölder continuous locally in t, then $v \in C^{1,2}$ and, hence, satisfies (3.1a).

Proof To solve the problem in this case, we use a trick to reduce our result to the previous case. We begin by observing that

$$c_s^t = \int_0^s c(t-r, B_r)\, dr$$

is continuous and locally of bounded variation. So Itô's formula, (10.2) in Chapter 2, implies that if $h \in C^1$ then

$$h(c_t^t) - h(c_0^t) = \int_0^t h'(c_s^t)\, dc_s^t$$

Taking $h(x) = e^{-x}$ we have

$$\exp(-c_t^t) - 1 = -\int_0^t \exp(-c_s^t) c(t-s, B_s)\, ds$$

Multiplying by $-\exp(c_t^t)$ gives

$$\exp(c_t^t) - 1 = \int_0^t c(t-s, B_s) \exp(c_t^t - c_s^t)\, ds$$

Plugging in the definitions of c_t^t and c_s^t we have

$$\exp\left(\int_0^t c(t-r, B_r)\, dr\right) = 1 + \int_0^t c(t-s, B_s) \exp\left(\int_s^t c(t-r, B_r)\, dr\right) ds$$

Multiplying by $f(B_t)$, taking expected values, and using Fubini's theorem, which is justified since everything is bounded, gives

$$v(t,x) = E_x f(B_t) + \int_0^t E_x \left\{ c(t-s, B_s) \exp\left(\int_s^t c(t-r, B_r)\, dr\right) f(B_t) \right\} ds$$

Conditioning on $\mathcal{F}_s$, noticing $c(t-s, B_s) \in \mathcal{F}_s$, and using the Markov property as in Exercise 2.7 of Chapter 1,

$$E_x\left(c(t-s, B_s) \exp\left(\int_s^t c(t-r, B_r)\, dr\right) f(B_t) \bigg| \mathcal{F}_s\right) = c(t-s, B_s) v(t-s, B_s)$$

Section 4.3 The Feynman-Kac Formula   141

Taking the expected value of the last equation and plugging into the previous one, we have

(*) $$v(t,x) = E_x f(B_t) + \int_0^t E_x\{c(t-s, B_s)v(t-s, B_s)\}ds$$
$$\equiv v_1(t,x) + v_2(t,x)$$

The first term on the right, $v_1(t,x)$, is $C^{1,2}$ by (1.6). The second term, $v_2(t,x)$, is of the form considered in the last section with $g(r,x) = c(r,x)v(r,x)$. If we start with the trivial observation that if c and f are bounded, then v is bounded on $[0,N] \times \mathbf{R}^d$, and apply (2.6b), we see that

$$|v_2(t,x) - v_2(t,y)| \leq C_N|x-y| \quad \text{whenever } t \leq N$$

To get a similar estimate for v_1 let $\bar{B}_t$ be a Brownian motion starting at 0 and observe that since f is Hölder continuous

$$|v_1(t,x) - v_1(t,y)| = |E\{f(x+\bar{B}_t) - f(y+\bar{B}_t)\}|$$
$$\leq E|f(x+\bar{B}_t) - f(y+\bar{B}_t)| \leq C|x-y|^\alpha$$

Combining the last two estimates we see that $v(r,x)$ is Hölder continuous locally in t. Since c and v are bounded and we have supposed that $c(r,x)$ is Hölder continuous locally in t the triangle inequality implies $g(r,x)$ is Hölder continuous locally in t. Using (2.6c) and (2.6d) now, we have $v_2(t,x) \in C^{1,2}$.   □

Unbounded c. As in the last section, we can generalize the results above to unbounded c's, but for simplicity we restrict our attention to the temporally homogeneous case $c(t,x) = h(x)$. Given the formula (*) above, which expresses v as a volume potential, it is perhaps not too surprising that the appropriate assumption is $c \in K_d$. The key to working in this generality is what Simon (1982) calls

(3.7) Khasminskii's lemma. Let $g \geq 0$ be a function on $\mathbf{R}^d$ with

$$\alpha \equiv \sup_x E_x \left(\int_0^t g(B_s)\,ds\right) < 1$$

Then

$$\sup_x E_x \exp\left(\int_0^t g(B_s)\,ds\right) \leq (1-\alpha)^{-1}$$

Proof The Markov property and the nonnegativity of g imply that

$$\sup_x E_x \int \cdots \int_{0 < s_1 < \cdots < s_n < t} ds_1 \ldots ds_n\, g(B_{s_1}) \ldots g(B_{s_n}) \leq \alpha^n$$

Since the integrand is symmetric in $(s_1, \ldots, s_n)$ we have

$$\int \cdots \int_{0<s_1<\cdots<s_n<t} ds_1 \ldots ds_n\, g(B_{s_1})\ldots g(B_{s_n})$$

$$= \frac{1}{n!} \int_{s_1=0}^{t} \cdots \int_{s_n=0}^{t} ds_1 \ldots ds_n\, g(B_{s_1})\ldots g(B_{s_n})$$

$$= \frac{1}{n!} \left(\int_{s=0}^{t} g(B_s)\, ds \right)^n$$

Summing on n now gives the desired formula. $\square$

From the last result, it should be clear why assuming $h \in K_d$ is natural in this context. This condition guarantees that

$$\sup_x E_x \left(\int_0^t |h(B_s)|\, ds \right) \to 0$$

and, hence, that

$$\sup_x E_x \exp \left(\int_0^t |h(B_s)|\, ds \right) \to 1$$

With these two results in hand, we can proceed with developing the theory much as we did in the case of bounded coefficients.

B. Elliptic Equations

In the next three sections of this chapter, we show how Brownian motion can be used to construct classical solutions of the following equations:

$$0 = \frac{1}{2}\Delta u$$

$$0 = \frac{1}{2}\Delta u + g$$

$$0 = \frac{1}{2}\Delta u + cu$$

in an open set G subject to the boundary condition: at each point of ∂G, u is continuous and $u = f$.

We will see that the solutions to these equations are (under suitable assumptions) given by

$$E_x f(B_\tau)$$

$$E_x \left(f(B_\tau) + \int_0^\tau g(B_s)\, ds \right)$$

$$E_x \left(f(B_\tau) \exp\left(\int_0^\tau c(B_s)\, ds \right) \right)$$

where $\tau = \inf\{t > 0 : B_t \notin G\}$. To see the similarities to the solutions given for the equations in Part A of this chapter think of those solutions in terms of space-time Brownian motions $\hat{B}_s = (t - s, B_s)$ run until time $\hat{\tau} = \inf\{s : \hat{B}_s \notin (0,\infty) \times \mathbf{R}^d\}$. Of course, $\hat{\tau} \equiv t$.

4.4. The Dirichlet Problem

In this section, we will consider the Dirichlet problem. That is, we will study

(4.1a) $u \in C^2$ and $\Delta u = 0$ in G.

(4.1b) At each point of ∂G, u is continuous and $u = f$.

To see what this means, note that if we let $h(t, x) = u(x)$, then h satisfies the heat equation

$$\frac{\partial h}{\partial t} = \frac{1}{2}\Delta h$$

Thus, u is an equilibrium temperature distribution when ∂G is held at a fixed temperature profile f.

As in the first three sections, the first step in solving (4.1) is to find a local martingale.

(4.2) Theorem. Let $\tau = \inf\{t > 0 : B_t \notin G\}$. If u satisfies (4.1a), then $M_t = u(B_t)$ is a local martingale on $[0, \tau)$.

Proof Applying Itô's formula gives

$$u(B_t) - u(B_0) = \int_0^t \nabla u(B_s) \cdot dB_s + \frac{1}{2}\int_0^t \Delta u(B_s) ds$$

for $t < \tau$. This proves (4.2), since $\Delta u = 0$ in G and the first term is a local martingale on $[0, \tau)$. □

The second step is a uniqueness theorem. For this we introduce

(†) $P_x(\tau < \infty) = 1$ for all x

which (4.7b) below will show is necessary for uniqueness. Note that (1.3) in Chapter 3 implies bounded sets G satisfy (†).

(4.3) Theorem. Suppose G satisfies (†). If there is a solution of (4.1) that is bounded, it must be

$$v(x) \equiv E_x f(B_\tau)$$

144   Chapter 4   Partial Differential Equations

Proof If u is bounded and satisfies (4.1) then $M_s, 0 \leq s < \tau$, is a bounded local martingale. Using (2.7) in Chapter 2, (†), and (4.1b), we have

$$M_\tau \equiv \lim_{s \uparrow \tau} M_s = f(B_\tau)$$
$$u(x) = E_x M_0 = E_x M_\tau = v(x) \qquad \square$$

(†) is not needed for the existence of solutions, but to drop (†) we need to modify the definition to take the possibility of $\tau = \infty$ into account. Let

$$\bar{v}(x) \equiv E_x \left(f(B_\tau) 1_{(\tau < \infty)} \right)$$

As in the first three sections, it is easy to show:

(4.4) Theorem. Let G be any open set and suppose f is bounded. If $\bar{v} \in C^2$, then it satisfies (4.1a).

Proof The Markov property implies that on $\{\tau > s\}$

$$E_x \left(f(B_\tau) 1_{(\tau < \infty)} \big| \mathcal{F}_s \right) = \bar{v}(B_s)$$

The left-hand side is a local martingale on $[0, \tau)$, so $\bar{v}(B_s)$ is also. If $\bar{v} \in C^2$, then repeating the calculation in the proof of (4.2) shows that, for $s \in [0, \tau)$,

$$\bar{v}(B_s) - \bar{v}(B_0) = \frac{1}{2} \int_0^s \Delta \bar{v}(B_r)\, dr + \text{ a local martingale}$$

The left-hand side is a local martingale on $[0, \tau)$, so the integral on the right-hand side is also. However, the integral is continuous and locally of bounded variation, so by (3.3) in Chapter 2 it must be $\equiv 0$. Since $\Delta \bar{v}$ is continuous in G, it follows that $\Delta \bar{v} \equiv 0$ in G. For if $\Delta \bar{v} \neq 0$ at some point we would have a contradiction. $\qquad \square$

Up to this point, everything has been the same as in Section 4.1. Differences appear when we consider the boundary condition (4.1b), since it is no longer sufficient for f to be bounded and continuous. The open set G must satisfy a regularity condition.

Definition. A point $y \in \partial G$ is said to be a **regular point** if $P_y(\tau = 0) = 1$.

(4.5) Theorem. Let G be any open set. Suppose f is bounded and continuous and y is a regular point of ∂G. If $x_n \in G$ and $x_n \to y$, then $\bar{v}(x_n) \to f(y)$.

Section 4.4 The Dirichlet Problem   145

Proof The first step is to show

(4.5a) Lemma. If $t > 0$, then $x \to P_x(\tau \le t)$ is lower semicontinuous. That is, if $x_n \to x$, then
$$\liminf_{n \to \infty} P_{x_n}(\tau \le t) \ge P_x(\tau \le t)$$

Proof By the Markov property
$$P_x(B_s \in G^c \text{ for some } s \in (\epsilon, t]) = \int p_\epsilon(x, y) P_y(\tau \le t - \epsilon) \, dy$$

Since $y \to P_y(\tau \le t - \epsilon)$ is bounded and measurable and
$$p_\epsilon(x, y) = (2\pi\epsilon)^{-d/2} e^{-|x-y|^2/2\epsilon}$$

it follows from the dominated convergence theorem that
$$x \to P_x(B_s \in G^c \text{ for some } s \in (\epsilon, t])$$

is continuous for each $\epsilon > 0$. Letting $\epsilon \downarrow 0$ shows that $x \to P_x(\tau \le t)$ is an increasing limit of continuous functions and, hence, lower semicontinuous. □

If y is regular for G and $t > 0$, then $P_y(\tau \le t) = 1$, so it follows from (4.5a) that if $x_n \to y$, then
$$\liminf_{n \to \infty} P_{x_n}(\tau \le t) \ge 1 \quad \text{for all } t > 0$$

With this established, it is easy to complete the proof. Since f is bounded and continuous, it suffices to show that if $D(y, \delta) = \{x : |x - y| < \delta\}$, then

(4.5b) Lemma. If y is regular for G and $x_n \to y$, then for all $\delta > 0$
$$P_{x_n}(\tau < \infty, B_\tau \in D(y, \delta)) \to 1$$

Proof Let $\epsilon > 0$ and pick t so small that
$$P_0\left(\sup_{0 \le s \le t} |B_s| > \frac{\delta}{2}\right) < \epsilon$$

Since $P_{x_n}(\tau \le t) \to 1$ as $x_n \to y$, it follows from the choices above that
$$\liminf_{n \to \infty} P_{x_n}(\tau < \infty, B_\tau \in D(y, \delta)) \ge \liminf_{n \to \infty} P_{x_n}\left(\tau \le t, \sup_{0 \le s \le t} |B_s - x_n| \le \frac{\delta}{2}\right)$$
$$\ge \liminf_{n \to \infty} P_{x_n}(\tau \le t) - P_0\left(\sup_{0 \le s \le t} |B_t| > \frac{\delta}{2}\right) > 1 - \epsilon$$

Since ϵ was arbitrary, this proves (4.5b) and, hence, (4.5). □

(4.5) shows that if every point of ∂G is regular, then $\bar{v}(x)$ will satisfy the boundary condition (4.1b) for any bounded continuous f. The next two exercises develop a converse to this result. The first identifies a trivial case.

Exercise 4.1. If $G \subset \mathbf{R}$ is open then each point of ∂G is regular.

Exercise 4.2. Let G be an open set in $\mathbf{R}^d$ with $d \geq 2$ and let $y \in \partial G$ have $P_y(\tau = 0) < 1$. Let f be a continuous function with $f(y) = 1$ and $f(z) < 1$ for all $z \neq y$. Show that there is a sequence of points $x_n \to y$ such that $\liminf_{n \to \infty} \bar{v}(x_n) < 1$.

From the discussion above, we see that for v to satisfy (4.1b) it is sufficient (and almost necessary) that each point of ∂G is a regular point. This situation raises two questions:

Do irregular points exist?

What are sufficient conditions for a point to be regular?

In order to answer the first question we will give two examples.

Example 4.1. Punctured Disc. Let $d \geq 2$ and let $G = D - \{0\}$, where $D = \{x : |x| < 1\}$. If we let $T_0 = \inf\{t > 0 : B_t = 0\}$, then $P_0(T_0 = \infty) = 1$ by (1.10) in Chapter 3, so 0 is not a regular point of ∂D. One can get an example with a bigger boundary in $d \geq 3$ by looking at $G = D - K$ where $K = \{x : x_1 = x_2 = 0\}$. In $d = 3$ this is a ball minus a line through the origin.

Example 4.2. Lebesgue's Thorn. Let $d \geq 3$ and let

$$G = (-1,1)^d - \cup_{1 \leq n \leq \infty}\{[2^{-n}, 2^{-n+1}] \times [-a_n, a_n]^{d-1}\}$$

where the $n = \infty$ term is the single point $\{0\}$.

Claim. If $a_n \downarrow 0$ sufficiently fast, then 0 is not a regular point of ∂G.

Proof $P_0((B_t^2, B_t^3) = (0,0)$ for some $t > 0) = 0$, so with probability 1, a Brownian motion B_t starting at 0 will not hit

$$I_n = \{x : x_1 \in [2^{-n}, 2^{-n+1}], x_2 = x_3 = \ldots = x_d = 0\}$$

Since B_t is transient in $d \geq 3$ it follows that for a.e. ω the distance between $\{B_s : 0 \leq s < \infty\}$ and I_n is positive. From the last observation, it follows immediately that if we let $T_n = \inf\{t : B_t \in [2^{-n}, 2^{-n+1}] \times [a_n, a_n]^{d-1}\}$ and pick a_n small enough, then $P_0(T_n < \infty) \leq 3^{-n}$. Now $\sum_{n=1}^{\infty} 3^{-n} = 3^{-1}(3/2) = 1/2$, so if we let $\tau = \inf\{t > 0 : B_t \notin G\}$ and $\sigma = \inf\{t > 0 : B_t \notin (-1,1)^d\}$, then we have

$$P_0(\tau < \sigma) \leq \sum_{n=1}^{\infty} P_0(T_n < \infty) \leq \frac{1}{2}$$

Thus $P_0(\tau > 0) \geq P_0(\tau = \sigma) \geq 1/2$ and 0 is an irregular point. □

The last two examples show that if G^c is too small near y, then y may be irregular. The next result shows that if G^c is not too small near y, then y is regular.

(4.5c) Cone Condition. If there is a cone V having vertex y and an $r > 0$ such that $V \cap D(y, r) \subset G^c$, then y is a regular point.

Proof The first thing to do is to define a cone with vertex y, pointing in direction v, with opening a as follows:

$$V(y, v, a) = \{x : x = y + \theta(v + z) \text{ where } \theta \in (0, \infty), z \perp v, \text{ and } |z| < a\}$$

Now that we have defined a cone, the rest is easy. Since the normal distribution is spherically symmetric,

$$P_y(B_t \in V(y, v, a)) = \epsilon_a > 0$$

where ϵ_a is a constant that depends only on the opening a. Let $r > 0$ be such that $V(y, v, a) \cap D(y, r) \subset G^c$. The continuity of Brownian paths implies

$$\lim_{t \to 0} P_y\left(\sup_{s \leq t} |B_s - y| > r\right) = 0$$

Combining the last two results with a trivial inequality we have

$$\epsilon_a \leq \liminf_{t \downarrow 0} P_y(B_t \in G^c) \leq \lim_{t \downarrow 0} P_y(\tau \leq t) \leq P_y(\tau = 0)$$

and it follows from Blumenthal's zero-one law that $P_y(\tau = 0) = 1$. □

(4.5c) is sufficient for most cases.

Exercise 4.3. Let $G = \{x : g(x) < 0\}$ where g is a C^1 function with $\nabla g(y) \neq 0$ for each $y \in \partial G$. Show that each point of ∂G is regular.

148 Chapter 4 Partial Differential Equations

However, in a pinch the following generalization can be useful.

Exercise 4.4. Define a flat cone $\bar{V}(y, v, a)$ to be the intersection of $V(y, v, a)$ with a $d - 1$ dimensional hyperplane that contains the line $\{y + \theta v : \theta \in \mathbf{R}\}$. Show that (4.5c) remains true if "cone" is replaced by "flat cone."

When $d = 2$ this says that if we can find a line segment ending at z that lies in the complement then z is regular. For example this implies that each point of the boundary of the slit disc $G = D - \{x : x_1 \geq 0, x_2 = 0\}$ is regular.

Having completed our discussion of the boundary condition, we now turn our attention to determining when $\bar{v}$ is smooth. As in Section 4.1, this is true under minimal assumptions on f.

(4.6) Theorem. Let G be any open set. If f is bounded, then $\bar{v} \in C^\infty$ and, hence, satisfies (4.1a).

Proof Let $x \in G$ and pick $\delta > 0$ so that $D(x, \delta) \subset G$. If we let $\sigma = \inf\{t : B_t \notin D(x, \delta)\}$, then the strong Markov property implies that (for more details see Example 3.2 in Chapter 1)

$$\bar{v}(x) = E_x\left(f(B_\tau)1_{(\tau<\infty)}\right) = E_x(\bar{v}(B_\sigma)) = \int_{\partial D(x,\delta)} \bar{v}(y)\,\pi(dy)$$

where π is surface measure on $D(x, \delta)$ normalized to be a probability measure. The last result is the "averaging property" of harmonic functions. Now a simple analytical argument takes over to show $\bar{v} \in C^\infty$.

(4.6a) Lemma. Let G be an open set and h be a bounded function which has the **averaging property**

$$h(x) = \int_{\partial D(x,\delta)} h(y)\,\pi(dy)$$

when $x \in G$ and $\delta < \delta_x$ where $\delta_x > 0$. Then $h \in C^\infty$ and $\Delta h = 0$ in G.

Proof Let ψ be a nonnegative infinitely differentiable function that vanishes on $[\delta^2, \infty)$ but is not $\equiv 0$. By repeatedly applying (1.7) it is routine to show that

$$g(x) = \int_{D(x,\delta)} \psi(|y - x|^2) h(y)\,dy \in C^\infty$$

Note that we use $|y-x|^2 = \sum_i(y_i - x_i)^2$ rather than $|y-x|$ which is not smooth at 0. Making a simple change of variables, then looking at things in polar coordinates, and using the averaging property we have

$$g(x) = \int_{D(0,\delta)} \psi(|z|^2) h(x+z)\, dz$$

$$= C\int_0^\delta dr\, r^{d-1}\psi(r^2)\left(\int_{\partial D(0,r)} h(x+z)\,\pi(dz)\right)$$

$$= C\left(\int_0^\delta dr\, r^{d-1}\psi(r^2)\right) h(x)$$

So h is a constant times g and hence C^∞.

To conclude now that $\Delta h = 0$ we note that the multivariate version of Taylor's theorem implies that if $|y-x| \leq r$ then

$$h(y) = h(x) + \sum_i (y_i - x_i) D_i h(x)$$
$$+ \sum_{ij}(y_i - x_i)(y_j - x_j) D_{ij} h(x) + \epsilon(y,x)$$

where $|\epsilon(y,x)| \leq C_3 r^3$. Integrating over $D(x,r)$ w.r.t. $dy/|D(0,r)|$ and using the averaging property we have

$$h(x) = h(x) + 0 + C_2 \Delta h(x) r^2 + O(r^3)$$

Subtracting $h(x)$ from each side, dividing by r^2, and letting $r \to 0$ it follows that $\Delta h(x) = 0$. □

Unbounded G. As in the previous three sections, our last topic is to discuss what happens when something becomes unbounded. This time we will focus on G and ignore f. Combining (4.3), (4.5), and (4.6) we have:

(4.7a) Theorem. Suppose that f is bounded and continuous and that each point of ∂G is regular. If for all $x \in G$, $P_x(\tau < \infty) = 1$, then v is the unique bounded solution of (4.1).

To see that there might be other unbounded solutions consider $G = (0,\infty)$, $f(0) = 0$, and note that $u(x) = cx$ is a solution. Conversely, we have

(4.7b) Theorem. Suppose that f is bounded and continuous and that each point of ∂G is regular. If for some $x \in G, P_x(\tau < \infty) < 1$, then the solution of (4.1) is not unique.

Proof Since $h(x) = P_x(\tau = \infty)$ has the averaging property given in (4.6a), it is C^∞ and has $\Delta h = 0$ in G. Since each point $y \in \partial G$ is regular, a trivial comparison and (4.5a) implies

$$\limsup_{x \to y} P_x(\tau = \infty) \le \limsup_{x \to y} P_x(\tau > 1) \le P_y(\tau > 1) = 0$$

The last two observations show that h is a solution of (4.1) with $f \equiv 0$, which completes the proof. □

By working a little harder, we can show that adding $\alpha P_x(\tau = \infty)$ to $\bar{v}(x)$ is the only way to produce new bounded solutions.

(4.7c) Theorem. Suppose that f is bounded and continuous and that each point of ∂G is regular. If u is bounded and satisfies (4.1) in G, then there is a constant α such that

$$u(x) = E_x(f(B_\tau); \tau < \infty) + \alpha P_x(\tau = \infty)$$

We will warm up for this by proving the following special case in which $G = \mathbf{R}^d$.

(4.7d) Theorem. If u is bounded and harmonic in $\mathbf{R}^d$ then u is constant.

Proof (4.2) above and (2.6) in Chapter 2 imply that $u(B_t)$ is a bounded martingale. So the martingale convergence theorem implies that as $t \to \infty$, $u(B_t) \to U_\infty$. Since U_∞ is measurable with respect to the tail σ-field, it follows from (2.12) in Chapter 1 that $P_x(a < U_\infty < b)$ is either $\equiv 0$ or $\equiv 1$ for any $a < b$. The last result implies that there is a constant c independent of x so that $P_x(U_\infty = c) \equiv 1$. Taking expected values it follows that $u(x) = E_x U_\infty \equiv c$. □

Proof of (4.7c) From the proof of (4.7d) we see that $u(B_t)$ is a bounded local martingale on $[0, \tau)$ so $U_\tau = \lim_{t \uparrow \tau} u(B_t)$. On $\{\tau < \infty\}$, we have $U_\tau = f(B_\tau)$ so what we need to show is that there is a constant α independent of the starting point B_0 so that $U_\tau = \alpha$ on $\{\tau = \infty\}$. Intuitively, this is a consequence of the triviality of the asymptotic σ-field, but the fact that $0 < P_x(\tau = \infty) < 1$ makes it difficult to extract this from (2.12) in Chapter 1.

To get around the difficulty identified in the previous paragraph, we will extend u to the whole space. The two steps in doing this are to

(a) Let $h(x) = u(x) - E_x(f(B_\tau); \tau < \infty)$. (4.6) and (4.5) imply that h is bounded and satisfies (4.1) with boundary function $f \equiv 0$.

(b) Let $M = \|h\|_\infty$ and look at

$$w(x) = \begin{cases} h(x) + M P_x(\tau = \infty) & x \in G \\ 0 & x \in G^c \end{cases}$$

To complete the proof now, we will show in four steps that $w(x) = \beta P_x(\tau = \infty)$, from which the desired result follows immediately.

(i) When restricted to G, w satisfies (4.1) with boundary function $f \equiv 0$.

The proof of (4.7b) implies that $MP_x(\tau < \infty)$ satisfies (4.1) with boundary function $f \equiv 0$. Combining this with (a) the desired result follows.

(ii) $w \geq 0$.

To do this we use the optional stopping theorem on the martingale $h(B_t)$ at time $\tau \wedge t$ and note $h(B_\tau) = 0$ to get

$$h(x) = E_x(h(B_t); \tau > t) \geq -MP_x(\tau > t)$$

Letting $t \to \infty$ proves $h(x) \geq -MP_x(\tau = \infty)$.

(iii) $w(B_t)$ is a submartingale.

Because of the Markov property it is enough to show that for all x and t we have $w(x) \leq E_x w(B_t)$. Since $w \geq 0$ this is trivial if $x \notin G$. To prove this for $x \in G$, we note that (i) implies $W_t = w(B_t)$ is a bounded local martingale on $[0, \tau)$ and $W_\tau = 0$ on $\{\tau < \infty\}$, so using the optional stopping theorem

$$w(x) = E_x(W_{\tau \wedge t}) = E_x(w(B_t); \tau > t) \leq E_x(w(B_t))$$

(iv) There is a constant β so that $w(x) = \beta P_x(\tau = \infty)$.

Since w is a bounded submartingale it follows that as $t \to \infty$, $w(B_t)$ converges to a limit W_∞. The argument in (4.7d) implies there is a constant β so that $P_x(W_\infty = \beta) = 1$ for all x. Letting $t \to \infty$ in

$$w(x) = E_x(w(B_t); \tau > t)$$

and using the bounded convergence gives (iv). $\square$

4.5. Poisson's Equation

In this section, we will see what happens when we add a function of x to the equation considered in the last section. That is, we will study:

(5.1a) $u \in C^2$ and $\frac{1}{2}\Delta u = -g$ in G.

(5.1b) At each point of ∂G, u is continuous and $u = 0$.

As in Section 4.2, we can add a solution of (4.1) to replace $u = 0$ in (5.1b) by $u = f$. As always, the first step in solving (5.1) is to find a local martingale.

(5.2) Theorem. Let $\tau = \inf\{t > 0 : B_t \notin G\}$. If u satisfies (5.1a), then

$$M_t = u(B_t) + \int_0^t g(B_s)\,ds$$

is a local martingale on $[0, \tau)$.

Proof Applying Itô's formula as we did in the last section gives

$$u(B_t) - u(B_0) = \int_0^t \nabla u(B_s) \cdot dB_s + \frac{1}{2}\int_0^t \Delta u(B_s)\,ds$$

for $t < \tau$. This proves (5.2), since $\frac{1}{2}\Delta u = -g$ and the first term on the right-hand side is a local martingale on $[0, \tau)$. $\square$

The next step is to prove a uniqueness result.

(5.3) Theorem. Suppose that G and g are bounded. If there is a solution of (5.1) that is bounded, it must be

$$v(x) \equiv E_x\left(\int_0^\tau g(B_t)\,dt\right)$$

Proof If u satisfies (5.1a) then M_t defined in (5.2) is a local martingale on $[0, \tau)$. If G is bounded, then (1.3) in Chapter 3 implies $E_x \tau < \infty$ for all $x \in G$. If u and g are bounded then for $t < \tau$

$$|M_t| \leq \|u\|_\infty + \tau\|g\|_\infty$$

Since the right-hand side is integrable, (2.7) in Chapter 2 and (5.1b) imply

$$M_\tau \equiv \lim_{t \uparrow \tau} M_t = \int_0^\tau g(B_t)\,dt$$
$$u(x) = E_x M_0 = E_x(M_\tau) = v(x) \quad \square$$

As usual, it is easy to show

(5.4) Theorem. Suppose that G is bounded and g is continuous. If $v \in C^2$, then it satisfies (5.1a).

Proof The Markov property implies that on $\{\tau > s\}$,

$$E_x\left(\int_0^\tau g(B_t)\,dt \,\bigg|\, \mathcal{F}_s\right) = \int_0^s g(B_t)\,dt + v(B_s)$$

The left-hand side is a local martingale on $[0, \tau)$, so the right-hand side is also. If $v \in C^2$, then repeating the calculation in the proof of (5.2) shows that for $s \in [0, \tau)$,

$$v(B_s) - v(B_0) + \int_0^s g(B_r)\,dr = \int_0^s \left(\frac{1}{2}\Delta v + g\right)(B_r)\,dr$$
$$+ \text{ a local martingale}$$

The left-hand side is a local martingale on $[0, \tau)$, so the integral on the right-hand side is also. However, the integral is continuous and locally of bounded variation, so by (3.3) in Chapter 2 it must be $\equiv 0$. Since $\frac{1}{2}\Delta v + g$ is continuous in G, it follows that it is $\equiv 0$ in G, for if it were $\neq 0$ at some point then we would have a contradiction. □

After the extensive discussion in the last section, the conditions needed to guarantee that the boundary conditions hold should come as no surprise.

(5.5) Theorem. Suppose that G and g are bounded. Let y be a regular point of ∂G. If $x_n \in G$ and $x_n \to y$, then $v(x_n) \to 0$.

Proof We begin by observing:

(i) It follows from (4.5a) that if $\epsilon > 0$, then $P_{x_n}(\tau > \epsilon) \to 0$.

(ii) If G is bounded, then (1.3) in Chapter 3 implies $C = \sup_x E_x \tau < \infty$ and, hence, $\|v\|_\infty \leq C\|g\|_\infty < \infty$.

Let $\epsilon > 0$. Beginning with some elementary inequalities then using the Markov property we have

$$|v(x_n)| \leq E_{x_n}\left(\int_0^{\tau \wedge \epsilon} |g(B_s)|\,ds\right) + E_{x_n}\left(\left|\int_\epsilon^\tau g(B_s)\,ds\right|; \tau > \epsilon\right)$$
$$\leq \epsilon\|g\|_\infty + E_{x_n}(|v(B_\epsilon)|; \tau > \epsilon) \leq \epsilon\|g\|_\infty + \|v\|_\infty P_{x_n}(\tau > \epsilon)$$

Letting $n \to \infty$, and using (i) and (ii) proves (5.5) since ϵ is arbitrary. □

Last, but not least, we come to the question of smoothness. For these developments we will assume that g is defined on $\mathbf{R}^d$ not just in G and has compact support. Recall we are supposing G is bounded and notice that the values of g on G^c are irrelevant for (5.1), so there will be no loss of generality if we later want to suppose that $\int g(x)\,dx = 0$. We begin with the case $d \geq 3$, because in this case (2.2) in Chapter 3 implies

$$\bar{w}(x) = E_x \int_0^\infty |g(B_t)|\,dt < \infty$$

154   Chapter 4   Partial Differential Equations

and, moreover, is a bounded function of x. This means we can define

$$w(x) = E_x \int_0^\infty g(B_t)\,dt$$

use the strong Markov property to conclude

$$w(x) = E_x \int_0^\tau g(B_t)\,dt + E_x w(B_\tau)$$

and change notation to get

(*) $$v(x) = w(x) - E_x w(B_\tau)$$

(4.6) tells us that the second term is C^∞ in G, so to verify that $v \in C^2$ we need only prove that w is, a task that is made simple by the fact that (2.4) and (2.3) in Chapter 3 gives the following explicit formula

$$w(x) = C_d \int |x-y|^{2-d} g(y)\,dy$$

The first derivative is easy.

(5.6a) **Theorem.** If g is bounded and has compact support, then w is C^1 and there is a constant C which only depends on d so that

$$|D_i w(x)| \le C\|g\|_\infty \int_{\{g \ne 0\}} \frac{dy}{|x-y|^{d-1}} < \infty$$

Proof   We will content ourselves to show that the expression we get by differentiating under the integral sign converges and leave it to the reader to apply (1.7) to make the argument rigorous. Now

$$D_i |x-y|^{2-d} = \left(\frac{2-d}{2}\right) \left(\sum_j (x_j - y_j)^2\right)^{-d/2} 2(x_i - y_i)$$

So differentiating under the integral sign

$$D_i w(x) = C_d(2-d) \int \frac{(x_i - y_i)}{|x-y|^d} g(y)\,dy$$

The integral on the right-hand side is convergent since

$$\int \left|\frac{(x_i - y_i)}{|x-y|^d} g(y)\right| dy \le \|g\|_\infty \int_{\{g \ne 0\}} \frac{dy}{|x-y|^{d-1}} < \infty \qquad \square$$

As in Section 4.2, trouble starts when we consider second derivatives. If $i \neq j$, then

$$D_{ij}|x-y|^{2-d} = (2-d)(-d)|x-y|^{-d-2}(x_i - y_i)(x_j - y_j)$$

In this case, the estimate used above leads to

$$|D_{ij}|x-y|^{2-d}| \leq C|x-y|^{-d}$$

which is (just barely) not locally integrable. As in Section 4.2, if g is Hölder continuous of order α, we can get an extra $|x-y|^\alpha$ to save the day. The details are tedious, so we will content ourselves to state the result.

(5.6b) **Theorem.** If g is Hölder continuous, then w is C^2.

The reader can find a proof either in Port and Stone (1978), pages 116-117, or in Gilbarg and Trudinger (1977), pages 53-55. Combining (*) with (5.6b) gives

(5.6) **Theorem.** Suppose that G is bounded. If g is Hölder continuous, then $v \in C^2$ and hence satisfies (5.1a).

The last result settles the question of smoothness in $d \geq 3$. To extend the result to $d \leq 2$, we need to find a substitute for (*). To do this, we let

$$w(x) = \int G(x,y) g(y) \, dy$$

where G is the potential kernel defined in (2.7) of Chapter 3, that is,

$$G(x,y) = \begin{cases} -\frac{1}{\pi} \log(|x-y|) & d=2 \\ -|x-y| & d=1 \end{cases}$$

G was defined as

$$\int_0^\infty \{p_t(x,y) - a_t\} \, dt$$

where the a_t were chosen to make the integral converge. So if $\int g \, dx = 0$, we see that

$$\int G(x,y) g(y) \, dy = \lim_{T \to \infty} E_x \int_0^T g(B_t) \, dt$$

Using this interpretation of w, we can easily show that (*) holds, so again our problem is reduced to proving that w is C^2, which is a problem in calculus. Once all the computations are done, we find that (5.6) holds in $d \leq 2$ and that in $d=1$, it is sufficient to assume that g is continuous. The reader can find

4.6. The Schrödinger Equation

In this section, we will consider what happens when we add cu to the left-hand side of the equation considered in Section 4.4. That is, we will study

(6.1a) $u \in C^2$ and $\frac{1}{2}\Delta u + cu = 0$ in G.

(6.1b) At each point of ∂G, u is continuous and $u = f$.

As always, the first step in solving (6.1) is to find a local martingale.

(6.2) Theorem. Let $\tau = \inf\{t > 0 : B_t \notin G\}$. If u satisfies (6.1a) then

$$M_t = u(B_t) \exp\left(\int_0^t c(B_s) ds\right)$$

is a local martingale on $[0, \tau)$.

Proof   Let $c_t = \int_0^t c(B_s) ds$. Applying Itô's formula gives

$$u(B_t)\exp(c_t) - u(B_0) = \int_0^t \exp(c_s)\nabla u(B_s) \cdot dB_s + \int_0^t u(B_s)\exp(c_s)\, dc_s$$
$$+ \frac{1}{2}\int_0^t \Delta u(B_s)\exp(c_s)\, ds$$

for $t < \tau$. This proves (6.2), since $dc_s = c(B_s)\, ds$, $\frac{1}{2}\Delta u + cu = 0$, and the first term on the right-hand side is a local martingale on $[0, \tau)$. □

At this point, the reader might expect that the next step, as it has been five times before, is to assume that everything is bounded and conclude that if there is a solution of (6.1) that is bounded, it must be

$$v(x) \equiv E_x(f(B_\tau)\exp(c_\tau))$$

We will not do this, however, because the following simple example shows that this result is false.

Example 6.1. Let $d = 1$, $G = (-a, a)$, $c \equiv \gamma$, and $f \equiv 1$. The equation we are considering is

$$\frac{1}{2}u'' + \gamma u = 0 \qquad u(a) = u(-a) = 1$$

The general solution is $A\cos bx + B\sin bx$, where $b = \sqrt{2\gamma}$. So if we want the boundary condition to be satisfied we must have

$$1 = A\cos ba + B\sin ba$$
$$1 = A\cos(-ba) + B\sin(-ba) = A\cos ba - B\sin ba$$

Adding the two equations and then subtracting them it follows that

$$2 = 2A\cos ba \qquad 0 = 2B\sin ba$$

From this we see that $B = 0$ always works and we may or may not be able to solve for A.

If $\cos ba = 0$ then there is no solution.

If $\cos ba \neq 0$ then $u(x) = \cos bx / \cos ba$ is a solution.

We will see later (in Example 9.1) that if $ab < \pi/2$ then

$$v(x) = \cos bx / \cos ba$$

However this cannot possibly hold for $ab > \pi/2$ since $v(x) \geq 0$ while the right-hand side is < 0 for some values of x. □

We will see below (again in Example 9.1) that the trouble with the last example is that if $ab > \pi/2$ then $c \equiv \gamma$ is too large, or to be precise, if we let

$$w(x) = E_x \exp\left(\int_0^\tau c(B_s)ds\right)$$

then $w \equiv \infty$ in $(-a,a)$. The rest of this section is devoted to showing that if $w \not\equiv \infty$, then "everything is fine." The development will require several stages. The first step is to show

(6.3a) Lemma. Let $\theta > 0$. There is a $\mu > 0$ so that if H is an open set with Lebesgue measure $|H| \leq \mu$ and $\tau_H = \inf\{t > 0 : B_t \notin H\}$ then

$$\sup_x E_x\left(\exp(\theta \tau_H)\right) \leq 2$$

Proof Pick $\gamma > 0$ so that $e^{\theta\gamma} \leq 4/3$. Clearly,

$$P_x(\tau_H > \gamma) \leq \int_H \frac{1}{(2\pi\gamma)^{d/2}} e^{-|x-y|^2/2\gamma} dy \leq \frac{|H|}{(2\pi\gamma)^{d/2}} \leq \frac{1}{4}$$

if we pick μ so that $\mu/(2\pi\gamma)^{d/2} \le 1/4$. Using the Markov property as in the proof of (1.2) in Chapter 3 we can conclude that

$$P_x(\tau_H > k\gamma) = E_x(P_{B_\gamma}(\tau_H > (k-1)\gamma); \tau_H > \gamma)$$
$$\le \frac{1}{4}\sup_y P_y(\tau_H > (k-1)\gamma)$$

So it follows by induction that for all integers $k \ge 0$ we have

$$\sup_x P_x(\tau_H > k\gamma) \le \frac{1}{4^k}$$

Since e^x is increasing, and $e^{\theta\gamma} \le 4/3$ by assumption, we have

$$E_x \exp(\theta\tau_H) \le \sum_{k=1}^{\infty} \exp(\theta\gamma k) P_x((k-1)\gamma < \tau_H \le k\gamma)$$
$$\le \sum_{k=1}^{\infty} \left(\frac{4}{3}\right)^k \frac{1}{4^{k-1}} = \frac{4}{3}\sum_{k=1}^{\infty} \frac{1}{3^{k-1}} = \frac{4}{3} \cdot \frac{1}{1-\frac{1}{3}} = 2$$

Careful readers may have noticed that we left $\tau_H = 0$ out of the expected value. However, by the Blumenthal 0-1 law either $P_x(\tau_H = 0) = 0$ in which case our computation is valid, or $P_x(\tau_H = 0) = 1$ in which case $E_x \exp(\theta\tau_H) = 1$. □

Let $c^* = \sup_x |c(x)|$. By (6.3a), we can pick r_0 so small that if $T_r = \inf\{t : |B_t - B_0| > r\}$ and $r \le r_0$, then $E_x \exp(c^* T_r) \le 2$ for all x.

(6.3b) Lemma. Let $2\delta \le r_0$. If $D(x, 2\delta) \subset G$ and $y \in D(x, \delta)$, then

$$w(y) \le 2^{d+2} w(x)$$

The reason for our interest in this is that it shows $w(x) < \infty$ implies $w(y) < \infty$ for $y \in D(x, \delta)$.

Proof If $D(y, r) \subset G$, and $r \le r_0$ then the strong Markov property implies

$$w(y) = E_y[\exp(c_{T_r})w(B(T_r))] \le E_y[\exp(c^* T_r)w(B(T_r))]$$
$$= E_y[\exp(c^* T_r)]\int_{\partial D(y,r)} w(z)\pi(dz) \le 2\int_{\partial D(y,r)} w(z)\pi(dz)$$

where π is surface measure on $\partial D(y, r)$ normalized to be a probability measure, since the exit time, T_r, and exit location, $B(T_r)$, are independent.

If $\delta \leq r_0$ and $D(y,\delta) \subset G$, multiplying the last inequality by r^{d-1} and integrating from 0 to δ gives

$$\frac{\delta^d}{d} w(y) \leq 2 \cdot \frac{1}{\sigma_d} \int_{D(y,\delta)} w(z)\, dz$$

where σ_d is the surface area of $\{x : |x| = 1\}$. Rearranging we have

(*) $$\int_{D(y,\delta)} w(z)\, dz \geq 2^{-1} \frac{\delta^d}{C_o} w(y)$$

where $C_o = d/\sigma_d$ is a constant that depends only on d.

Repeating the first argument in the proof with $y = x$ and using the fact that $c_{T_r} \geq -c^* T_r$ gives

$$w(x) = E_x[\exp(c_{T_r}) w(B(T_r))] \geq E_x[\exp(-c^* T_r) w(B(T_r))]$$
$$= E_x[\exp(-c^* T_r)] \int_{\partial D(x,r)} w(z) \pi(dz)$$

Since $1/x$ is convex, Jensen's inequality implies

$$E_x[\exp(-c^* T_r)] \geq 1/E_x[\exp(c^* T_r)] \geq 1/2$$

Combining the last two displays, multiplying by r^{d-1}, and integrating from 0 to 2δ we get the lower bound

$$\frac{(2\delta)^d}{d} w(x) \geq 2^{-1} \cdot \frac{1}{\sigma_d} \int_{D(x,2\delta)} w(z)\, dz$$

Rearranging and using $D(x,2\delta) \supset D(y,\delta)$, $w \geq 0$ we have

(**) $$w(x) \geq 2^{-1} \frac{C_o}{(2\delta)^d} \int_{D(y,\delta)} w(z)\, dz$$

where again $C_o = d/\sigma_d$. Combining (**) and (*) it follows that

$$w(x) \geq 2^{-1} \frac{C_o}{(2\delta)^d} \int_{D(y,\delta)} w(z)\, dz$$
$$\geq 2^{-1} \frac{C_o}{(2\delta)^d} \cdot 2^{-1} \frac{\delta^d}{C_o} w(y) = 2^{-(d+2)} w(y) \qquad \square$$

(6.3b) and a simple covering argument lead to

160   Chapter 4   Partial Differential Equations

(6.3c) Theorem. Let G be a connected open set. If $w \not\equiv \infty$ then
$$w(x) < \infty \quad \text{for all } x \in G$$

Proof From (6.3b), we see that if $w(x) < \infty$, $2\delta \le r_0$, and $D(x, 2\delta) \subset G$, then $w < \infty$ on $D(x, \delta)$. From this result, it follows that $G_0 = \{x : w(x) < \infty\}$ is an open subset of G. To argue now that G_0 is also closed (when considered as a subset of G) we observe that if $2\delta < r_0$, $D(y, 3\delta) \subset G$, and we have $x_n \in G_0$ with $x_n \to y \in G$ then for n sufficiently large, $y \in D(x_n, \delta)$ and $D(x_n, 2\delta) \subset G$, so $w(y) < \infty$. □

Before we proceed to the uniqueness result, we want to strengthen the last conclusion.

(6.3d) Theorem. Let G be a connected open set with finite Lebesgue measure, $|G| < \infty$. If $w \not\equiv \infty$ then
$$\sup_x w(x) < \infty$$

Proof Let $K \subset G$ be compact so that $|G - K| < \mu$ the constant in (6.3a) for $\theta = c^*$. For each $x \in K$ we can pick a δ_x so that $2\delta_x \le r_0$ and $D(x, 2\delta_x) \subset G$. The open sets $D(x, \delta_x)$ cover K so there is a finite subcover $D(x_i, \delta_{x_i})$, $1 \le i \le I$. Clearly,
$$\sup_{1 \le i \le I} w(x_i) < \infty$$
(6.3b) implies $w(y) \le 2^{d+2} w(x_i)$ for $y \in D(x_i, \delta_{x_i})$, so
$$M = \sup_{y \in K} w(y) < \infty$$
If $y \in H = G - K$, then $E_y(\exp(c^* \tau_H)) \le 2$ by (6.3a) so using the strong Markov property
$$w(y) = E_y \left(\exp(c_{\tau_H}); \tau_H = \tau\right) + E_y \left(\exp(c_{\tau_H}) w(B_{\tau_H}); B_{\tau_H} \in K\right)$$
$$\le 2 + M E_y \left(\exp(c_{\tau_H}); B_{\tau_H} \in K\right) \le 2 + 2M \qquad □$$

With (6.3d) established, we are now more than ready to prove our uniqueness result. To simplify the statements of the results that follow we will now list the assumptions that we will make for the rest of the section.

(A1) G is a bounded connected open set.

(A2) f and c are bounded and continuous.

(A3) $w \not\equiv \infty$.

(6.3) Theorem. If there is a solution of (6.1) that is bounded, it must be

$$v(x) \equiv E_x(f(B_\tau)\exp(c_\tau))$$

Proof (6.2) implies that $M_s = u(B_s)\exp(c_s)$ is a local martingale on $[0,\tau)$. Since f, c, and u are bounded, letting $s \uparrow \tau \wedge t$ and using the bounded convergence theorem gives

$$u(x) = E_x(f(B_\tau)\exp(c_\tau); \tau \leq t) + E_x(u(B_t)\exp(c_t); \tau > t)$$

Since f is bounded and $w(x) = E_x \exp(c_\tau) < \infty$, the dominated convergence theorem implies that as $t \to \infty$, the first term converges to $E_x(f(B_\tau)\exp(c_\tau))$. To show that the second term $\to 0$, we begin with the observation that since $\{\tau > t\} \in \mathcal{F}_t$, the definition of conditional expectation and the Markov property imply

$$E_x(u(B_t)\exp(c_\tau); \tau > t) = E_x(E_x(u(B_t)\exp(c_\tau)|\mathcal{F}_t); \tau > t)$$
$$= E_x(u(B_t)\exp(c_t)w(B_t); \tau > t)$$

Now we claim that for all $y \in G$

$$w(y) \geq \exp(-c^*)P_y(\tau \leq 1) \geq \epsilon > 0$$

The first inequality is trivial. The last two follow easily from (A1). See the first display in the proof of (1.2) in Chapter 3.

Replacing $w(B_t)$ by ϵ,

$$E_x(|u(B_t)|\exp(c_t); \tau > t) \leq \epsilon^{-1} E_x(|u(B_t)|\exp(c_\tau); \tau > t)$$
$$\leq \epsilon^{-1}\|u\|_\infty E_x(\exp(c_\tau); \tau > t) \to 0$$

as $t \to \infty$, by the dominated convergence theorem since $w(x) = E_x \exp(c_\tau) < \infty$ and $P_x(\tau < \infty) = 1$. Going back to the first equation in the proof, we have shown $u(x) = v(x)$ and the proof is complete. □

This completes our consideration of uniqueness. The next stage in our program, fortunately, is as easy as it always has been. Recall that here and in what follows we are assuming (A1)–(A3).

(6.4) Theorem. If $v \in C^2$, then it satisfies (6.1a) in G.

Proof The Markov property implies that on $\{\tau > s\}$,

$$E_x(\exp(c_\tau)f(B_\tau)|\mathcal{F}_s) = \exp(c_s)E_{B_s}(\exp(c_\tau)f(B_\tau))$$
$$= \exp(c_s)v(B_s)$$

The left-hand side is a local martingale on $[0,\tau)$, so the right-hand side is also. If $v \in C^2$, then repeating the calculation in the proof of (6.2) shows that for $s \in [0,\tau)$,

$$v(B_s)\exp(c_s) - v(B_0) = \int_0^s \left(\frac{1}{2}\Delta v + cv\right)(B_r)\exp(c_r)dr$$
$$+ \text{ a local martingale}$$

The left-hand side is a local martingale on $[0,\tau)$, so the integral on the right-hand side is also. However, the integral is continuous and locally of bounded variation, so by (3.3) in Chapter 2 it must be $\equiv 0$. Since $v \in C^2$ and c is continuous, it follows that $\frac{1}{2}\Delta v + cv \equiv 0$, for if it were $\neq 0$ at some point then we would have a contradiction. □

Having proved (6.4), the next step is to consider the boundary condition. As in the last two sections, we need the boundary to be regular.

(6.5) Theorem. v satisfies (6.1b) at each regular point of ∂G.

Proof Let y be a regular point of ∂G. We showed in (4.5a) and (4.5b) that if $x_n \to y$, then $P_{x_n}(\tau \leq \delta) \to 1$ and $P_{x_n}(B_\tau \in D(y,\delta)) \to 1$ for all $\delta > 0$. Since c is bounded and f is bounded and continuous, the bounded convergence theorem implies that

$$E_{x_n}(\exp(c_\tau)f(B_\tau); \tau \leq 1) \to f(y)$$

To control the contribution from the rest of the space, we observe that if $|c| \leq M$ then using the Markov property and the boundedness of w established in (6.3d) we have

$$E_{x_n}(\exp(c_\tau)f(B_\tau); \tau > 1) \leq e^M \|f\|_\infty E_{x_n}(w(B_1); \tau > 1)$$
$$\leq e^M \|f\|_\infty \|w\|_\infty P_{x_n}(\tau > 1) \to 0 \qquad \square$$

This brings us finally to the problem of determining when v is smooth enough to be a solution. We use the same trick used in Section 4.3 to reduce to the previous case. We begin with the identity established there, which holds for all t and Brownian paths ω and, hence, holds when $t = \tau(\omega)$

$$\exp\left(\int_0^\tau c(B_s)ds\right) = 1 + \int_0^\tau c(B_s)\exp\left(\int_s^\tau c(B_r)dr\right)ds$$

Multiplying by $f(B_\tau)$ and taking expected values gives

$$v(x) = E_x f(B_\tau) + \int_0^\infty E_x\left(c(B_s)\exp\left(\int_s^\tau c(B_r)dr\right) f(B_\tau)1_{(s<\tau)}\right)ds$$

Conditioning on $\mathcal{F}_s$ and using the Markov property, we can write the above as

$$v(x) = E_x f(B_\tau) + \int_0^\infty E_x(c(B_s)v(B_s); \tau > s)\, ds$$
$$\equiv v_1(x) + v_2(x)$$

The first term, $v_1(x)$, is C^∞ by (4.6). The second term is

$$v_2(x) = E_x\left(\int_0^\tau c(B_s)v(B_s)\,ds\right)$$

so if we let $g(x) = c(x)v(x)$ then we can apply results from the last section. If c and f are bounded and $w \not\equiv \infty$, then v is bounded by (6.3d), so it follows from results in the last section that v_2 is C^1 and has a bounded derivative.

Since $v_1 \in C^\infty$ and G is bounded, it follows that v is C^1 and has a bounded derivative. If c is Hölder continuous, then $g(x) = c(x)v(x)$ is Hölder continuous, and we can use (5.6b) from the last section to conclude $v_2 \in C^2$ and hence

(6.6) Theorem. If in addition to (A1)–(A3), c is Hölder continuous, then $v \in C^2$ and, hence, satisfies (6.1a).

C. Applications to Brownian Motion

In the next three sections we will use the p.d.e. results proved in the last three to derive some formulas for Brownian motion. The first two sections are closely related but the third can be read independently.

4.7. Exit Distributions for the Ball

In this section, we will use results for the Dirichlet problem proved in Section 4.4 to find the exit distributions for $D = \{x : |x| < 1\}$. Our main result is

(7.1) Theorem. If f is bounded and measurable, then

$$(*) \qquad E_x f(B_\tau) = \int_{\partial D} \frac{1 - |x|^2}{|x - y|^d} f(y) \pi(dy)$$

where π is surface measure on ∂D normalized to be a probability measure.

Proof An application of the monotone class theorem shows that if $(*)$ holds for $f \in C^\infty$ it is valid for bounded measurable f. In view of (4.3), we can prove $(*)$ for $f \in C^\infty$ by showing that if $k_y(x) = (1 - |x|^2)/|x - y|^d$ and

$$v(x) = \begin{cases} \int_{\partial D} k_y(x) f(y) \pi(dy) & x \in D \\ f(x) & x \in \partial D \end{cases}$$

then v solves the Dirichlet problem (4.1):

(7.2a) In D, $v \in C^2$ and $\Delta v = 0$.

(7.2b) At each point of ∂D, v is continuous and $v(x) = f(x)$.

The first, somewhat painful, step in doing this is to show

(7.3) Lemma. If $y \in \partial D$ then $\Delta k_y = 0$ in D.

Proof To warm up for this, we observe

$$D_i |x - y|^p = D_i \left(\sum_j (x_j - y_j)^2 \right)^{p/2} = p|x - y|^{p-2}(x_i - y_i)$$

so taking $p = -d$ we have

$$D_i k_y(x) = (-2x_i) \cdot \frac{1}{|x - y|^d} + (1 - |x|^2) \cdot \frac{-d(x_i - y_i)}{|x - y|^{d+2}}$$

Differentiating again and using our fact with $p = -(d+2)$ gives

$$D_{ii} k_y(x) = (-2) \cdot \frac{1}{|x - y|^d} + 2 \cdot (-2x_i) \cdot \frac{-d(x_i - y_i)}{|x - y|^{d+2}}$$
$$+ (1 - |x|^2) \left\{ \frac{d(d+2)(x_i - y_i)^2}{|x - y|^{d+4}} - \frac{d}{|x - y|^{d+2}} \right\}$$

Section 4.7 Exit Distributions for the Ball   165

Summing the last expression on i gives

$$\Delta k_y(x) = \frac{-2d}{|x-y|^d} + 4d \cdot \frac{|x|^2 - x \cdot y}{|x-y|^{d+2}}$$
$$+ (1 - |x|^2) \left\{ \frac{d^2 + 2d}{|x-y|^{d+2}} - \frac{d^2}{|x-y|^{d+2}} \right\}$$
$$= \frac{-2d|x-y|^2}{|x-y|^{d+2}} + \frac{4d(|x|^2 - x \cdot y)}{|x-y|^{d+2}} + \frac{(1-|x|^2) \cdot 2d}{|x-y|^{d+2}}$$

If we replace the 1 by $|y|^2$, the expression collapses

$$\Delta k_y(x) = \frac{2d}{|x-y|^{d+2}}(-|x-y|^2 + 2|x|^2 - 2x \cdot y + |y|^2 - |x|^2) = 0$$

since $|x-y|^2 = (x-y) \cdot (x-y) = |x|^2 - 2x \cdot y + |y|^2$.

Inside the open set D, v is a linear combination of the k_y. So bringing the differentiation under the integral (and leaving it to the reader to justify this using (1.7)) gives

$$\Delta v(x) = \int_{\partial D} \pi(dy) f(y) \Delta k_y(x) = 0$$

Thus, v satisfies (7.2a). To check (7.2b), the first step is to show

$$I(x) = \int_{\partial D} \frac{1-|x|^2}{|x-y|^d} \pi(dy) \equiv 1$$

This is just "calculus," but we prefer to use a soft noncomputational approach instead. We begin by observing that $I(0) = 1$, I is invariant under rotations, and $\Delta I = 0$ in D. (For the last conclusion apply the result for Δv with $f \equiv 1$.) To conclude $I \equiv 1$ now, let $x \in D$ with $|x| = r < 1$, and let $\tau = \inf\{t : |B_t| > r\}$. Applying (1.2) of Chapter 3 with $G = D(0, r)$ now shows

$$I(0) = E_0 I(B_\tau) = I(x)$$

where the second equality follows from invariance under rotations.

To show that $v(x) \to f(y)$ as $x \to y \in \partial D$, we observe that if $z \neq y$,

$$k_z(x) = \frac{1-|x|^2}{|x-z|^d} \to \frac{0}{|y-z|^d} \quad \text{as } x \to y$$

From the last calculation it is clear that if $\delta > 0$, then the convergence is uniform for $z \in B_0(\delta) = \partial D - D(y,\delta)$. Thus, if we let $B_1(\delta) = \partial D - B_0(\delta)$, then

$$\int_{B_0(\delta)} k_z(x)\pi(dz) \to 0 \quad \text{and} \quad \int_{B_1(\delta)} k_z(x)\pi(dz) \leq 1$$

since $I(x) = 1$. To prove that $v(x) \to f(y)$ now we observe

$$|v(x) - f(y)| = \left| \int k_z(x) f(z) \pi(dz) - f(y) \int k_z(x) \pi(dz) \right|$$

$$\leq 2\|f\|_\infty \int_{B_0(\delta)} k_z(x) \pi(dz) + \sup_{z \in B_1(\delta)} |f(z) - f(y)|$$

The first term $\to 0$ as $x \to y$, while the sup in the second is small if δ is. This shows that (7.2b) holds and completes the proof of (7.1). □

The derivation given above is a little unsatisfying, since it starts with the answer and then verifies it, but it is simpler than messing around with Kelvin's transformations (see pages 100-103 in Port and Stone (1978)). It also has the merit of explaining why $k_y(x)$ is the probability density of exiting at y: k_y is a nonnegative harmonic function that has $k_y(0) = 1$ and $k_y(x) \to 0$ when $x \to z \in \partial D$ and $z \neq y$.

Exercise 7.1. The point of this exercise is to apply the reasoning of this section to $\tau = \inf\{t : B_t \notin H\}$ where $H = \{(x,y) : x \in \mathbf{R}^{d-1}, y > 0\}$. For $\theta \in \mathbf{R}^{d-1}$ let

$$h_\theta(x,y) = \frac{C_d y}{(|x-\theta|^2 + y^2)^{d/2}}$$

where C_d is chosen so that $\int d\theta\, h_\theta(0,1) = 1$ and let

$$u(x,y) = \int d\theta\, h_\theta(x,y) f(\theta, 0)$$

where f is bounded and continuous.

(a) Show that $\Delta h_\theta = 0$ in H and use (1.7) to conclude $\Delta u = 0$ in H.

(b) Show $I(x,y) = \int d\theta\, h_\theta(x,y) \equiv 1$.

(c) Show that if $x_n \to x$, $y_n \to 0$ then $u(x_n, y_n) \to f(x, 0)$.

(d) Conclude that $E_{(x,y)} f(B_\tau) = \int d\theta\, h_\theta(x,y) f(\theta, 0)$.

4.8. Occupation Times for the Ball

In the last section we considered how B_t leaves $D = \{x : |x| < 1\}$. In this section we will investigate how it spends its time before it leaves. Let $\tau = \inf\{t : B_t \notin D\}$ and let $G(x,y)$ be the potential kernel defined in (2.7) of Chapter 3. That is,

$$G(x,y) = \begin{cases} -|x-y| & d=1 \\ -\frac{1}{\pi}\log|x-y| & d=2 \\ C_d|x-y|^{2-d} & d \geq 3 \end{cases}$$

where $C_d = \Gamma(d/2 - 1)/2\pi^{d/2}$.

(8.1) Theorem. If g is bounded and measurable then

$$E_x \int_0^\tau g(B_t)\,dt = \int G_D(x,y)g(y)\,dy$$

where

$$G_D(x,y) = G(x,y) - \int \frac{1-|x|^2}{|x-z|^d} G(z,y)\pi(dz)$$

Proof Combine (*) in Section 4.5 with (7.1). □

We think of $G_D(x,y)$ as the expected amount of time a Brownian motion starting at x spends at y before exiting G. To be precise, if $A \subset G$ then the expected amount of time B_t spends in A before exiting G is $\int_A G_D(x,y)\,dy$. Our task in this section is to compute $G_D(x,y)$. In $d=1$, where $D = (-1,1)$, (8.1) tells us that

$$G_D(x,y) = G(x,y) - \frac{x+1}{2}G(1,y) - \frac{1-x}{2}G(-1,y)$$
$$= -|x-y| + \frac{x+1}{2}(1-y) + \frac{1-x}{2}(y+1)$$
$$= -|x-y| + 1 - xy$$

Considering the two cases $x \geq y$ and $x \leq y$ leads to

(8.2) $$G_D(x,y) = \begin{cases} (1-x)(1+y) & -1 \leq y \leq x \leq 1 \\ (1-y)(1+x) & -1 \leq x \leq y \leq 1 \end{cases}$$

Geometrically, if we fix y then $x \to G_D(x,y)$ is determined by the conditions that it is linear on $[0,y]$ and on $[y,1]$ with $G_D(0,y) = G_D(1,y) = 0$, and $G_D(y,y) = 1 - y^2$.

168 Chapter 4 Partial Differential Equations

In $d \geq 2$, (8.1) works well when $y = 0$ for then $G(z,0)$ is constant on $\{|z| = 1\}$ and the expression in (8.1) reduces to

$$(8.3) \qquad G_D(x,0) = G(x,0) - G(e_1,0) = \begin{cases} C_d(|x|^{2-d} - 1) & d \geq 3 \\ (-1/\pi) \log |x| & d = 2 \end{cases}$$

To get an explicit formula for $G_D(x,y)$ in $d \geq 2$ for $y \neq 0$ we will cheat and look up the answer. This weakness of character has the advantage of making the point that $G_D(x,y)$ is nothing more than the Green's function for D with Dirichlet boundary conditions. Folland (1976), page 109, defines the Green's function for D to be the function $K(x,y)$ on $\bar{D} \times \bar{D}$ determined by the following properties:

(A) For each $y \in D$, $K(\cdot, y) - G(\cdot, y)$ is C^2 and harmonic in D.

(B) For each $y \in D$, if $x_n \to x \in \partial D$, $K(x_n, y) \to 0$.

Remark. For convenience, we have changed Folland's notation to conform to ours and interchanged the roles of x and y. This interchange makes no difference, since the Green's function is symmetric, that is, $K(x,y) = K(y,x)$. (See Folland (1976), page 110.)

(8.4) Lemma. The occupation time density G_D is equal to the Green's function K defined above.

Proof Using (8.1) and (7.1) we have

$$G_D(x,z) - G(x,z) = -\int_{\partial D} \frac{1 - |x|^2}{|x-y|^d} G(y,z)\pi(dy) = -E_x G(B_\tau, y)$$

This is harmonic in D by (4.6). (4.5) implies if $x_n \to x \in \partial D$, then

$$E_{x_n} G(B_\tau, y) \to G(x,y) \qquad \square$$

Having made the connection in (8.4), we can now find G_D by "guessing and verifying" functions with properties (A) and (B). To "guess," we turn to page 123 in Folland (1976) and find

(8.5) Theorem. In $d \geq 3$, if $0 < |y| < 1$ then

$$G_D(x,y) = G(x,y) - |y|^{2-d} G(x, y/|y|^2)$$

Proof To check (A) and (B), we observe that if we let Δ_x denote the Laplacian acting in the x variable for fixed y then

(*) $$\Delta_x G(x,y) = 0 \quad \text{for } x \neq y$$

(a) Since $y/|y|^2 \notin D$, (*) implies that the second term is harmonic for $x \in D$.

(b) Let $x \in \partial D$. Clearly, the right-hand side is continuous at x. To see that it vanishes, we note

$$G(x,y) - |y|^{2-d} G(x, y/|y|^2) = \frac{C_d}{|x-y|^{d-2}} - \frac{C_d}{|y|^{d-2}} \left| x - \frac{y}{|y|^2} \right|^{-(d-2)}$$
$$= \frac{C_d}{|x-y|^{d-2}} - \frac{C_d}{|x|y| - y|y|^{-1}|^{d-2}} = 0 \qquad \square$$

The last equality follows from a fact useful for the next proof:

(8.6) Lemma. If $|x| = 1$ then $||x|y| - y|y|^{-1}| = |x - y|$.

Proof Using $|z|^2 = z \cdot z$ and then $|x|^2 = 1$ we have

$$||x|y| - y|y|^{-1}|^2 = |x|^2 |y|^2 - 2x \cdot y + 1$$
$$= |y|^2 - 2x \cdot y + |x|^2 = |x - y|^2 \qquad \square$$

Again turning to page 123 of Folland (1976) we "guess"

(8.7) Theorem. In $d = 2$ if $0 < |y| < 1$ then

$$G_D(x,y) = \frac{-1}{\pi} \left(\ln|x-y| - \ln||x|y| - y|y|^{-1}| \right)$$

Proof Again we need to check (A) and (B).

(a) $$G_D(x,y) - G(x,y) = \frac{1}{\pi} \left(\ln \left| x - \frac{y}{|y|^2} \right| + \ln|y| \right)$$

Again the first term is harmonic by (*) since $y/|y|^2 \notin D$. The second term does not depend on x and hence is trivially harmonic. (b) Let $x \in \partial D$. Clearly, the right-hand side is continuous at x. The fact that it vanishes follows immediately from (8.6). $\qquad \square$

4.9. Laplace Transforms, Arcsine Law

In this section we will apply results from Section 4.6 to do three things:

(a) Complete the discussion of Example 6.1.

(b) Prove a remarkable observation of Ciesielski and Taylor (1962) that for Brownian motion starting at 0, the distribution of the exit time from $D = \{x : |x| < 1\}$ in dimension d is the same as that of the total occupation time of D in dimension $d+2$.

(c) Prove Lévy's arcsine law for the occupation time of $[0, \infty)$.

The third topic is independent of the first two.

Example 9.1. Take $d = 1$, $G = (-a, a)$, $c(x) \equiv \gamma > 0$, and $f \equiv 1$ in the problem considered in Section 4.6.

(9.1) Theorem. Let $\tau = \inf\{t : B_t \notin (-a, a)\}$. If $0 < \gamma < \pi^2/8a^2$, then

$$E_x e^{\gamma \tau} = \frac{\cos(x\sqrt{2\gamma})}{\cos(a\sqrt{2\gamma})}$$

If $\gamma \geq \pi^2/8a^2$ then $E_x e^{\gamma \tau} \equiv \infty$.

Proof By Example 6.1,

$$u(x) = \cos(x\sqrt{2\gamma})/\cos(a\sqrt{2\gamma})$$

is a nonnegative solution of

$$\frac{1}{2}u'' + \gamma u = 0 \qquad u(-a) = u(a) = 1$$

To check that $w \not\equiv \infty$, we observe that (6.2) implies that $M_t = u(B_t)e^{\gamma t}$ is a local martingale on $[0, \tau)$. If we let T_n be a sequence of stopping times that reduces M_t, then

$$u(x) = E_x\left(u(B_{T_n \wedge n})e^{\gamma(T_n \wedge n)}\right)$$

Letting $n \to \infty$ and noting $T_n \wedge n \uparrow \tau$, it follows from Fatou's lemma that

$$u(x) \geq E_x e^{\gamma \tau}$$

The last equation implies $w \not\equiv \infty$, so (6.3) implies the result for $\gamma < \pi^2/8a^2$.

To see that $w(x) \equiv \infty$ when $\gamma \geq \pi^2/8a^2$, suppose not. In this case (6.3) implies $v(x) = E_x(f(B_\tau)\exp(c_\tau))$ is the unique solution of our equation but computations in Example 6.1 imply that in this case there is no nonnegative solution. □

Exercise 9.1. Show that if $\beta \geq 0$ then

$$E_x \exp(-\beta \tau) = \frac{\cosh(x\sqrt{2\beta})}{\cosh(a\sqrt{2\beta})}$$

Since $\cos(z) = (e^{iz} + e^{-iz})/2$ this is what we get if we set $\gamma = -\beta$ in (9.1).

Example 9.2. Our second topic is the observation of Ciesielski and Taylor (1962). The proof of the general result, see Getoor and Sharpe (1979), sections 5 and 8, or Knight (1981), pages 88-89, requires more than a little familiarity with Bessel functions, so we will only show that the distribution of the exit time from $(-1,1)$ in one dimension starting from 0 is the same as that of the total occupation time of $D = \{x : |x| < 1\}$ in three dimensions starting from the origin.

Exercise 9.1 gives the Laplace transform of the exit time from $(-1,1)$ so we need only compute

$$v(x) = E_x \exp\left(-\beta \int_0^\infty 1_D(B_t)\, dt\right)$$

Of course, we only have to compute $v(0)$ but as in Example 9.1, and Exercise 9.1, we will do this by using a differential equation to compute $v(x)$ for all x.

Several properties of v are immediately obvious (in any dimension $d \geq 3$):

(i) Spherical symmetry implies $v(x) = f(|x|)$.

(ii) Using the strong Markov property at the hitting time of D and (1.12) from Chapter 3 gives for $r > 1$

(a) $\qquad f(r) = r^{2-d} f(1) + (1 - r^{2-d}) \cdot 1 = 1 + r^{2-d} \cdot (f(1) - 1)$

(iii) The strong Markov property and (6.3) imply that v is the unique solution of

$$\frac{1}{2}\Delta v - \beta v = 0 \qquad \text{for } x \in D$$
$$v(x) = f(1) \qquad \text{for } x \in \partial D$$

To express the last equation in terms of f we note that if $v(x) = f(|x|)$ then

$$D_i v(x) = f'(|x|) \frac{x_i}{|x|}$$

$$D_{ii} v(x) = f'(|x|)\left\{\frac{1}{|x|} - \frac{x_i^2}{|x|^3}\right\} + f''(|x|)\frac{x_i^2}{|x|^2}$$

172   Chapter 4   Partial Differential Equations

Summing over i gives

$$\frac{1}{2}\Delta v = \frac{1}{2}f''(|x|) + \frac{d-1}{2|x|}f'(|x|)$$

Multiplying by 2 we see that f satisfies

(b) $\qquad f''(r) + \frac{d-1}{r}f'(r) - 2\beta f(r) = 0 \quad \text{for } r < 1$

To tie equations (a) and (b) together, we note that by applying the reasoning in (iii) to $D(0,q)$ with $q > 1$ and using the proof of (6.6) (which refers us to (5.6a)) we can conclude

(iv) v is C^1 and hence $f'(r)$ is continuous at $r = 1$.

Facts (ii)–(iv) give us the information we need to solve for f at least in principle: (b) is a second order ordinary differential equation, so if we specify $f(0) = C$ and $f'(0) = 0$, the latter from (i), there is a unique solution f_C on $[0,1]$. Given $f_C(1)$, (ii) gives us the solution on $[1,\infty)$. We then pick C so that (iv) holds.

To carry out our plan, we begin with the "well known" fact that the only solutions to (b) which stay bounded near 0 have the form $cr^{1-(d/2)}I_{(d/2)-1}(r\sqrt{2\beta})$ where

$$I_\nu(z) = \sum_{m=0}^{\infty} \frac{(z/2)^{\nu+2m}}{m!\Gamma(\nu+m+1)}$$

is one of the happy families of Bessel functions. Letting $a = \sqrt{2\beta}$ and recalling $\Gamma(x) = (x-1)\Gamma(x-1)$, we see that when $d = 3$ and hence $\nu = 1/2$ the solution has a simple form

$$C(a)\sum_{m=0}^{\infty}\frac{(ar)^{2m}}{(2m+1)!} = \frac{C(a)}{ar}\sinh(ar)$$

Having "found" the solution we want, we can now forget about its origins and simply check that it works. Differentiating we have

$$f'(r) = \frac{C(a)}{r}\cosh(ar) - \frac{C(a)}{ar^2}\sinh(ar)$$

$$f''(r) = \frac{C(a)a}{r}\sinh(ar) - \frac{2C(a)}{r^2}\cosh(ar) + \frac{2C(a)}{ar^3}\sinh(ar)$$

$$f''(r) + \frac{2}{r}f'(r) = \frac{C(a)a}{r}\sinh(ar) = a^2 f(r)$$

which shows that (b) holds.

To complete the solution, we have to pick $C(a)$ to make $f \in C^1$. Using formulas for $f'(r)$ for $r < 1$ from the previous display, for $r > 1$ from (a), and recalling $d = 3$, we have

$$f'(1-) = C(a)\cosh(a) - C(a)a^{-1}\sinh(a)$$
$$f'(1+) = -\{f(1) - 1\} = 1 - C(a)a^{-1}\sinh(a)$$

Solving gives $C(a) = 1/\cosh(a)$. Since $a = \sqrt{2\beta}$ this matches the result in Exercise 9.1 when $x = 0$. □

Example 9.3. Our third and final topic is Kac's (1951) derivation of

(9.2) Lévy's arcsine law. Let $H_t = |\{s \in [0,t] : B_s \geq 0\}|$. If $0 \leq \theta \leq 1$ then

$$P_0(H_t \leq \theta t) = \frac{1}{\pi}\int_0^\theta \frac{dr}{\sqrt{r(1-r)}} = \frac{2}{\pi}\arcsin(\sqrt{\theta})$$

Remark. The reader should note that by scaling the distribution of H_t/t does not depend on t, so the fraction of time in $[0,t]$ that a Brownian gambler is ahead does not converge to $1/2$ in probability as $t \to \infty$. Indeed $r = 1/2$ is the value for which the probability density $1/\pi\sqrt{r(1-r)}$ is smallest.

Proof For the last equality see the other arcsine law, (4.2) in Chapter 1. To prove the first we start with

(9.3) Lemma. Let $c(x) = -\alpha - \beta 1_{[0,\infty)}(x)$ with $\alpha, \beta \geq 0$. Suppose v is bounded, C^1, and satisfies

$$\frac{1}{2}\Delta v + cv = -1$$

for all $x \neq 0$. Then

$$v(x) = \int_0^\infty dt\, e^{-\alpha t} E_x e^{-\beta H_t}$$

Proof Our assumptions about v imply that $v''(x) = -2(1 + c(x)v(x))\,dx$ in the sense of distribution. So two results from Chapter 2, the Meyer-Tanaka formula, (11.4), and (11.7) imply

$$v(B_t) - v(B_0) = \int_0^t v'(B_s)\,dB_s - \int_0^t (1 + c(B_s)v(B_s))\,ds$$

Letting $c_t = \int_0^t c(B_s)\,ds$ and using the integration by parts formula, (10.1) in Chapter 2, with $X_t = v(B_t)$ and $Y_t = \exp(c_t)$, which is locally of bounded variation, we have

$$v(B_t)\exp(c_t) - v(B_0) = \int_0^t \exp(c_s)v'(B_s)\,dB_s$$
$$- \int_0^t \exp(c_s)\{1 + c(B_s)v(B_s)\}\,ds$$
$$+ \int_0^t v(B_s)\exp(c_s)\,dc_s$$

So $M_t = v(B_t)\exp(c_t) + \int_0^t \exp(c_s)\,ds$ is a local martingale. Since v is bounded and $c_t \leq 0$, M_t is bounded. As $t \to \infty$, $\exp(c_t) \leq e^{-\alpha t} \to 0$, so using the martingale and bounded convergence theorems gives

$$v(x) = E_x M_0 = E_x M_\infty = E_x \int_0^\infty \exp(c_s)\,ds$$

Plugging in the definition of $c(x)$ now and using Fubini's theorem leads to the formula given above. □

(9.3) tells us that we want to find a bounded C^1 function v with

$$(\alpha + \beta)v = \frac{1}{2}v'' + 1 \qquad x > 0$$
$$\alpha v = \frac{1}{2}v'' + 1 \qquad x < 0$$

To solve the equation $\gamma v = \frac{1}{2}v'' + 1$ we write $v = v_0 + v_1$ where $v_1(x) \equiv 1/\gamma$ and note that

$$\gamma v_1 = \frac{1}{2}v_1'' + 1$$
$$\gamma v_0 = \frac{1}{2}v_0''$$

so we have $v_0(x) = Ce^{\pm x\sqrt{2\gamma}}$. Picking the signs to keep v bounded, we have

$$v(x) = \begin{cases} Ae^{-x\sqrt{2(\alpha+\beta)}} + \frac{1}{\alpha+\beta} & x > 0 \\ Be^{x\sqrt{2\alpha}} + \frac{1}{\alpha} & x < 0 \end{cases}$$

To find A and B we note that we want v to be C^1 so

$$A + \frac{1}{\alpha+\beta} = B + \frac{1}{\alpha}$$
$$-A\sqrt{2(\alpha+\beta)} = B\sqrt{2\alpha}$$

Section 4.9 Laplace Transforms, Arcsine Law

It is a little tedious to solve these equations but it is easy to check that

$$A = \frac{\sqrt{\alpha+\beta} - \sqrt{\alpha}}{(\alpha+\beta)\sqrt{\alpha}} \qquad B = \frac{\sqrt{\alpha} - \sqrt{\alpha+\beta}}{\alpha\sqrt{\alpha+\beta}}$$

gives a solution. What we are interested in is

(9.4) $$v(0) = A + \frac{1}{\alpha+\beta} = \frac{1}{\sqrt{\alpha(\alpha+\beta)}}$$

To go backwards from here to the answer written in (9.2), we warm up by changing variables $x = \sqrt{2\gamma t}$, $dx = (1/2)\sqrt{2\gamma/t}\,dt$ to get

$$\int_0^\infty \frac{e^{-\gamma t}}{\sqrt{t}}\,dt = \int_0^\infty e^{-x^2/2}\sqrt{2/\gamma}\,dx$$
$$= \sqrt{\frac{2}{\gamma}} \cdot \frac{1}{2} \cdot \sqrt{2\pi} = \sqrt{\pi/\gamma}$$

This identity and Fubini's theorem imply

$$\frac{1}{\sqrt{\alpha+\beta}}\frac{1}{\sqrt{\alpha}} = \frac{1}{\pi}\int_0^\infty \frac{e^{-(\alpha+\beta)s}}{\sqrt{s}}\int_s^\infty \frac{e^{-\alpha(t-s)}}{t-s}\,dt\,ds$$
$$= \int_0^\infty e^{-\alpha t}\frac{1}{\pi}\int_0^t \frac{e^{-\beta s}}{\sqrt{s(t-s)}}\,ds\,dt$$

From (9.3), (9.4), and the uniqueness of Laplace transforms it follows that

$$E_0\exp(-\beta H_t) = \frac{1}{\pi}\int_0^t \frac{e^{-\beta s}}{\sqrt{s(t-s)}}\,ds\,dt$$

and invoking the uniqueness again we have the desired result. □

5 Stochastic Differential Equations

5.1. Examples

In this chapter we will be interested in solving stochastic differential equations (or SDE's) that are written informally as

$$dX_s = b(X_s)\,ds + \sigma(X_s)\,dB_s$$

or more formally, as an integral equation:

$$(\star) \qquad X_t = X_0 + \int_0^t b(X_s)\,ds + \int_0^t \sigma(X_s)\,dB_s$$

We will give a precise formulation of $(\star)$ at the beginning of Section 5.2.

In $(\star)$, B_t is an m dimensional Brownian motion, σ is a $d \times m$ matrix, and to make the matrix operations work out right we will think of X, b and B as column vectors. That is, they are matrices of size $d \times 1$, $d \times 1$, and $m \times 1$ respectively. Writing out the coordinates $(\star)$ says that for $1 \le i \le d$

$$X_t^i = X_0^i + \int_0^t b_i(X_s)\,ds + \sum_{j=1}^m \int_0^t \sigma_{ij}(X_s)\,dB_s^j$$

Note that the number m of Brownian motions used to "drive" the equation may be more or less than d. This generalization does not cause any additional difficulty, is useful in some situations (see Example 1.5), and will help us distinguish between $\sigma\sigma^T$ which is $d \times d$ and $\sigma^T\sigma$ which is $m \times m$. Here σ^T denotes the transpose of the matrix σ.

To explain what we have in mind when we write down $(\star)$ and why we want to solve it, we will now consider some examples. In some of the later examples we will use the term **diffusion process**. This is customarily defined to be a "strong Markov process with continuous paths." However, in this section, the term will be used to mean "a family of solutions of the SDE, one for each starting point." (4.6) will show that under suitable conditions the family of solutions defines a diffusion process.

Example 1.1. Exponential Brownian Motion. Let $X_t = X_0 \exp(\mu t + \sigma B_t)$ where X_0 is a real number and B_t is a standard one dimensional Brownian motion. Using Itô's formula

$$(1.1) \qquad X_t = X_0 + \int_0^t \mu X_s \, ds + \int_0^t \sigma X_s \, dB_s + \frac{1}{2} \int_0^t \sigma^2 X_s \, ds$$

so X_t solves $(\star)$ with $b(x) = (\mu + (\sigma^2/2))x$ and $\sigma(x) = \sigma x$. Exponential Brownian motion is often used as a simple model for stock prices because it stays nonnegative and fluctuations in the price are proportional to the price of the stock.

To explain the last phrase, we return to the general equation $(\star)$ and suppose that b and σ are continuous. If $X_0 = x$ then stopping at $t \wedge T_n$ for suitable stopping times $T_n \uparrow \infty$, taking expected values in $(\star)$ and letting $n \to \infty$ we have

$$EX_t^i = x^i + E \int_0^t b_i(X_s) \, ds$$

So if b_i is continuous

$$\frac{d}{dt} EX_t^i \Big|_{t=0} = b_i(x)$$

Because of the last equation, b is called the **infinitesimal drift**.

To give the meaning of the other coefficient we note that if

$$a(x) = \sigma(x)\sigma^T(x)$$

then the formula for the covariance of two stochastic integrals, (8.7) in Chapter 2, implies

$$\langle X^i, X^j \rangle_t = \sum_k \int_0^t \sigma_{ik}(X_s)\sigma_{jk}(X_s) \, ds = \int_0^t a_{ij}(X_s) \, ds$$

Using the integration by parts formula

$$X_t^i X_t^j = x^i x^j + \int_0^t X_s^j \, dX_s^i + \int_0^t X_s^i \, dX_s^j + \langle X^i, X^j \rangle_t$$

Substituting

$$dX_s^k = b_k(X_s) \, ds + \sum_j \sigma_{kj}(X_s) \, dB_s^j$$

using the associative law, and taking expected values

$$E(X_t^i X_t^j) = x^i x^j + E\int_0^t X_s^j b_i(X_s)\,ds$$
$$+ E\int_0^t X_s^i b_j(X_s)\,ds + E\int_0^t a_{ij}(X_s)\,ds$$

If a_{ij} is also continuous, differentiating gives

$$\left.\frac{d}{dt}E(X_t^i X_t^j)\right|_{t=0} = x^j b_i(x) + x^i b_j(x) + a_{ij}(x)$$

Using $EX_t^i = x^i + E\int_0^t b_i(X_s)\,ds$ and differentiating again we get

$$\left.\frac{d}{dt}\left(EX_t^i EX_t^j\right)\right|_{t=0} = x^j b_i(x) + x^i b_j(x)$$

Subtracting the last two equations gives

$$\left.\frac{d}{dt}\left(E(X_t^i X_t^j) - EX_t^i EX_t^j\right)\right|_{t=0} = a_{ij}(x)$$

justifying the name **infinitesimal covariance**.

When $d = 1$, we call $a(x)$ the **infinitesimal variance** and $\sigma(x) = \sqrt{a(x)}$ the **infinitesimal standard deviation**. In Example 1.1, $\sigma(x) = \sigma x$ is proportional to the stock price x. The infinitesimal drift $b(x) = (\mu + (\sigma^2/2))x$ is also proportional to x. Note, however, that in addition to the drift μ built in to the exponential there is a drift $\sigma^2 x/2$ that comes from the Brownian motion. More precisely it comes from the fact that e^x is convex and hence $\exp(\sigma B_t)$ is a submartingale.

Example 1.2. The Bessel Processes. Let W_t be a d-dimensional Brownian motion with $d > 1$, and $X_t = |W_t|$. Differentiating gives

$$D_i|x| = \frac{1}{2}\cdot\frac{2x_i}{|x|} \qquad D_{ii}|x| = \frac{1}{|x|} - \frac{x_i^2}{|x|^3} \qquad \Delta|x| = \frac{d-1}{|x|}$$

So using Itô's formula

$$X_t - X_0 = \sum_i \int_0^t \frac{W_s^i}{|W_s|}\,dW_s^i + \frac{1}{2}\int_0^t \frac{d-1}{|W_s|}\,ds$$

We will use B_t to denote the first term on the right-hand side, since B_t is a local martingale and $\langle B \rangle_t = t$, i.e., B_t is a Brownian motion by Lévy's theorem, (4.1) in Chapter 3. Changing notation now we have

$$(1.2) \qquad X_t - X_0 = B_t + \int_0^t \frac{d-1}{2X_s}\, ds$$

so $(\star)$ holds with $b(x) = (d-1)/2|x|$ and $\sigma(x) = 1$.

Example 1.3. The Ornstein Uhlenbeck Process. Let B_t be a one dimensional Brownian motion and consider

$$(1.3) \qquad dX_t = -\alpha X_t\, dt + \sigma\, dB_t$$

which describes one component of the velocity of a Brownian particle which is slowed by friction (i.e., experiences an acceleration $-\alpha$ times its velocity X_t). This is one case where we can find the solution explicitly.

$$(1.4) \qquad X_t = e^{-\alpha t}\left(X_0 + \int_0^t e^{\alpha s}\sigma\, dB_s\right)$$

To check this informally, we note that differentiating the first term in the product gives $-\alpha X_t\, dt$ and differentiating the second gives $\sigma\, dB_t$. To check this formally, we use Itô's formula with

$$f(u,v) = e^{-\alpha u} v \qquad U_t = t \qquad V_t = X_0 + \int_0^t e^{\alpha s}\sigma\, dB_s$$

to conclude

$$X_t - X_0 = \int_0^t -\alpha X_s\, ds + \int_0^t e^{-\alpha s}(e^{\alpha s}\sigma)\, dB_s$$

which is (1.3) in integral form.

From the representation in (1.4) we get the following useful fact

(1.5) Theorem. When $X_0 = x$, the distribution of X_t is normal with mean $xe^{-\alpha t}$ and variance $\sigma^2 \int_0^t e^{-2\alpha r}\, dr$.

Proof To see that the integral has a normal distribution with mean 0 and the indicated variance we use Exercise 6.7 in Chapter 2. Adding $e^{-\alpha t}X_0 = e^{-\alpha t}x$ now we have the desired result. □

Example 1.4. The Kalman-Bucy Filter. Consider

$$(1.6) \qquad \begin{aligned} dY_t &= V_t\, dt + \alpha\, dW_t \\ dV_t &= -cV_t\, dt + \sigma\, dB_t \end{aligned}$$

Here V_t is an Ornstein-Uhlenbeck velocity process, and Y_t is an observation of the position process subject to observation error described by $\alpha\, dW_t$, where W is a Brownian motion independent of B. The fundamental problem here is: how does one best estimate the present velocity V_t from the observation $\{Y_s : s \leq t\}$? This is a problem of important practical significance but not one that we will treat here. See for example, Rogers and Williams (1987), p. 327–329, or Øksendal (1992), Chapter VI.

Example 1.5. Brownian Motion on the Circle. Let B_t be a one dimensional Brownian motion, let $X_t = \cos B_t$ and $Y_t = \sin B_t$. Since $X_t^2 + Y_t^2 = 1$, (X_t, Y_t) always stays on the unit circle. Itô's formula implies that

(1.7)
$$dX_t = -Y_t\, dB_t - \frac{1}{2} X_t\, dt$$
$$dY_t = X_t\, dB_t - \frac{1}{2} Y_t\, dt$$

To write this in the form $(\star)$ we take

$$b(x,y) = \begin{pmatrix} -x/2 \\ -y/2 \end{pmatrix} \qquad \sigma(x,y) = \begin{pmatrix} -y \\ x \end{pmatrix}$$

Note that σ is always a unit vector tangent to the circle but we need the drift $b(x,y)$, which is perpendicular to the circle and points inward, to keep (X_t, Y_t) from flying off.

Example 1.6. Feller's Branching Diffusion. This $X_t \geq 0$ and satisfies

(1.8) $$dX_t = \beta X_t\, dt + \sigma \sqrt{X_t}\, dB_t$$

To explain where this equation comes from, we recall that in a branching process, each individual in generation m has an independent and identically distributed number of offspring in generation $m + 1$. Consider a sequence of branching processes $\{Z_m^n, m \geq 0\}$ in which the probability of k children is p_k^n and suppose

(A1) the mean of p_k^n is $1 + (\beta_n/n)$ with $\beta_n \to \beta$

(A2) the variance of p_k^n is σ_n^2 with $\sigma_n^2 \to \sigma^2 > 0$

(A3) for any $\delta > 0$, $\sum_{k > \delta n} k^2 p_k^n \to 0$.

Since the individuals in generation 0 have an independent and identically distributed number of children

$$E(Z_1^n \,|\, Z_0^n = nx) = nx \cdot \left(1 + \frac{\beta}{n}\right)$$
$$\operatorname{var}(Z_1^n \,|\, Z_0^n = nx) = nx \cdot \sigma_n^2$$

So if we let $X_t^n = Z_{[nt]}^n/n$ then

$$E\left(X_{1/n}^n - x \,\Big|\, X^n(0) = x\right) = x\beta_n \cdot \frac{1}{n}$$

$$\mathrm{var}\left(X_{1/n}^n \,\Big|\, X^n(0) = x\right) = x\sigma_n^2 \cdot \frac{1}{n}$$

The last two equations say that the infinitesimal mean and variance of the rescaled process X_t^n are $x\beta_n$ and $x\sigma_n^2$. These quantities obviously converge to those of the process X_t in (1.8). In Example 8.2 of Chapter 8, we will show that under (A1)–(A3) the processes $\{X_t^n, t \geq 0\}$ converge weakly to $\{X_t, t \geq 0\}$.

Example 1.7. Wright-Fisher Diffusion. This $X_t \in [0,1]$ and satisfies

(1.9) $\qquad dX_t = (-\alpha X_t + \beta(1-X_t))\,dt + \sqrt{X_t(1-X_t)}\,dB_t$

The motivation for this model comes from genetics, but to avoid the details of that subject we will suppose instead we have an urn with n balls in it that may be labelled A or a. To build up the urn at time $m+1$ we sample with replacement from the urn at time m. However, with probability α/n we ignore the draw and place an a in, and with probability β/n we ignore the draw and place an A in. In genetics terms the last two events correspond to mutations. Let Z_m^n be the number of A's in the urn at time m. Now when the fraction of A's in the urn at time 0 is x, the probability of drawing an A on a given trial is

$$p_n = x \cdot \left(1 - \frac{\alpha}{n}\right) + (1-x) \cdot \frac{\beta}{n}$$

Since we are sampling with replacement,

$$E\left(Z_1^n \,|\, Z_0^n = nx\right) = np_n$$
$$\mathrm{var}\left(Z_1^n \,|\, Z_0^n = nx\right) = np_n(1-p_n)$$

If we let $X_t^n = Z_{[nt]}^n/n$ then

$$E\left(X_{1/n}^n - x \,\Big|\, X^n(0) = x\right) = \{-\alpha x + \beta(1-x)\} \cdot \frac{1}{n}$$

$$\mathrm{var}\left(X_{1/n}^n \,\Big|\, X^n(0) = x\right) = p_n(1-p_n) \cdot \frac{1}{n}$$

Now $p_n \to x$ as $n \to \infty$ so again the infinitesimal mean and variance of X_t^n converge to those of X_t in (1.9) but the rigorous connection has to wait until Example 8.3 of Chapter 8. This is just one of many of the diffusions that can

5.2. Itô's Approach

We begin this section by giving some essential definitions and introducing a counterexample that will explain the need for some of the formalities. We will then proceed to our main business: the first existence and uniqueness result for SDE, which was proved by K. Itô long before the subject became entangled in the complications of the general theory of processes.

To finally become precise, a solution to our SDE $(\star)$ is a triple $(X_t, B_t, \mathcal{F}_t^\star)$ where X_t and B_t are continuous processes adapted to a filtration $\mathcal{F}_t^\star$ so that

(i) B_t is a Brownian motion with respect to $\mathcal{F}_t^\star$, i.e., for each s and t, $B_{t+s} - B_t$ is independent of $\mathcal{F}_t^\star$ and is normally distributed with mean 0 and variance s.

(ii) X_t satisfies

$$(\star) \qquad X_t = X_0 + \int_0^t b(X_s)\,ds + \int_0^t \sigma(X_s)\,dB_s$$

We will always assume that σ and b are measurable and locally bounded so the integral exists.

It is easy to see that if the SDE holds for some $\mathcal{F}_t^\star$ it will hold for $\mathcal{F}_t^{X,B}$ the filtration generated by (X_t, B_t). When X_t is adapted to $\mathcal{F}_t^B$ the filtration generated by the Brownian motion, then we can take $\mathcal{F}_t^\star = \mathcal{F}_t^B$ and X_t is called a **strong solution**. It is difficult to imagine how X_t could satisfy $(\star)$ without being adapted to $\mathcal{F}_t^B$ but there is a famous example due to H. Tanaka showing that this can happen.

Example 2.1. Let W_t be a one dimensional Brownian motion with $W_0 = 0$ and let

$$B_t = \int_0^t \operatorname{sgn}(W_s)\,dW_s$$

where $\operatorname{sgn}(x) = 1$ if $x \geq 0$, $\operatorname{sgn}(x) = -1$ if $x < 0$. Since B_t is a local martingale with $\langle B \rangle_t = t$, (4.2) in Chapter 3 implies that B_t is a Brownian motion with respect to $\mathcal{F}_t^W$, the filtration generated by W_t. Since $\operatorname{sgn}(x)^2 = 1$, the associative law implies that

$$\int_0^t \operatorname{sgn}(W_s)\,dB_s = \int_0^t dW_s = W_t$$

So W is a solution of $(\star)$ with $b \equiv 0$, $\sigma(x) = \operatorname{sgn}(x)$, and $\mathcal{F}_t^\star = \mathcal{F}_t^W = \mathcal{F}_t^{W,B}$.

To see that W is a not a strong solution we apply the Meyer-Tanaka formula, (11.4) in Chapter 2, with $X_t = W_t$ and $f(x) = |x|$ to get

$$|W_t| - L_t^0 = \int_0^t \text{sgn}(W_s)\, dW_s = B_t$$

where L_t^0 is the local time of $|W_t|$ at 0. The occupation time density formula in (11.7) of Chapter 2, and the continuity of the local time established in (11.8) of Chapter 2, imply that

$$\frac{1}{2\epsilon}\int_0^t 1_{\{|W_s|\le \epsilon\}}\, ds = \frac{1}{2\epsilon}\int_{-\epsilon}^\epsilon L_t^a\, da \to L_t^0 \quad \text{as } \epsilon \to 0$$

So L_t^0 and hence $B_t = |W_t| - L_t^0$ is measurable with respect to the filtration generated by $|W_t|$, $t \ge 0$. However, W_t is clearly not adapted to the filtration generated by $|W_t|$. □

As usual when we have an equation, we are interested in having a unique solution. For SDE there are several notions of uniqueness. Our first, **pathwise uniqueness** holds if whenever X_t and X_t' are solutions of $(\star)$ with $X_0 = X_0' = x$ driven by the same Brownian motion B_t, then with probability one $X_t = X_t'$ for all $t \ge 0$. Here, the filtrations $\mathcal{F}_t$ and $\mathcal{F}_t'$ are allowed to be different.

(2.1) Theorem. Pathwise uniqueness does not hold in Example 2.1

Proof We observed in Example 2.1 that W was a solution. To show that $-W$ is also a solution, we note that $\text{sgn}(-x) = -\text{sgn}(x)$ except when $x = 0$, in which case the left side is 1 and the right side is -1. So

$$\int_0^t \text{sgn}(-W_s)\, dB_s = -\int_0^t \text{sgn}(W_s)\, dB_s + 2\int_0^t 1_{\{W_s = 0\}}\, dB_s$$

To prove that the second integral is 0 (and hence the right-hand side is $= -W_t$) recall that W_t is a Brownian motion. Example 3.1 in Chapter 2 implies that $\{s : W_s = 0\}$ has measure 0, and hence

$$E\left(\int_0^t 1_{\{W_s=0\}}\, dB_s\right)^2 = E\int_0^t 1_{\{W_s=0\}}\, ds = E|\{s \le t : W_s = 0\}| = 0 \quad \square$$

Turning to positive results:

Section 5.2 Itô's Approach

(2.2) Theorem. If for all i, j, x, and y we have $|\sigma_{ij}(x) - \sigma_{ij}(y)| \le K|x-y|$ and $|b_i(x) - b_i(y)| \le K|x-y|$, then the stochastic differential equation

$$(\star) \qquad X_t = x + \int_0^t \sigma(X_s)\, dB_s + \int_0^t b(X_s)\, ds$$

has a strong solution and pathwise uniqueness holds.

Proof We construct the solution by a successive approximation scheme that reduces to the classical Picard iteration method for solving ordinary differential equations in the special case $\sigma \equiv 0$. We begin by introducing the notation

$$\tilde{X}_t = X_0 + \int_0^t \sigma(X_s)\, dB_s + \int_0^t b(X_s)\, ds$$

to describe one iteration. In words, we use the process $\{X_s, t \ge 0\}$ as input for the coefficients σ and b, and the output is the process $\{\tilde{X}_t, t \ge 0\}$. To solve $(\star)$ we begin with the initial guess $X_s^0 \equiv x$ and define the successive guess by iteration:

$$X_s^{n+1} = \tilde{X}_s^n \quad \text{for } n \ge 0$$

The rest of this section is devoted to proving that this procedure constructs a strong solution and that pathwise uniqueness holds. To help the reader digest the argument we have divided it into sections. One of the claims is easy. It is clear by induction that X_t^n is adapted to $\mathcal{F}_t^B$.

The basic estimate which is the key to the proof is:

(2.3) Lemma. Let τ be a stopping time with respect to $\mathcal{F}_t^\star$, let $T < \infty$ and let $B = (4Td + 16d^2)K^2$. Then

$$E\left(\sup_{0 \le t \le T \wedge \tau} |\tilde{Y}_t - \tilde{Z}_t|^2\right) \le 2E|Y_0 - Z_0|^2 + BE\int_0^{T \wedge t} |Y_s - Z_s|^2\, ds$$

Proof The inequality is trivial if the right-hand side is infinite so we can suppose that each term on the right is finite. To begin we recall that $(a+b)^2 \le 2a^2 + 2b^2$. Using this twice

$$(a+b+c)^2 \le 2a^2 + 2(b+c)^2 \le 2a^2 + 4b^2 + 4c^2$$

So the left-hand side of (2.3) is

(a)
$$\le 2E|Y_0 - Z_0|^2 + 4E \sup_{0 \le t \le T} \left| \int_0^{t \wedge \tau} \sigma(Y_s) - \sigma(Z_s)\, dB_s \right|^2$$
$$+ 4E \sup_{0 \le t \le T} \left| \int_0^{t \wedge \tau} b(Y_s) - b(Z_s)\, ds \right|^2$$

To bound the third term in (a), we observe that the Cauchy-Schwarz inequality implies that

$$\left(\int_0^{t\wedge\tau} b_i(Y_s) - b_i(Z_s)\,ds\right)^2 \le \int_0^{t\wedge\tau} (b_i(Y_s) - b_i(Z_s))^2\,ds \cdot \int_0^{t\wedge\tau} 1\,ds$$

Taking $\sup_{0\le t\le T}$ and using

(b) $$\sup_{0\le t\le T} |v(t)|^2 \le \sum_{i=1}^{d} \sup_{0\le t\le T} v_i^2(t)$$

with the Lipschitz continuity assumption gives

(c)
$$4E \sup_{0\le t\le T} \left|\int_0^{t\wedge\tau} b(Y_s) - b(Z_s)\,ds\right|^2$$
$$\le 4E \sum_{i=1}^{d} \sup_{0\le t\le T} \left(\int_0^{t\wedge\tau} b_i(Y_s) - b_i(Z_s)\,ds\right)^2$$
$$\le 4T \cdot E \sum_{i=1}^{d} \int_0^{T\wedge\tau} (b_i(Y_s) - b_i(Z_s))^2\,ds$$
$$\le 4TdK^2 \cdot E \int_0^{T\wedge\tau} |Y_s - Z_s|^2\,ds$$

To bound the second term in (a), let σ_i be the ith row of σ, let

$$M_t^i = \int_0^{t\wedge\tau} (\sigma_i(Y_s) - \sigma_i(Z_s)) \cdot dB_s$$

and let $\tau_n = \inf\{t : |M_t^i| \ge n\} \wedge \tau$. The L^2 maximal inequality and the formula for the variance of a stochastic integral imply that

$$E \sup_{0\le t\le T\wedge\tau_n} (M_t^i)^2 \le 4E\left(M_{T\wedge\tau_n}^i\right)^2 \le 4E\int_0^{T\wedge\tau} |\sigma_i(Y_s) - \sigma_i(Z_s)|^2\,ds$$

Letting $n \to \infty$, then using (b) and Lipschitz continuity we have

(d)
$$4E \sup_{0\le t\le T} \left|\int_0^{t\wedge\tau} (\sigma(Y_s) - \sigma(Z_s))\,dB_s\right|^2$$
$$\le 4E \sum_{i=1}^{d} \sup_{0\le t\le T} \left|\int_0^{t\wedge\tau} (\sigma_i(Y_s) - \sigma_i(Z_s)) \cdot dB_s\right|^2$$
$$\le 16E \sum_{i=1}^{d} \int_0^{T\wedge\tau} |\sigma_i(Y_s) - \sigma_i(Z_s)|^2\,ds$$
$$\le 16d^2K^2 E \int_0^{T\wedge\tau} |Y_s - Z_s|^2\,ds$$

Combining inequalities (a), (c), and (d) proves (2.3). □

Convergence of the sequence of approximations. Let

$$\Delta_n(t) = E\left(\sup_{0\le s\le t} |X_s^n - X_s^{n-1}|^2\right)$$

and observe that since $X_n^0 = X_{n-1}^0$, (2.3) implies that for $n \ge 1$

(2.4) $$\Delta_{n+1}(T) \le B\int_0^T \Delta_n(s)\,ds$$

To get started we have to estimate $|\Delta_1(s)|$. Our earlier inequality for squares implies that for norms $|a+b|^2 \le 2|a|^2 + 2|b|^2$. So we have

$$|X_s^1 - X_s^0|^2 \le 2\left(|\sigma(x)B_s|^2 + |b(x)s|^2\right)$$

Using the fact that

$$\sup_{0\le s\le t} |\sigma(x)B_s| \stackrel{d}{=} t^{1/2} \sup_{0\le s\le 1} |\sigma(x)B_s|$$

and that the right-hand side has finite expected value we have

$$\Delta_1(t) \le C(t + t^2)$$

Combining this with (2.4) and using induction we have

(2.5) $$\Delta_n(t) \le B^{n-1}C\left(\frac{t^n}{n!} + \frac{2t^{n+1}}{(n+1)!}\right)$$

From (2.5) we easily get the existence of a limit. Chebyshev's inequality shows that

$$P\left(\sup_{0\le t\le T} |X_t^n - X_t^{n-1}| > 2^{-n}\right) \le 2^{2n}\Delta_n(T)$$

(2.5) implies the right-hand side is summable, so the Borel-Cantelli gives

$$P\left(\sup_{0\le t\le T} |X_t^n - X_t^{n-1}| > 2^{-n} \text{ i.o.}\right) = 0$$

Since $\sum_n 2^{-n} < \infty$ it follows that with probability 1, $X_t^n \to$ a limit X_t^∞ uniformly on $[0,T]$.

188   Chapter 5   Stochastic Differential Equations

The limit is a solution. We begin by showing that convergence also occurs uniformly in L^2.

(2.6) Lemma. For all $0 \leq m < n \leq \infty$,

$$E\left(\sup_{0 \leq s \leq T} |X_s^m - X_s^n|^2\right) \leq \left(\sum_{k=m+1}^{n} \Delta_k(T)^{1/2}\right)^2$$

Proof Let $\|Z\|_2 = (EZ^2)^{1/2}$. If $n < \infty$, then it follows from monotonicity of expected value and the triangle inequality for the norm $\|\cdot\|_2$ that

$$\left\|\sup_{0 \leq s \leq T} |X_s^m - X_s^n|\right\|_2 \leq \left\|\sum_{k=m+1}^{n} \sup_{0 \leq s \leq T} |X_s^k - X_s^{k-1}|\right\|_2$$

$$\leq \sum_{k=m+1}^{n} \left\|\sup_{0 \leq s \leq T} |X_s^k - X_s^{k-1}|\right\|_2 = \sum_{k=m+1}^{n} \Delta_k(T)^{1/2}$$

Letting $n \to \infty$ and using Fatou's lemma proves the result for $n = \infty$. □

To see that X_t^∞ is a solution, we let $Y_t = X_t^n$ and $Z_t = X_t^\infty$ in (2.3) to get

$$E\left(\sup_{0 \leq t \leq T} |X_t^{n+1} - \tilde{X}_t^\infty|^2\right) \leq BE \int_0^T |X_s^n - X_s^\infty|^2 \, ds$$

$$\leq BT \cdot E\left(\sup_{0 \leq s \leq T} |X_s^n - X_s^\infty|^2\right) \to 0$$

by (2.6), so $\tilde{X}_t^\infty = \lim X_t^{n+1} = X_t^\infty$.

Uniqueness. The key to the proof is

(2.7) Gronwall's Inequality. Suppose $\varphi(t) \leq A + B \int_0^t \varphi(s) \, ds$ for all $t \geq 0$ and $\varphi(t)$ is continuous. Then $\varphi(t) \leq Ae^{Bt}$.

Proof If we let $\psi(t) = (A + \epsilon)e^{Bt}$ then $\psi'(t) = B\psi(t)$ so

$$\psi(t) = A + \epsilon + B \int_0^t \psi(s) \, ds$$

Let $\tau = \inf\{t : \varphi(t) \geq \psi(t)\}$. Since ψ and φ are continuous, it follows that if $\tau < \infty$ then $\varphi(\tau) = \psi(\tau)$, but this is impossible since

$$\psi(\tau) = A + \epsilon + B \int_0^\tau \psi(s) \, ds > A + B \int_0^\tau \varphi(s) \, ds \geq \varphi(t)$$

The last contradiction implies that $\varphi(t) \leq (A+\epsilon)e^{Bt}$, but $\epsilon > 0$ is arbitrary, so the desired result follows. □

To prove pathwise uniqueness now, let $(Y_t, B_t, \mathcal{F}_t^1)$ and $(Z_t, B_t, \mathcal{F}_t^2)$ be two solutions of $(\star)$. Replacing the $\mathcal{F}_t^i$ by $\mathcal{F}_t = \mathcal{F}_t^1 \vee \mathcal{F}_t^2$ we can suppose $\mathcal{F}_t^1 = \mathcal{F}_t^2$. To prepare for developments in the next section, we will use a more general formulation than we need now.

(2.8) Lemma. Let Y_t and Z_t be continuous processes adapted to $\mathcal{F}_t$ with $Y_0 = Z_0 = x$ and let $\tau(R) = \inf\{t : |Y_t| \text{ or } |Z_t| > R\}$. Suppose that B_t is Brownian motion w.r.t. $\mathcal{F}_t$ and that Y_t and Z_t satisfy $(\star)$ on $[0, \tau(R)]$. Then $Y_t = Z_t$ on $[0, \tau(R)]$.

Proof Let

$$\varphi(t) = E\left(\sup_{0 \leq s \leq t \wedge \tau(R)} |Y_s - Z_s|^2\right)$$

Since Y_t and Z_t are solutions, (2.3) implies

$$\varphi(t) \leq BE \int_0^{t \wedge \tau(R)} |Y_s - Z_s|^2 \, ds$$

$$\leq BE \int_0^t |Y_{s \wedge \tau} - Z_{s \wedge \tau}|^2 \, ds \leq B \int_0^t \varphi(s) \, ds$$

so (2.7) holds with $A = 0$ and it follows that $\varphi \equiv 0$. □

The last result implies that two solutions must agree until the first time one of them leaves the ball of radius R. (Of course this implies that the two solutions must exit at the same time.) Since solutions are by definition continuous functions from $[0, \infty) \to \mathbf{R}^d$, we must have $\tau(R) \uparrow \infty$ as $R \uparrow \infty$ and the desired result follows. □

Temporal Inhomogeneity. In some applications it is important to allow the coefficients b and σ to depend on time, that is, to consider

$$(\star\star) \qquad X_t = X_0 + \int_0^t b(s, X_s) \, ds + \int_0^t \sigma(s, X_s) \, dB_s$$

In many cases it is straightforward to generalize proofs from $(\star)$ to $(\star\star)$. For example, by repeating the argument above with some minor modifications one can show

(2.9) Theorem. If for all i, j, x, y, and $T < \infty$ there is a constant K_T so that for $t \leq T$, $|\sigma_{ij}(t,0)| \leq K_T$, $|b_i(t,0)| \leq K_T$,

$$|\sigma_{ij}(t,x) - \sigma_{ij}(t,y)| \leq K_T|x-y| \quad \text{and} \quad |b_i(t,x) - b_i(t,y)| \leq K_T|x-y|$$

then the stochastic differential equation (★★) has a strong solution and pathwise uniqueness holds.

Comparing with (2.2) one sees that the new result simply assumes Lipschitz continuity locally uniformly in t but needs new conditions: $|\sigma_{ij}(t,0)| \leq K_T$ and $|b_i(t,0)| \leq K_T$ that are automatic in the temporally homogeneous case. The dependence of the coefficients on t usually does not introduce any new difficulties but it does make the statements and proofs uglier. (Here it would be impolite to point to Karatzas and Shreve (1991) and Stroock and Varadhan (1979).) So with the exception of (5.1) below, where we need to prove the result for time-dependent coefficients, we will restrict our attention to the temporally homogeneous case.

5.3. Extensions

In this section we will (a) extend Itô's result to coefficients that are only locally Lipschitz, (b) prove a result for one dimensional SDE due to Yamada and Watanabe, and (c) introduce examples to show that the results in (2.2) and (3.3) are fairly sharp.

a. Locally Lipschitz Coefficients

In this subsection we will extend (2.2) in the following way.

(3.1) Theorem. Suppose (i) for any $n < \infty$ we have

$$|\sigma_{ij}(x) - \sigma_{ij}(y)| \leq K_n|x-y| \qquad |b_i(x) - b_i(y)| \leq K_n|x-y|$$

when $|x|, |y| \leq n$ and (ii) there is a constant $A < \infty$ and a function $\varphi(x) \geq 0$ so that if X_t is a solution of (★) then $e^{-At}\varphi(X_t)$ is a local supermartingale. Then (★) has a strong solution and pathwise uniqueness holds.

(3.2) Theorem. Let $a = \sigma\sigma^T$ and suppose

$$\sum_{i=1}^{d} \{2x_i b_i(x) + a_{ii}(x)\} \leq B(1 + |x|^2)$$

then (ii) in (3.1) holds with $A = B$ and $\varphi(x) = 1 + |x|^2$.

We begin with the easier proof.

Proof of (3.2) Using Itô's formula with $X_t^0 = t$ and
$$f(x_0, \ldots, x_d) = e^{-Bx_0}\varphi(x_1, \ldots, x_d)$$
we have

$$e^{-At}\varphi(X_t) - \varphi(X_0) = -B\int_0^t e^{-Bs}\varphi(X_s)\,ds$$

$$+ \sum_{i=1}^d \int_0^t e^{-Bs} 2X_s^i\, dX_s^i + \frac{1}{2}\sum_{i=1}^d \int_0^t e^{-Bs} 2\,d\langle X^i\rangle_s$$

$$= \text{local martingale}$$

$$+ \int_0^t e^{-Bs}\left(-B\varphi(X_s) + \sum_{i=1}^d \{2X_s^i b_i(X_s) + a_{ii}(X_s)\}\right) ds$$

Our assumption implies that the last term is ≤ 0 so $e^{-At}\varphi(X_t)$ is a local supermartingale. □

Proof of (3.1) Let $R < \infty$, and introduce σ^R and b^R with

(a) $\sigma^R(x) = \sigma(x)$ and $b^R(x) = b(x)$ for $|x| \leq R$
(b) $\sigma^R(x) = 0$ and $b^R(x) = 0$ for $|x| \geq 2R$
(c) $|\sigma_{ij}^R(x) - \sigma_{ij}^R(y)| \leq K'|x - y|$, $|b_i^R(x) - b_i^R(y)| \leq K'|x - y|$

For an explicit construction do

Exercise 3.1. Suppose $|h(x) - h(y)| \leq C|x - y|$ when $|x| \leq R$. Let

$$h(x) = \frac{2R - |x|}{R} \cdot h(Rx/|x|) \quad \text{for } R \leq |x| \leq 2R$$

and $h(x) = 0$ for $|x| \geq 2R$. Show that h is Lipschitz continuous on $\mathbf{R}^d$ with constant $C = 2C_1 + R^{-1}|h(0)|$.

Fix a Brownian motion and let X_t^n be the unique strong solution of

$(\star_n)$ $$X_t = X_0 + \int_0^t b^n(X_s)\,ds + \int_0^t \sigma^n(X_s)\,dB_s$$

Let $T_n = \inf\{t : |X_t| \geq n\}$. (2.8) implies that if $m < n$ then $X_t^m = X_t^n$ for $t \leq T_m$. From this it follows that we can define a process X_t^∞ so that

$$X_t^\infty = X_t^n \quad \text{for } t \leq T_n$$

and X_t^∞ will be a solution of $(\star)$ for $t < T_\infty = \lim_{n \uparrow \infty} T_n$. To complete the proof now it suffices to show that $T_\infty = \infty$ a.s. If $X_0 = x$ then using the optional stopping theorem at time $t \wedge T_n$, which is valid since the local supermartingale is bounded before that time, we have

$$\varphi(x) \geq E\left(e^{-A(t \wedge T_n)} \varphi(X_{t \wedge T_n})\right) \geq e^{-At} P(T_n \leq t) \inf_{y:|y|=n} \varphi(y)$$

Rearranging gives

$$P(T_n \leq t) \leq e^{At} \varphi(x) \bigg/ \inf_{y:|y|=n} \varphi(y)$$

Since we have supposed $\varphi(y) \to \infty$ as $y \to \infty$ it follows that $P(T_n \leq t) \to 0$ as $n \to \infty$ which proves the desired result. □

In words, what we have shown is that for locally Lipschitz coefficients the solution is unique up to T_n the exit time from the ball of radius n. If $T_n \uparrow \infty$ we get a solution defined for all time. If not then the solution reaches infinity in finite time and we say that an **explosion** has occurred.

Intuitively (3.2) says there is no explosion if, when $|x|$ is large, the part of the drift that points out, $b(x) \cdot x/|x|$, is smaller than $C|x|$ and the variance of each component is smaller than $C|x|^2$. We do not need conditions on the off diagonal elements of a_{ij} since the Kunita-Watanabe inequality implies

$$|\langle X^i, X^j \rangle_t|^2 \leq \langle X^i \rangle_t \langle X^j \rangle_t$$

Note that since our condition concerns a sum, a drift toward the origin can compensate for a larger variance.

Example 3.1. Consider $d = 1$ and suppose $b(x) = -x^3$. Then (3.2) holds if $a(x) \leq B(1 + x^2) + 2x^4$.

The next two examples show that when considered separately the conditions on b and a are close to the best possible.

Example 3.2. Consider $d = 1$ and let

$$\sigma(x) = 0, \qquad b(x) = (1 + |x|)^\delta \quad \text{with } \delta > 1$$

Suppose $X_0 = 0$. $b > 0$ so $t \to X_t$ is strictly increasing. Let $T_n = \inf\{t : X_t = n\}$. While $X_t \in (n-1, n)$ its velocity $b(X_t) \geq n^\delta$ so $T_n - T_{n-1} \leq n^{-\delta}$. Letting $T_\infty = \lim T_n$ and summing we have $T_\infty \leq \sum_{n=1}^\infty n^{-\delta} < \infty$ if $\delta > 1$.

We like the last proof since it only depends on the asymptotic behavior of the coefficients. One can also solve the equation explicitly, which has the advantage of giving the exact explosion time. Let $Y_t = 1 + X_t$. $dY_t = Y_t^\delta \, dt$ so if we guess $Y_t = (1 - at)^{-p}$ then

$$dY_t/dt = \frac{ap}{(1-at)^{p+1}}$$

Setting $a = 1/p$ and then $p = 1/(\delta - 1)$ to have $p + 1 = \delta p$ we get a solution that explodes at time $1/a = p = 1/(\delta - 1)$.

Example 3.3. Suppose $b = 0$, and $\sigma = (1 + |x|)^\delta I$. (2.2) implies there is no explosion when $\delta \leq 1$. Later we will show (see Example 6.2 and (6.3)) that

in $d \geq 3$ explosion occurs for $\delta > 1$
in $d \leq 2$ no explosion for any $\delta < \infty$

b. Yamada and Watanabe in d = 1

In one dimension one can get by with less smoothness in σ.

(3.3) Theorem. Suppose that (i) there is a strictly increasing continuous function ρ with $|\sigma(x) - \sigma(y)| \leq \rho(|x - y|)$ where $\rho(0) = 0$ and for all $\epsilon > 0$

$$\int_0^\epsilon \rho^{-2}(u) \, du = \infty$$

(ii) there is a strictly increasing and concave function κ with $|b(x) - b(y)| \leq \kappa(|x - y|)$ where $\kappa(0) = 0$ and for all $\epsilon > 0$

$$\int_0^\epsilon \kappa^{-1}(u) \, du = \infty$$

Then pathwise uniqueness holds.

Remark. Note that there is no mention here of existence of solutions. That is taken care of by a result in Section 8.4.

Proof The first step is to define a sequence φ_n of smooth approximations of $|x|$. Let $a_n \downarrow 0$ be defined by $a_0 = 1$ and

$$\int_{a_n}^{a_{n-1}} \rho^{-2}(u) \, du = n$$

194 Chapter 5 Stochastic Differential Equations

Let $0 \le \psi_n(u) \le 2\rho^{-2}(u)/n$ be a continuous function with support in (a_n, a_{n-1}) and
$$\int_{a_n}^{a_{n-1}} \psi_n(u)\,du = 1$$

Since the upper bound in the definition of ψ_n integrates to 2 over (a_n, a_{n-1}) such a function exists. Let
$$\varphi_n(x) = \int_0^{|x|} dy \int_0^y dz\, \psi_n(z)$$

By definition $\varphi_n(x) = \varphi_n(-x)$, and $\varphi_n'(x) = \int_0^x \psi_n(u)\,du$, for $x > 0$, so
$$|\varphi_n'(x)| \text{ is } \begin{cases} = 0 & \text{for } |x| \le a_n \\ \le 1 & \text{for } a_n \le |x| \le a_{n-1} \\ = 1 & \text{for } a_{n-1} \le |x| \end{cases}$$

and it follows that $\varphi_n(x) \uparrow |x|$ as $n \uparrow \infty$.

Suppose now that X_t^1 and X_t^2 are two solutions with $X_0^1 = X_0^2 = x$ driven by the same Brownian motion, and let $\Delta_t = X_t^1 - X_t^2$.
$$\Delta_t = \int_0^t \{\sigma(X_s^1) - \sigma(X_s^2)\}\,dB_s + \int_0^t \{b(X_s^1) - b(X_s^2)\}\,ds$$

so Itô's formula gives

(a)
$$\begin{aligned}\varphi_n(\Delta_t) &= \int_0^t \varphi_n'(\Delta_s)\,d\Delta_s + \frac{1}{2}\int_0^t \varphi_n''(\Delta_s)\,d\langle\Delta\rangle_s \\ &= \int_0^t \varphi_n'(\Delta_s)\{\sigma(X_s^1) - \sigma(X_s^2)\}\,dB_s \\ &\quad + \int_0^t \varphi_n'(\Delta_s)\{b(X_s^1) - b(X_s^2)\}\,ds \\ &\quad + \frac{1}{2}\int_0^t \varphi_n''(\Delta_s)\{\sigma(X_s^1) - \sigma(X_s^2)\}^2\,ds \\ &\equiv I_0(t) + I_1(t) + I_2(t)\end{aligned}$$

$I_0(t)$ is a local martingale so there are stopping times $T_m \uparrow \infty$ with

(b)
$$EI_0(t \wedge T_m) = 0$$

To deal with I_2 in (a) we note that $\varphi_n''(x) = \psi_n(|x|) \le 2\rho^{-2}(|x|)/n$ when $|x| \in (a_n, a_{n-1})$ and is 0 otherwise. So using (i)

(c)
$$|I_2(t)| \le \frac{1}{2}\int_0^t \frac{2\rho^{-2}(|\Delta_s|)}{n}\rho^2(|\Delta_s|)\,ds \le \frac{t}{n}$$

For I_1 in (a) we observe $|\varphi_n'(x)| \leq 1$, so using (ii) and the concavity of κ we have

(d)
$$E|I_1(t)| \leq \int_0^t E|b(X_s^1) - b(X_s^2)|\, ds$$
$$\leq \int_0^t E\kappa(|\Delta_s|)\, ds \leq \int_0^t \kappa(E|\Delta_s|)\, ds$$

Combining (a)–(d) we have

$$E\varphi_n(\Delta_{t \wedge T_m}) \leq \frac{t}{n} + \int_0^t \kappa(E|\Delta_s|)\, ds$$

Letting $m \to \infty$ and using Fatou's lemma, then letting $n \to \infty$ and using the monotone convergence theorem (recall $\varphi_n(x) \geq 0$ and $\varphi_n(x) \uparrow |x|$) we have

$$E|\Delta_t| \leq \int_0^t \kappa(E|\Delta_s|)\, ds$$

To finish the proof now we prove a slight improvement of Gronwall's inequality.

(3.4) Lemma. Suppose $f(t) \leq \int_0^t \kappa(f(s))\, ds$ where f is continuous, $\kappa(x) > 0$ is increasing for $x > 0$, and has $\int_0^\epsilon \kappa^{-1}(x)\, dx = \infty$ for any $\epsilon > 0$. Then $f(t) \equiv 0$.

Proof Let $g(t)$ be the solution of $g'(t) = \kappa(g(t))$ with $g(0) = \epsilon$. It follows from the proof of (2.7) that $f(t) \leq g(t)$. To show that g is small when ϵ is small note that g is strictly increasing and let h be its inverse. Since $g(h(x)) = x$, we have $h'(x) = 1/g'(h(x)) = 1/\kappa(g(h(x))) = 1/\kappa(x)$ and hence if $a < b$

$$h(b) - h(a) = \int_a^b 1/\kappa(x)\, dx$$

In words the right-hand side gives the amount time it takes g to climb from a to b. Taking $a = \epsilon$ and $b = \delta > 0$, the assumed divergence of the integral implies that the amount of time to climb to δ approaches ∞ as $\epsilon \to 0$ and the desired result follows. □

c. Examples

We will now give examples to show that for bounded coefficients the combination of (2.2) and (3.3) is fairly sharp. In the first case the counterexamples will have to wait until Section 5.6.

Example 3.4. Consider $b(x) = 0$ and $\sigma(x) = |x|^\delta \wedge 1$. (2.2) and (3.3) imply that

$$\text{pathwise uniqueness holds for } \begin{cases} \delta \geq 1/2 & \text{in } d = 1 \\ \delta \geq 1 & \text{in } d > 1 \end{cases}$$

Example 6.1 will show

$$\text{pathwise uniqueness fails if } \begin{cases} \delta < 1/2 & \text{in } d = 1 \\ \delta < 1 & \text{in } d > 1 \end{cases}$$

Example 3.5. Consider $\sigma(x) = 0$, $b(x) = |x|^\delta \wedge 1$, and $X_0 = 0$. If $\delta \geq 1$ then (2.2) implies that there is a unique solution: $X_s \equiv 0$. However, if $\delta < 1$ there are others. To find one, we guess $X_t = Ct^p$. Then

$$dX_s = Cps^{p-1}\, ds \qquad b(X_s) = C^\delta s^{p\delta}$$

so to make $dX_s = b(X_s)\, ds$ we set

$$(p-1) = p\delta \quad \text{i.e., } p = 1/(1-\delta)$$
$$Cp = C^\delta \quad \text{i.e., } C = p^{-1/(1-\delta)}$$

If $\delta < 1$ then $p > 0$ and we have created a second solution starting at 0. Once we have two solutions we can construct infinitely many. Let $a > 0$ and let

$$X_t = \begin{cases} 0 & t \leq a \\ C(t-a)^p & t \geq a \end{cases}$$

Exercise 3.2. The main reason for interest in (3.3) is the weakening of Lipschitz continuity of σ. However, the condition on b is also an improvement, and further sharpens the line between theorems and counterexamples. (a) Show that (ii) in (3.3) holds when $C > 0$ and

$$\kappa(x) = \begin{cases} Cx \log(1/x) & \text{for } x \leq e^{-2} \\ C(e^{-2} + x) & \text{for } x \geq e^{-2} \end{cases}$$

(b) Consider $g(t) = \exp(-1/t^p)$ with $p > 0$ and show that $g'(t) = \psi_p(g(t))$ where $\psi_p(y) = py\{\log(1/y)\}^{(p+1)/p}$.

5.4. Weak Solutions

Intuitively, a strong solution corresponds to solving the SDE for a given Brownian motion, while in producing a weak solution we are allowed to construct the Brownian motion and the solution at the same time. In signal processing applications one must deal with the noisy signal that one is given, so strong solutions are required. However, if the aim is to construct diffusion processes then there is nothing wrong with a weak solution.

Now if we do not insist on staying on the original space $(\Omega, \mathcal{F}, P)$ then we can use the map $\omega \to (X_t(\omega), B_t(\omega))$ to move from the original space $(\Omega, \mathcal{F})$ to $(C \times C, \mathcal{C} \times \mathcal{C})$ where we can use the filtration generated by the coordinates $\omega_1(s), \omega_2(s)$, $s \le t$. Thus a weak solution is completely specified by giving the joint distribution (X_t, B_t).

Reflecting our interest in constructing the process X_t and sneaking in some notation we will need in a minute, we say that there is **uniqueness in distribution** if whenever (X, B) and (X', B') are solutions of SDE(b, σ) with $X_0 = X'_0 = x$ then X and X' have the same distribution, i.e., $P(X \in A) = P(X' \in A)$ for all $A \in \mathcal{C}$. Here a solution of SDE(b, σ) is something that satisfies $(\star)$ in the sense defined in Section 5.2.

Example 4.1. Since sgn$(x)^2 = 1$, any solution to the equation in Example 2.1 is a local martingale with $\langle X \rangle_t \equiv t$. So Lévy's theorem, (4.1) in Chapter 3, implies that X_t is a Brownian motion. The last result asserts that uniqueness in distribution holds, but (2.1) shows that pathwise uniqueness fails. The next result, also due to Yamada and Watanabe (1971), shows that the other implication is true.

(4.1) Theorem. If pathwise uniqueness holds then there is uniqueness in distribution.

Remark. Pathwise uniqueness also implies that every solution is a strong solution. However that proof is more complicated and the conclusion is not important for our purposes so we refer the reader to Revuz and Yor (1991), p. 340–341.

Proof Suppose (X^1, B^1) and (X^2, B^2) are two solutions with $X_0^1 = X_0^2 = x$. (As remarked above the solutions are specified by giving the joint distributions (X^i, B^i).) Since $(C, \mathcal{C})$ is a complete separable metric space, we can find regular conditional distributions $Q^i(\omega, A)$ defined for $\omega \in C$ and $A \in \mathcal{C}$ (see Section 4.1 in Durrett (1995)) so that

$$P(X^i \in A | B^i) = Q^i(B^i, A)$$

Let P_0 be the measure on $(C, \mathcal{C})$ for the standard Brownian motion and define a measure π on $(C^3, \mathcal{C}^3)$ by

$$\pi(A_0 \times A_1 \times A_2) = \int_{A_0} dP_0(\omega_0) Q_1(\omega_0, A_1) Q_2(\omega_0, A_2)$$

If we let $Y_t^i(\omega_0, \omega_1, \omega_2) = \omega_i(t)$ then it is clear that

$$(X^1, B^1) \stackrel{d}{=} (Y^1, Y^0) \quad \text{and} \quad (X^2, B^2) \stackrel{d}{=} (Y^2, Y^0)$$

Take $A_2 = C$ or $A_1 = C$ respectively in the definition. Thus, we have two solutions of the equation driven by the same Brownian motion. Pathwise uniqueness implies $Y^1 = Y^2$ with probability one. (i) follows since

$$X^1 \stackrel{d}{=} Y^1 = Y^2 \stackrel{d}{=} X^2 \qquad \square$$

Let $a(x) = \sigma \sigma^T(x)$ and recall from Section 5.1 that

$$\langle X^i, X^j \rangle_t = \int_0^t a_{ij}(X_s)\, ds$$

We say that X is a **solution to the martingale problem** for b and a, or simply X solves MP(b, a), if for each i and j,

$$X_t^i - \int_0^t b_i(X_s)\, ds \quad \text{and} \quad X_t^i X_t^j - \int_0^t a_{ij}(X_s)\, ds$$

are local martingales. The second condition is, of course, equivalent to

$$\langle X^i, X^j \rangle_t = \int_0^t a_{ij}(X_s)\, ds$$

To be precise in the definition above we need to specify a filtration $\mathcal{F}_t$ for the local martingales. In posing the martingale problem we are only interested in the process X_t, so we can assume without loss of generality that the underlying space is $(C, \mathcal{C})$ with $X_t(\omega) = \omega_t$ and the filtration is the one generated by X_t.

In the formulation above b and $a = \sigma \sigma^T$ are the basic data for the martingale problem. This presents us with the problem of obtaining σ from a. To solve this problem we begin by introducing some properties of a. Since $\langle X^i, X^j \rangle_t = \langle X^j, X^i \rangle_t$, a is **symmetric**, that is, $a_{ij} = a_{ji}$. Also, if $z \in \mathbf{R}^d$ then

$$\sum_{i,j} z_i a_{ij} z_j = \sum_{i,j,k} z_i \sigma_{i,k} \sigma_{j,k} z_j = \sum_k \left(\sum_i z_i \sigma_{ik} \right)^2 \geq 0$$

So a is **nonnegative definite** and linear algebra provides us with the following useful information.

(4.2) Lemma. For any symmetric nonnegative definite matrix a, we can find an orthogonal matrix U (i.e., its rows are perpendicular vectors of length 1) and a diagonal matrix Λ with diagonal entries $\lambda_1 \geq \lambda_2 \ldots \geq \lambda_d \geq 0$ so that $a = U^T \Lambda U$. a is invertible if and only if $\lambda_d > 0$.

(4.2) tells us $a = U^T \Lambda U$, so if we want $a = \sigma \sigma^T$ we can take $\sigma = U^T \sqrt{\Lambda}$, where $\sqrt{\Lambda}$ is the diagonal matrix with entries $\sqrt{\lambda_i}$. (4.2) tells us how to find the square root of a given a. The usual algorithms for finding U are such that if we apply them to a measurable $a(x)$ then the resulting square root $\sigma(x)$ will also be measurable. It takes more sophistication to start with a smooth $a(x)$ and construct a smooth $\sigma(x)$. As in the case of (4.2), we will content ourselves to just state the result. For detailed proofs of (4.3) and (4.4) see Stroock and Varadhan (1979), Section 5.2.

(4.3) Theorem. If $\theta^T a(x) \theta \geq \alpha |\theta|^2$ and $\|a(x) - a(y)\| \leq C|x - y|$ for all x, y then $\|a^{1/2}(x) - a^{1/2}(y)\| \leq (C/2\alpha^{1/2})|x - y|$ for all x, y.

When a can degenerate we need to suppose more smoothness to get a Lipschitz continuous square root.

(4.4) Theorem. Suppose $a_{ij} \in C^2$ and
$$\max_{1 \leq i \leq d} |\theta^T D_{ii} a(x) \theta| \leq C|\theta|^2$$
then $\|a^{1/2}(x) - a^{1/2}(y)\| \leq d(2C)^{1/2}|x - y|$ for all x, y.

To see why we need $a \in C^2$ to get a Lipschitz $a^{1/2}$, consider $d = 1$ and $a(x) = |x|^\lambda$. In this case $\sigma(x) = |x|^{\lambda/2}$ is Lipschitz continuous at 0 if and only if $\lambda \geq 2$.

Returning to probability theory, our first step is to make the connection between solutions to the martingale problem and solutions of the SDE.

(4.5) Theorem. Let X be a solution to $\mathrm{MP}(b, a)$ and σ a measurable square root of a. Then there is a Brownian motion B_t, possibly defined on an enlargement of the original probability space, so that (X, B) solves $\mathrm{SDE}(b, \sigma)$.

Proof If σ is invertible at each X this is easy. Let $Y_t^i = X_t^i - \int_0^t b_i(X_s)\, ds$ and let

(a) $$B_t^i = \sum_j \int_0^t \sigma_{ij}^{-1}(X_s)\, dY_s^j$$

From the last two definitions and the associative law it is immediate that

(b) $$\int_0^t \sigma(X_s)\, dB_s = Y_t - Y_0 = X_t - X_0 - \int_0^t b(X_s)\, ds$$

so (⋆) holds. The B_t^i are local martingales with $\langle B^i, B^j \rangle_t = \delta_{ij} t$ so Exercise 4.1 in Chapter 3 implies that B is a Brownian motion.

To deal with the general case in one dimension, we let

$$I_s = \begin{cases} 1 & \text{if } \sigma(X_s) \neq 0 \\ 0 & \text{otherwise} \end{cases}$$

let $J_s = 1 - I_s$, let W_s be an independent Brownian motion (to define this we may need to enlarge the probability space) and let

(a') $$B_s = \int_0^t \frac{I_s}{\sigma(X_s)} dY_s + \int_0^t J_s \, dW_s$$

The two integrals are well defined since their variance processes are $\leq t$. B_t is a local martingale with $\langle B \rangle_t = t$ so (4.1) in Chapter 3 implies that B_t is a Brownian motion.

To extract the SDE now, we observe that the associative law and the fact that $J_s \sigma(X_s) \equiv 0$ imply

$$\int_0^t \sigma(X_s) \, dB_s = \int_0^t I_s \, dY_s$$
$$= Y_t - Y_0 - \int_0^t J_s \, dY_s$$

In view of (b) the proof will be complete when we show $\int_0^t J_s \, dY_s \equiv 0$. To do this we note that

$$E\left(\int_0^t J_s \, dY_s\right)^2 = E \int_0^t J_s^2 \, d\langle Y \rangle_s$$
$$= E \int_0^t J_s^2 \sigma(X_s) \, ds = 0$$

To handle higher dimensions we let $U(x)$ be the orthogonal matrices constructed in (4.2) and let

$$Z_t = \int_0^t U^T(X_s) \, dY_s$$

Introducing Kronecker's $\delta_{ij} = 1$ if $i = j$ and 0 otherwise, we have

$$\langle Z^i, Z^j \rangle_t = \int_0^t \sum_{k,\ell} U_{i,k}(X_s) a_{k,\ell}(X_s) U_{\ell,j}^T(X_s) \, ds = \int_0^t \delta_{ij} \lambda_i(X_s) \, ds$$

Our result for one dimension implies that the Z_t^i may be realized as stochastic integrals with respect to independent Brownian motions. Tracing back through the definitions gives the desired representation. For more details see Ikeda and Watanabe (1981), p. 89–91, Revuz and Yor (1991), p. 190–191, or Karatzas and Shreve (1991), p. 315–317. □

(4.5) shows that there is a 1-1 correspondence between distributions of solutions of the SDE and solutions of the martingale problem, which by definition are distributions on $(C,\mathcal{C})$. Thus there is uniqueness in distribution for the SDE if and only if the **martingale problem has a unique solution**. Our final topic is an important reason to be interested in uniqueness.

(4.6) **Theorem.** Suppose a and b are locally bounded functions and MP(b,a) has a unique solution. Then the strong Markov property holds.

Proof For each $x \in \mathbf{R}^d$ let P_x be the probability measure on $(C,\mathcal{C})$ that gives the distribution of the unique solution starting from $X_0 = x$. If you talk fast the proof is easy. If T is a stopping time then the conditional distribution of $\{X_{T+s}, s \geq 0\}$ given $\mathcal{F}_T$ is a solution of the martingale problem starting from X_T and so by uniqueness it must be $P_{X(T)}$. Thus, if we let θ_T be the random shift defined in Section 1.3, then for any bounded measurable $Y : C \to \mathbf{R}$ we have the strong Markov property

$$(4.7) \qquad E(Y \circ \theta_T \,|\, \mathcal{F}_T) = E_{X(T)} Y$$

To begin to turn the last paragraph into a proof, fix a starting point $X_0 = x_0$. To simplify this rather complicated argument, we will follow the standard practice (see Stroock and Varadhan (1979), p. 145–146, Karatzas and Shreve (1991), p. 321–322, or Rogers and Williams (1987), p. 162–163) of giving the details only for the case in which the coefficients are bounded and hence

$$M_t^{i0} = X_t^i - \int_0^t b_i(X_s)\,ds$$

$$M_t^{ij} = X_t^i X_t^j - \int_0^t a_{ij}(X_s)\,ds$$

are martingales. We have added an extra index in the first case so we can refer to all the martingales at once by saying "the M_t^{ij}."

Consider a bounded stopping time T and let $Q(\omega, A)$ be a regular conditional distribution for $\{X_{T+t}, t \geq 0\}$ given $\mathcal{F}_T$. We want to claim that

(4.8) **Lemma.** For P_{x_0} a.e. ω, all the

$$M_{T+t}^{ij} - M_T^{ij} = M_t^{ij} \circ \theta_T$$

are martingales under $Q(\omega, \cdot)$.

To prove this we consider rational $s < t$ and $B \in \mathcal{F}_s$ and note that the optional stopping theorem implies that if $A \in \mathcal{F}_T$ then

$$E_{x_0}\left(\{(M_t^{ij} - M_s^{ij})1_B\} \circ \theta_T\} 1_A\right) = 0$$

Letting B run through a countable collection that is closed under intersection and generates $\mathcal{F}_s$ it follows that P_{x_0} a.s., the expected value of $(M_t^{ij} - M_s^{ij})1_B$ is 0, which implies (4.8). Using uniqueness now gives (4.7) and completes the proof. □

5.5. Change of Measure

Our next step is to use Girsanov's formula from Section 2.12 to solve martingale problems or more precisely to change the drift in an existing solution. To accomodate Example 5.2 below we must consider the case in which the added drift b depends on time as well as space.

(5.1) Theorem. Suppose X_t is a solution of $MP(\beta, a)$, which, for concreteness and without loss of generality, we suppose is defined on the canonical probability space $(C, \mathcal{C}, P)$ with $\mathcal{F}_t$ the filtration generated by $X_t(\omega) = \omega_t$. Suppose $a^{-1}(x)$ exists for all x, and that $b(s, x)$ is measurable and has $|b^T a^{-1} b(s, x)| \leq M$. We can define a probability measure Q locally equivalent to P, so that under Q, X_t is a solution to $MP(\beta + b, a)$.

Proof Let $\bar{X}_t = X_t - \int_0^t \beta(X_s)ds$, let $c(s, x) = a^{-1}(x)b(s, x)$ and let

$$Y_t = \int_0^t c(s, X_s) \cdot d\bar{X}_s$$

which exists since

$$\langle Y \rangle_t = \int_0^t \sum_{ij} c_i(s, X_s)c_j(s, X_s) \, d\langle \bar{X}^i, \bar{X}^j \rangle_s$$

$$= \int_0^t c^T ac(s, X_s) \, ds = \int_0^t b^T a^{-1} b(s, X_s) \, ds \leq Mt$$

by our assumption that $|b^T a^{-1} b(s, x)| \leq M$. Letting Q_t and P_t be the restrictions of Q and P to $\mathcal{F}_t$ we define our Radon-Nikodym derivative to be

$$\frac{dQ_t}{dP_t} = \alpha_t = \exp\left(Y_t - \frac{1}{2}\langle Y \rangle_t\right)$$

Since $\langle Y \rangle_t \leq Mt$, (3.7) in Chapter 3 implies that α_t is a martingale, and invoking (12.3) in Chapter 2 we have defined Q.

To apply Girsanov's formula ((12.4) from Chapter 2), we have to compute $A_t^j = \int_0^t \alpha_s^{-1} d\langle \alpha, \bar{X}^j \rangle_s$. Itô's formula and the associative law imply that

$$\alpha_t - 1 = \int_0^t \alpha_s \, dY_s = \int_0^t \alpha_s \, c(s, X_s) \cdot d\bar{X}_s$$

So by the formula for the covariance of stochastic integrals

$$\langle \alpha, \bar{X}^j \rangle_t = \sum_i \int_0^t \alpha_s \, c_i(s, X_s) \, d\langle \bar{X}^i, \bar{X}^j \rangle_s$$

$$= \int_0^t \alpha_s (c^T a)_j (s, X_s) \, ds = \int_0^t \alpha_s \, b_j(s, X_s) \, ds$$

since $c(s, x) = a^{-1}(x) b(s, x)$. It follows that

$$A_t^j = \int_0^t \alpha_s^{-1} d\langle \alpha, \bar{X}^j \rangle_s = \int_0^t b_j(s, X_s) \, ds$$

Using Girsanov's formula now, it follows that

$$\bar{X}_t^j - \int_0^t b_j(s, X_s) \, ds = X_t^j - \int_0^t \beta_j(X_s) + b_j(s, X_s) \, ds$$

is a local martingale/Q.

This is half of the desired conclusion but the other half is easy. We note that under P

$$\langle X^i, X^j \rangle_t = \int_0^t a_{ij}(X_s) \, ds$$

and (12.6) in Chapter 2 implies that the covariance of X^i and X^j is the same under Q. □

Exercise 5.1. To check your understanding of the formulas, consider the case in which b does not depend on s, start with the Q constructed above, change to a measure R to remove the added drift b, and show that $dR_t/dQ_t = 1/\alpha_t$ so $R = P$.

Example 5.1. Suppose X_t is a solution of MP$(0, I)$, i.e., a standard Brownian motion and consider the special situation in which $b(x) = \nabla U(x)$. In this case recalling the definition of Y_t in the proof of (5.1) and applying Itô's formula to $U(X_t)$ we have

$$Y_t = \int_0^t \nabla U(X_s) \cdot X_s = U(X_t) - U(X_0) - \frac{1}{2}\int_0^t \Delta U(X_s)\,ds$$

Thus, we can get rid of the stochastic integral in the definition of the Radon-Nikodym derivative:

$$\frac{dQ_t}{dP_t} = \exp\left(U(X_t) - U(X_0) - \frac{1}{2}\int_0^t \Delta U(X_s) + |\nabla U(X_s)|^2\,ds\right)$$

In the special case of the Ornstein-Uhlenbeck process, $b(x) = -2\alpha x$, where α is a positive real number, we have $U(x) = -\alpha|x|^2$ $\Delta U(x) = -2d\alpha$, and the expression above simplifies to

$$\frac{dQ_t}{dP_t} = \exp\left(-\alpha|X_t|^2 + \alpha|X_0|^2 + d\alpha t - \int_0^t 2\alpha^2|X_s|^2\,ds\right)$$

In the trivial case of constant drift, i.e., $b(x) = \mu = \nabla U(x)$ we have $U(x) = \mu \cdot x$ and $\Delta U(x) = 0$ so

$$\frac{dQ_t}{dP_t} = \exp\left(\mu \cdot X_t - \mu \cdot X_0 - \frac{1}{2}|\mu|^2 t\right)$$

It is comforting to note that when $X_0 = x$, the one dimensional density functions satisfy

$$Q_t(X_t = y) = \exp\left(\mu \cdot y - \mu \cdot x - \frac{1}{2}|\mu t|^2\right) P_t(X_t = y)$$
$$= (2\pi t)^{-d/2}\exp(-|y - x - \mu t|^2/2t)$$

as it should be since under Q, X_t has the same distribution as $B_t + \mu t$. □

(5.1) deals with existence of solutions, our next result with uniqueness.

(5.2) Theorem. Under the hypotheses of (5.1) there is a 1-1 correspondence between solutions of $\mathrm{MP}(\beta, a)$ and $\mathrm{MP}(\beta + b, a)$.

Proof Let P_1 and P_2 be solutions of $\mathrm{MP}(\beta, a)$. By change of measure we can produce solutions Q_1 and Q_2 of $\mathrm{MP}(\beta + b, a)$. We claim that if $P_1 \neq P_2$ then $Q_1 \neq Q_2$. To prove this we consider two cases:

CASE 1. Suppose P_1 and P_2 are not locally equivalent measures. Then since Q_i is locally equivalent to P_i, Q_1 and Q_2 are not locally equivalent and cannot be equal.

CASE 2. Suppose P_1 and P_2 are locally equivalent. In the definition of α_t, $\langle Y \rangle_t$ and Y_t are independent of P_i by (12.6) and (12.7) in Chapter 2, so $dQ_1/dP_1 = dQ_2/dP_2$ and $Q_1 \neq Q_2$. □

One can considerably relax the boundedness condition in (5.1). The next result will cover many examples. If it does not cover yours, note that the main ingredients of the proof are that the conditions of (5.1) hold locally, and we know that solutions of MP(b, a) do not explode.

(5.3) Theorem. Suppose X is a solution of MP$(0, a)$ constructed on $(C, \mathcal{C})$. Suppose that $a^{-1}(x)$ exists, $b^T a^{-1} b(x)$ is locally bounded and measurable, and the quadratic growth condition from (3.2) holds. That is,

$$\sum_{i=1}^{d} \{2x_i b_i(x) + a_{ii}(x)\} \leq A(1 + |x|^2)$$

Then there is a 1-1 correpondence between solutions to MP(b, a) and MP$(0, a)$.

Remark. Combining (5.3) with (3.1) and (4.1) we see that if in addition to the conditions in (5.3) we have $a = \sigma\sigma^T$ where σ is locally Lipschitz, then MP(b, a) has a unique solution. By using (3.3) instead of (3.1) in the previous sentence we can get a better result in one dimension.

Proof Let Z and α be as in the proof of (5.1) and let $T_n = \inf\{t : |X_t| > n\}$. From the proof of (5.1) it follows that $\alpha(t \wedge T_n)$ is a martingale. So if we let P_t^n be the restriction of P to $\mathcal{F}_{t \wedge T_n}$ and let $dQ_t^n/dP_t^n = \alpha(t \wedge T_n)$ then under Q^n, X_t is a solution of MP(b, a) up to time T_n.

The quadratic growth condition implies that solutions of MP(b, a) do not explode. We will now show that in the absence of explosions, α_t is a martingale. First, to show α_t is integrable, we note that Fatou's lemma implies

(a) $$E\alpha_t \leq \liminf_{n \to \infty} E(\alpha(t \wedge T_n)) = 1$$

where E denotes expected value with respect to P. Next we note that

(b) $$E|\alpha_t - \alpha_{t \wedge T_n}| = E(|\alpha_t - \alpha(T_n)|; T_n \leq t)$$
$$\leq E(\alpha(T_n); T_n \leq t) + E\alpha_t - E(\alpha_t; T_n > t)$$

The absence of explosions implies that as $n \to \infty$

(c) $$E(\alpha(T_n); T_n \leq t) = Q^n(T_n \leq t) \to 0$$

This and the fact that $\alpha(t \wedge T_n)$ is a martingale implies

(d) $\qquad E(\alpha_t; T_n > t) = 1 - E(\alpha(T_n); T_n \leq t) \to 1$

as $n \to \infty$. (b), (c), and (d) imply that $\alpha(t \wedge T_n) \to \alpha_t$ in L^1. It follows (see e.g., (5.6) in Chapter 4 of Durrett (1995)) that

$$\alpha(t \wedge T_n) = E\left(\alpha_t \mid \mathcal{F}_{t \wedge T_n}\right)$$

and from this that α_s, $0 \leq s \leq t$ is a martingale. The rest of the proof is the same as that of (5.1) and (5.2). □

Our last chore in this section is to complete the proof of (2.10) in Chapter 1 by showing

(5.4) Theorem. Let B_t be a Brownian motion starting at 0, let g be a continuous function with $g(0) = 0$, and let $\epsilon > 0$. Then

$$P\left(\sup_{0 \leq t \leq 1} |B_t - g(t)| < \epsilon\right) > 0$$

Proof We can find a C^1 function h with $h(0) = 0$ and $|h(s) - g(s)| < \epsilon/2$ for $s \leq t$, so we can suppose without loss of generality that $g \in C^1$ and let $b_s = g'(s)$. If we let $Y_t = \int_0^t b_s \, dB_s$ and define Q_t by

$$\frac{dQ_t}{dP_t} = \exp\left(\int_0^t b_s \, dB_s - \frac{1}{2} \int_0^t |b_s|^2 \, dB_s\right)$$

then it follows from (5.1) that under Q, $B_t - g(t)$ is a standard Brownian motion. Let $G = \{|B_s - g(s)| < \epsilon \text{ for } s \leq t\}$. Exercise 2.11 in Chapter 1 implies that $Q_t(G) > 0$. It follows from the Cauchy-Schwarz inequality that

$$Q_t(G) = \int \frac{dQ_t}{dP_t} \cdot 1_G \, dP_t \leq \left(\int \left(\frac{dQ_t}{dP_t}\right)^2 dP_t\right)^{1/2} P_t(G)^{1/2}$$

To complete the proof now we let E denote expectation with respect to P_t and observe that if $|b_s| \leq M$ for $s \leq t$ then

$$\int \left(\frac{dQ_t}{dP_t}\right)^2 dP_t = E \exp\left(\int_0^t 2b_s \, dB_s - \int_0^t |b_s|^2 \, ds\right)$$

$$= \exp\left(\int_0^t |b_s|^2 \, ds\right) \leq e^{M^2 t}$$

since $\exp\left(\int_0^t 2b_s\, dB_s - (1/2)\int_0^t |2b_s|^2\, ds\right)$ is a martingale and hence has expected value 1. For this step it is important to remember that $b_s = g'(s)$ is a non-random vector. □

5.6. Change of Time

In the last section we learned that we could alter the b in the SDE by changing the measure. However, we could not alter σ with this technique because the quadratic variation is not affected by absolutely continuous change of measure. In this section we will introduce a new technique, change of time, that will allow us to alter the diffusion coefficient. To prepare for developments in Section 6.1 we will allow for the possibility that our processes are defined on a random time interval.

(6.1) Theorem. Let X_t be a solution of MP(b, a) for $t < \zeta$. That is, for each i and j,

$$X_t^i - \int_0^t b_i(X_s)\, ds \quad \text{and} \quad X_t^i X_t^j - \int_0^t a_{ij}(X_s)\, ds$$

are local martingales on $[0, \zeta)$. Let g be a positive function and suppose that

$$\sigma_t = \int_0^t g(X_s)\, ds < \infty \quad \text{for all } t < \zeta$$

Define the inverse of σ by $\gamma_s = \inf\{t : \sigma_t > s \text{ or } t \geq \zeta\}$ and let $Y_s = X(\gamma_s)$ for $s < \xi = \sigma_\zeta$. Then Y_s is a solution of MP($b/g, a/g$) for $s < \xi$.

Proof We begin by noting that $X_t = Y(\sigma_t)$, then in the second step change variables $r = \sigma_s$, $dr = g(X_s)\, ds$, $X_s = X_{\gamma(r)} = Y_r$, to get

$$X_t^i - \int_0^t b_i(X_s)\, ds = Y_{\sigma_t}^i - \int_0^t b_i(Y_{\sigma_s})\, ds$$
$$= Y_{\sigma_t}^i - \int_0^{\sigma_t} b_i(Y_r)/g(Y_r)\, dr$$

Changing variables $t = \gamma_u$ gives

$$Y_u^i - \int_0^u b_i(Y_r)/g(Y_r)\, dr = X_{\gamma_u}^i - \int_0^{\gamma_u} b_i(X_s)\, ds$$

Since the γ_u are stopping times, the right-hand side is a local martingale on $[0, \xi)$ and we have checked one of the two conditions to be a solution of MP($b/g, a/g$).

Repeating the last computation we have

$$X_t^i X_t^j - \int_0^t a_{ij}(X_s)\,ds = Y_{\sigma_t}^i Y_{\sigma_t}^j - \int_0^t a_{ij}(Y_{\sigma_s})\,ds$$
$$= Y_{\sigma_t}^i Y_{\sigma_t}^j - \int_0^{\sigma_t} a_{ij}(Y_r)/g(Y_r)\,dr$$

So changing variables $t = \gamma_u$ gives

$$Y_u^i Y_u^j - \int_0^u a_{ij}(Y_r)/g(Y_r)\,dr = X_{\gamma_u}^i X_{\gamma_u}^j - \int_0^{\gamma_u} a_{ij}(X_s)\,ds$$

Again the right-hand side is a local martingale on $[0,\xi)$ and this completes the proof. □

We will now use (6.1) to prove the claims made in Examples 3.4 and 3.3. There we start with a Brownian motion defined for all time so $\zeta = \infty$. In addition, we want the time changed process defined for all time so we will check that

$$\xi \equiv \sigma_\infty \equiv \int_0^\infty g(X_s)\,ds = \infty$$

where the first two $\equiv$'s are definitions and the $=$ is what we need to check.

Example 6.1. To construct a solution to MP$(0, (|x|^{2\delta} \wedge 1)I)$ which by (4.5) will give rise to a solution of SDE$(0, (|x|^\delta \wedge 1)I)$ we start with $X_t = B_t$ a Brownian motion, which solves MP$(0, I)$, and take $g(x) = |x|^{-2\delta} \vee 1$. Since $g(x) \geq 1$, it is trivial that $\int_0^\infty g(X_s)\,ds = \infty$. To check that $\int_0^t g(B_s)\,ds < \infty$ we note that integrating in polar coordinates,

$$E_0|B_1|^{-2\delta} = C\int_0^\infty r^{-2\delta} e^{-r^2/2} r^{d-1}\,dr < \infty$$

when $-2\delta + d - 1 > -1$, that is, $\delta < d/2$. Using the Brownian scaling relationship $B_s \stackrel{d}{=} s^{1/2} B_1$ we have

$$E_0 \int_0^t |B_s|^{-2\delta}\,ds = E_0|B_1|^{-2\delta} \int_0^t s^{-\delta}\,ds < \infty$$

when $\delta < 1$. Combining the two results we have a nontrivial solution when $\delta < \min(d/2, 1)$. That is, when $\delta < 1/2$ in $d = 1$ and when $\delta < 1$ in $d \geq 2$.

Exercise 6.1. We know from the result of Yamada and Watanabe in (3.3) that the construction in Example 6.1 must fail in $d = 1$ when $\delta = 1/2$. To show this directly prove that if $t > 0$ then $\int_0^{T_\epsilon} |B_s|^{-1} = \infty$ P_0-a.s.

We now prove the first claim made in Example 3.3.

Example 6.2. Suppose $d \geq 3$. To construct a solution to $MP(0,(1+|x|)^{2\delta}I)$ which by (4.5) will give rise to a solution of $SDE(0,(1+|x|)^{\delta}I)$ we start with $X_t = B_t$ a Brownian motion, which solves $MP(0, I)$ and take $g(x) = (1+|x|)^{-2\delta}$. Since $g(x) \leq 1$, it is trivial that $\int_0^t g(B_s)\,ds < \infty$. To give conditions for $\xi = \int_0^\infty g(B_s)\,ds < \infty$ we take expected value and use (2.3) in Chapter 3 to conclude

$$E_0 \xi = \int \frac{C_d}{|y|^{d-2}} \cdot \frac{1}{(1+|y|)^{2\delta}}\,dy$$

$$= C_d' \int_0^\infty r^{-(d-2)}(1+r)^{-2\delta} r^{d-1}\,dr$$

$$= C_d' \int_0^\infty r(1+r)^{-2\delta}\,dr < \infty$$

since $\delta > 1$. From the last computation we see that the time change maps the Brownian motion $X_t = B_t$, $0 \leq t < \infty$ into Y_t, $0 \leq t < \xi$ where $\xi < \infty$. Since $\lim_{t\uparrow\infty} |B_t| = \infty$ we have $\lim_{t\uparrow\xi} |Y_t| = \infty$ and there is no way to continue beyond time ξ. The next result shows that every solution suffers this fate. Again with developments in Section 6.1 in mind it is a little more general than what we need now.

(6.2) Theorem. Let G be an open subset of $\mathbf{R}^d$, let $\tau_G = \inf\{t : X_t \notin G\}$ and suppose $a(x) = h(x)I$ where for any compact $K \subset G$ we have

$$0 < \epsilon_K \leq a(x) \leq C_K < \infty \quad \text{for } x \in K$$

Then uniqueness in distribution holds for $MP(0,a)$ on $[0, \tau_G)$.

Proof (6.1) implies that if we start with a solution of $MP(0,a)$ on $[0, \tau_G)$ with distribution Q and take $g = h$ then the time changed process is a solution of $MP(0, I)$ on $[0, \tau_G)$, which by (2.8) must be a Brownian motion run until it exits G.

Let P_G denote the distribution of the Brownian motion stopped when it leaves G. Since h is bounded away from 0 and ∞ on compact $K \subset G$, the map Ψ that takes sample paths $\omega(\cdot)$ of the solution $MP(0, I)$ on $[0, \tau_G)$ to sample paths $\omega(\gamma(\cdot))$ of the solution of $MP(0, a)$ on $[0, \tau_G)$ is 1-1 and measurable. From this it follows that $Q = P_G \circ \Psi^{-1}$ and the proof is complete. □

Next we prove the second claim made in Example 3.3.

(6.3) Theorem. Suppose $d \leq 2$ and $a(x) = h(x)I$ where for all R,

$$0 < \epsilon_R \leq h(y) \leq C_R < \infty \quad \text{when } |y| \leq R$$

Then MP(0, a) is **well posed**, i.e., uniqueness in distribution holds and there is no explosion.

Proof We start with $X_t = B_t$, a Brownian motion, which solves MP(0, I) and take $g(x) = h(x)^{-1}$. Since $g(B_t) \leq \epsilon_R^{-1}$ for $t \leq T_R = \inf\{t : |B_t| > R\}$ we have $\int_0^t g(B_s)\, ds < \infty$ for any t. On the other hand

$$\int_0^\infty g(B_s)\, ds \geq C_1^{-1} |\{t : |B_t| \leq 1\}| = \infty$$

since Brownian motion is recurrent, so applying (6.1) we have a solution of MP(0, a). Uniqueness follows from (6.2) so the proof is complete. □

Combining (6.3) with (5.2) we can construct solutions of MP(b, a) in one dimension for very general b and a.

(6.4) Theorem. Consider dimension $d = 1$ and suppose

(i) $0 < \epsilon_R \leq a(y) \leq C_R < \infty$ when $|y| \leq R$

(ii) $b(x)$ is locally bounded

(iii) $x \cdot b(x) + a(x) \leq A(1 + x^2)$

Then MP(b, a) is well posed.

6 One Dimensional Diffusions

6.1. Construction

In this section we will take an approach to constructing solutions of

(⋆) $$dX_t = b(X_t)\,dt + \sigma(X_t)\,dB_t$$

that is special to one dimension, but that will allow us to obtain a very detailed understanding of this case. To obtain results with a minimum of fuss and in a generality that encompasses most applications we will assume:

(1D) b and σ are continuous and $a(x) = \sigma^2(x) > 0$ for all x

The purpose of this section is to answer the following questions: Does a solution exist when (1D) holds? Is it unique? Does it explode? Our approach may be outlined as follows:

(i) We define a function φ so that if X_t is a solution of the SDE on $[0, \xi)$ then $Y_t = \varphi(X_t)$ is a local martingale on $[0, \xi)$.

(ii) Our Y_t has $\langle Y \rangle_t = \int_0^t h(Y_s)\,ds$ so construct Y_t to be a solution of MP$(0, h)$ by time changing a Brownian motion.

(iii) We define $X_t = \varphi^{-1}(Y_t)$ and check that X_t is a solution of the SDE.

To begin to carry out our plan, suppose that X_t is a solution of MP(b, a) for $t < \xi$. If $f \in C^2$, Itô's formula implies that for $t < \xi$

(1.1) $$f(X_t) - f(X_0) = \int_0^t f'(X_s)\,dX_s + \frac{1}{2}\int_0^t f''(X_s)\,d\langle X \rangle_s$$

$$= \text{local mart.} + \int_0^t Lf(X_s)\,ds$$

where

(1.2) $$Lf(x) = \frac{1}{2}a(x)f''(x) + b(x)f'(x)$$

and $a(x) = \sigma^2(x)$. From this we see that $f(X_t)$ is a local martingale on $[0, \zeta)$ if and only if $Lf(x) \equiv 0$. Setting $Lf = 0$ and noticing (1D) implies $a(x) > 0$ gives a first order differential equation for f'

(1.3) $$(f')' = \frac{-2b}{a}f'$$

Solving this equation we find,

$$f'(y) = B \exp\left(\int_0^y -\frac{2b(z)}{a(z)}\,dz\right)$$

and it follows that

$$f(x) = A + \int_0^x B \exp\left(\int_0^y -\frac{2b(z)}{a(z)}\,dz\right) dy$$

Any of these functions can be called the **natural scale**. We will usually take $A = 0$ and $B = 1$ to get

(1.4) $$\varphi(x) = \int_0^x \exp\left(\int_0^y -\frac{2b(z)}{a(z)}\,dz\right) dy$$

However, in some situations it will be convenient to replace the 0's at the lower limits by some other point. Note that our assumption (1D) implies φ is C^2 so our use of Itô's formula in (1.1) is justified.

Since $Y_t = \varphi(X_t)$ is a local martingale, results in Section 3.4 imply that it is a time change of Brownian motion. To find the time change function we note that (1.1) and the formula for the variance of a stochastic integral imply

(1.5) $$\langle Y \rangle_t = \int_0^t \varphi'(X_s)^2 d\langle X \rangle_s$$
$$= \int_0^t \varphi'(X_s)^2 a(X_s)\,ds = \int_0^t h(Y_s)\,ds$$

where

$$h(y) = \{\varphi'(\varphi^{-1}(y))\}^2 a(\varphi^{-1}(y)) > 0 \quad \text{is continuous}$$

So if we let $\tau_t = \inf\{s : \langle Y \rangle_s > t\}$ then $W_t = Y_{\tau_t}$ is a Brownian motion run for an amount of time $\langle Y \rangle_\xi$.

Section 6.1 Construction

To construct solutions of MP(b,a) we will reverse the calculations above: we will use a time change of Brownian motion to construct a Y_t which solves MP($0,h$) and then let $X_t = \varphi^{-1}(Y_t)$. With Examples 1.6 and 1.7 of Chapter 5 in mind we generalize our set-up to allow the coefficients b and σ to be defined on an open interval (α, β) with $\alpha < 0 < \beta$. Since $\varphi'(x) > 0$ for all x, the image of (α, β) under φ is an open interval (ℓ, r) with $-\infty \leq \ell < 0 < r \leq \infty$.

Letting W_t be a Brownian motion, $\zeta = \inf\{t : W_t \notin (\ell, r)\}$, $g = 1/h$,

$$\sigma_t = \int_0^t g(W_s)\,ds \text{ for } t < \zeta \quad \text{and} \quad \gamma_s = \inf\{t : \sigma_t > s \text{ or } t \geq \zeta\}$$

Using (6.1) in Chapter 5, we see that $Y_s = W(\gamma_s)$ is a solution of MP($0,h$) for $s < \xi \equiv \sigma_\zeta$.

To define X_t now, we let ψ be the inverse of φ and let $X_t = \psi(Y_t)$. To check that X solves MP(b,a) until it exits from (α, β) at time ξ, we differentiate $\varphi(\psi(x)) = x$ and rearrange, then differentiate again to get

$$\psi'(x) = \frac{1}{\varphi'(\psi(x))}$$

$$\psi''(x) = \frac{-1}{\{\varphi'(\psi(x))\}^2}\varphi''(\psi(x))\psi'(x)$$

Using the first equality and $\varphi''(y) = -(2b(y)/a(y))\varphi'(y)$, which follows from (1.3), we have

$$\psi''(x) = \frac{1}{\{\varphi'(\psi(x))\}^2} \cdot \frac{2b(\psi(x))}{a(\psi(x))}$$

These calculations show $\psi \in C^2$ so Itô's formula, (1.1), and (1.5) imply that for $t < \xi$

$$\psi(Y_t) - \psi(Y_0) = \int_0^t \psi'(Y_s)\,dY_s + \frac{1}{2}\int_0^t \psi''(Y_s)h(Y_s)\,ds$$

For the second term we note that combining the formulas for ψ'' and h gives (recall $\psi = \varphi^{-1}$)

$$\frac{1}{2}\psi''(y)h(y) = b(\psi(y))$$

To deal with the first term observe that Y_t is a solution of MP($0,h$) up to time ξ so by (4.5) in Chapter 5 there is a Brownian motion B_s with $dY_s = \sqrt{h(Y_s)}\,dB_s$. Since $\psi'(y)\sqrt{h(y)} = \sigma(\psi(y))$, letting $X_t = \psi(Y_t)$ we have for $t < \xi$

$$X_t - X_0 = \int_0^t \sigma(X_s)\,dB_s + \int_0^t b(X_s)\,ds$$

The last computation shows that if Y_t is a solution of MP$(0,h)$ then $X_t = \psi(Y_t)$ is a solution of MP(b,a), while (1.1) shows that if X_t is a solution of MP(b,a) then $Y_t = \varphi(X_t)$ is a solution of MP$(0,h)$. This establishes there is a 1-1 correspondence between solutions of MP$(0,h)$ and MP(b,a). Using (6.2) of Chapter 5 now, we see that

(1.6) **Theorem.** Consider $(C,\mathcal{C})$ and let $X_t(\omega) = \omega_t$. Let $\alpha < \beta$ and $\tau_{(\alpha,\beta)} = \inf\{X_t(\omega) \notin (\alpha,\beta)\}$. Under (1D), uniqueness in distribution holds for MP(b,a) on $[0,\tau_{(\alpha,\beta)})$.

Using the uniqueness result with (4.6) of Chapter 5 we get

(1.7) **Theorem.** For each $x \in (\alpha,\beta)$ let P_x be the law of X_t, $t < \tau_{(\alpha,\beta)}$ from (1.6). If we set $X_t = \Delta$ for $t \geq \tau_{(\alpha,\beta)}$, where Δ is the cemetery state of Section 1.3, then the resulting process has the strong Markov property.

6.2. Feller's Test

In the previous section we showed that under (1D) there were unique solutions to MP(b,a) on $[0,\tau_{(\alpha,\beta)})$ where $\tau_{(\alpha,\beta)} = \inf\{t : X_t \notin (\alpha,\beta)\}$. The next result gives necessary and sufficient conditions for no explosions, i.e., $\tau_{(\alpha,\beta)} = \infty$ a.s. Let

$$T_y = \inf\{t : X_t = y\} \quad \text{for } y \in (\alpha,\beta)$$
$$T_\alpha = \lim_{y \downarrow \alpha} T_y \quad \text{and} \quad T_\beta = \lim_{y \uparrow \beta} T_y$$

In stating (2.1) we have without loss of generality supposed $0 \in (\alpha,\beta)$. If you are confronted by an (α,β) that does not have this property pick your favorite γ in the interval and translate the system by $-\gamma$. Changing variables in the integrals we see that (2.1) holds when all of the 0's are replaced by γ's.

(2.1) **Feller's test.** Let $\varphi(x)$ be the natural scale defined by (1.4) and let $m(x) = 1/(\varphi'(x)a(x))$.

(a) $P_x(T_\beta < T_0)$ is positive for some (all) $x \in (0,\beta)$ if and only if

$$\int_0^\beta dx\, m(x)\, (\varphi(\beta) - \varphi(x)) < \infty$$

(b) $P_x(T_\alpha < T_0)$ is positive for some (all) $x \in (\alpha,0)$ if and only if

$$\int_\alpha^0 dx\, m(x)\, (\varphi(x) - \varphi(\alpha)) < \infty$$

Section 6.2 Feller's Test

(c) If both integrals are infinite then $P_x(\tau_{(\alpha,\beta)} = \infty) = 1$ for all $x \in (\alpha, \beta)$.

Remarks. If $\varphi(\beta) = \infty$ then the first integrand is $\equiv \infty$ and the integral is ∞. Similarly the second integral is ∞ if $\varphi(\alpha) = -\infty$. The statement in (a) means that the following are equivalent

(a1) $P_x(T_\beta < T_0) > 0$ for some $x \in (0, \beta)$

(a2) $\int_0^\beta dx\, m(x)\, (\varphi(\beta) - \varphi(x)) < \infty$

(a3) $P_x(T_\beta < T_0) > 0$ for all $x \in (0, \beta)$

Proof The key to the proof of our sufficient condition for no explosions given in (3.1) and (3.2) of Chapter 5 was the fact that if A is sufficiently large

$$S_t = (1 + |X_t|^2)e^{-At} \quad \text{is a supermartingale}$$

To get the optimal result on explosions we have to replace $1 + |x|^2$ by a function that is tailor-made for the process. A natural choice that gives up nothing is a function $g \geq 0$ so that

$$e^{-t}g(X_t) \quad \text{is a local martingale}$$

To find such a g it is convenient to look at things on the natural scale, i.e., let $Y_t = \varphi(X_t)$. Let $\ell = \lim_{x \downarrow \alpha} \varphi(x)$ and let $r = \lim_{x \uparrow \beta} \varphi(x)$. We will find a C^2 function $f(x)$ so that f is decreasing on $(\ell, 0)$, f is increasing on $(0, r)$ and

$$e^{-t}f(Y_t) \quad \text{is a local martingale}$$

Denouement. Once we have f the conclusion of the argument is easy, so to explain our motivations we begin with the end. Let

$$f(\ell) = \lim_{y \downarrow \ell} f(y) \qquad f(r) = \lim_{y \uparrow r} f(y)$$

(2.6) and the calculations at the end of the proof will show that

(2.2a) The integral in (a) is finite if and only if $f(r) < \infty$.

(2.2b) The integral in (b) is finite if and only if $f(\ell) < \infty$.

Once these are established the conclusions of (2.1) follow easily. Let

$$\bar{T}_y = \inf\{t : Y_t = y\} \quad \text{for } y \in (\ell, r)$$
$$\bar{T}_r = \lim_{y \uparrow r} \bar{T}_y \quad \text{and} \quad \bar{T}_\ell = \lim_{y \downarrow \ell} \bar{T}_y$$
$$\bar{\tau}_{(\ell,r)} = \bar{T}_\ell \wedge \bar{T}_r$$

Proof of (c) Suppose $f(\ell) = f(r) = \infty$. Let $a_n < 0 < b_n$ be chosen so that $f(a_n) = f(b_n) = n$ and let $\tau_n = \bar{T}_{a_n} \wedge \bar{T}_{b_n}$. Since $e^{-s}f(Y_s)$ is bounded before time $\tau_n \wedge t$ we can apply the optional stopping theorem at that time to conclude that if $y \in (a_n, b_n)$, then

$$P_y(\tau_n < t) \le \frac{e^t f(y)}{n} \to 0 \quad \text{as } n \to \infty$$

Letting $t \to \infty$ we conclude $0 = P_y(\bar{\tau}_{(\ell,r)} < \infty) = P_{\varphi^{-1}(y)}(\tau_{(\alpha,\beta)} < \infty)$.

Proof of (a) Let $0 < y < r$, let $b_n \uparrow r$ with $b_1 > y$, and let $\tau_n = \bar{T}_0 \wedge \bar{T}_{b_n}$.

If $f(r) = \infty$, applying the optional stopping theorem at time $\tau_n \wedge t$, we conclude that

$$P_y(\bar{T}_{b_n} < \bar{T}_0 \wedge t) \le \frac{e^t f(y)}{f(b_n)} \to 0 \quad \text{as } n \to \infty$$

Letting $t \to \infty$ we conclude $0 = P_y(\bar{T}_r < \bar{T}_0) = P_{\varphi^{-1}(y)}(T_\beta < T_0)$. This shows that if (a2) is false then (a1) is false, i.e., (a1) implies (a2).

If $f(r) < \infty$, applying the optional stopping theorem at time τ_n (which is justified since $e^{-t}f(Y_t) \le f(b_n)$ for $t \le \tau_n$) we conclude that

$$1 < f(y) = E_y\left(e^{-\tau_n} f(Y_{\tau_n})\right)$$
$$\le 1 + f(r) E_y\left(e^{-\bar{T}_{b_n}}; \bar{T}_{b_n} < \bar{T}_0\right)$$

Rearranging gives

$$E_y\left(e^{-\bar{T}_{b_n}}; \bar{T}_{b_n} < \bar{T}_0\right) \ge \frac{f(y) - 1}{f(r)} > 0$$

Noting $\bar{T}_r = \bar{T}_0 < \infty$ is impossible, and letting $n \to \infty$ we have

$$E_y(e^{-\bar{T}_{b_n}}; \bar{T}_{b_n} < \bar{T}_0) \downarrow E_y(e^{-\bar{T}_r}; \bar{T}_r < \bar{T}_0)$$

which is a contradiction, unless $0 < P_y(\bar{T}_r < \bar{T}_0) = P_{\varphi^{-1}(y)}(T_\beta < T_0)$. This shows that (a2) implies (a3). (a3) implies (a1) is trivial so the proof of (a) is complete.

Proof of (b) is identical to that of (a).

The search for f. To find a function f so that $e^{-t}f(Y_t)$ is a local martingale, we apply Itô's formula to $f(x_1, x_2) = e^{-x_1} f(x_2)$ with $X_t^1 = t$, $X_t^2 = Y_t$, and recall Y_t is a local martingale with variance process given by (1.5), so

$$e^{-t}f(Y_t) - f(Y_0) = \int_0^t \left\{-e^{-s}f(Y_s) + \frac{1}{2}e^{-s}f''(Y_s)h(Y_s)\right\}ds$$
$$+ \text{ local mart.}$$

so we want

(2.3) $$\frac{1}{2}h(x)f''(x) - f(x) = 0$$

To solve this equation we let $f_0 \equiv 1$ and define for $n \geq 1$

(2.4) $$f_n(x) = \int_0^x dy \int_0^y dz \, \frac{2f_{n-1}(z)}{h(z)}$$

(1.8) in Chapter 4 implies that if we set $f(x) = \sum_{n=0}^\infty f_n(x)$ and

(2.5) $$\sum_{n=0}^\infty \sup_{|x| \leq R} |f_n(x)| < \infty \quad \text{for any } R < \infty$$

then $f''(x) = \sum_{n=0}^\infty f_n''(x)$. Using $f_n''(x) = 2f_{n-1}(x)/h(x)$ for $n \geq 1$ and $f_0''(x) = 0$ it follows that

$$f''(x) = \sum_{n=0}^\infty f_n''(x) = \sum_{n=1}^\infty \frac{2f_{n-1}(x)}{h(x)} = \frac{2f(x)}{h(x)}$$

i.e., f satisfies (2.3).

To prove (2.5), we begin with some simple properties of the f_n.

(2.6a) $f_n \geq 0$

(2.6b) f_n is convex with $f_n'(0) = 0$

(2.6c) f_n is increasing on $(0, r)$ and decreasing on $(\ell, 0)$

Proof Since $h(z) > 0$ (2.6a) follows easily by induction from the definition in (2.4). Using (2.6a) in the definition, (2.6b) follows. (2.6c) is an immediate consequence of (2.6b). □

Our next step is to use induction to show

(2.7) **Lemma.** $f_n(x) \leq (f_1(x))^n/n!$, (2.5) holds, and

$$1 + f_1 \leq f \leq \exp(f_1)$$

Proof Using (2.6c) and (2.6a) the second and third conclusions are immediate consquences of the first one. The inequality $f_n(x) \leq (f_1(x))^n/n!$ is obvious if

$n=1$. Using (i) the definition of f_n in (2.4), (ii) (2.6c), (iii) the definition of f_1 and f_0, (iv) the result for $n-1$, and (v) doing a little calculus, we have

$$f_n(x) = \int_0^x dy \int_0^y dz \, \frac{2 f_{n-1}(z)}{h(z)}$$

$$\leq \int_0^x dy \, f_{n-1}(y) \int_0^y dz \, \frac{2}{h(z)}$$

$$= \int_0^x dy \, f_{n-1}(y) f_1'(y)$$

$$\leq \int_0^x dy \, \frac{(f_1(y))^{n-1}}{(n-1)!} f_1'(y) = \frac{(f_1(x))^n}{n!} \qquad \square$$

To complete the proof now we only have to prove (2.2a) and (2.2b). In view of (2.7) it suffices to prove these results with f replaced by f_1. Changing variables $z = \varphi(v)$, $dz = \varphi'(v) \, dv$ then $y = \varphi(u)$, $dy = \varphi'(u) \, du$ we have

$$f_1(x) = \int_0^x dy \int_0^y dz \, \frac{2}{\varphi'(\psi(z))^2 a(\psi(z))}$$

$$= \int_0^x dy \int_0^{\psi(y)} dv \, \frac{2}{\varphi'(v) a(v)}$$

$$= \int_0^{\psi(x)} du \, \varphi'(u) \int_0^u dv \, \frac{2}{\varphi'(v) a(v)}$$

Letting $x \uparrow r$ and using Fubini's theorem

$$f_1(r) = \int_0^\beta dv \, \frac{2}{\varphi'(v) a(v)} \int_v^\beta du \, \varphi'(u)$$

$$= 2 \int_0^\beta dv \, m(v) \, (\varphi(\beta) - \varphi(v))$$

A similar argument shows that the second expression in (2.1) is (except for a factor of 2) $f_1(\ell)$ and the proof is complete. $\square$

To see what (2.1) says in a concrete case we consider

Example 2.1. Let $\sigma(x) = 1$, $b(x) = (1+|x|)^\delta/2$ where $\delta > 0$. When $\delta \leq 1$ the coefficients are Lipschitz continuous, so there is no explosion. We will now use (2.1) to reprove that result and show that explosion occurs when $\delta > 1$. When $y < 0$, $\varphi'(y) \geq 1$, so $\varphi(-\infty) = -\infty$ and the second integral in (2.1) is ∞.

To evaluate the first integral we note that if $y > 0$

$$\varphi'(y) = \exp\left(\int_0^y -(1+z)^\delta \, dz\right) = \exp\left(\frac{-(1+y)^{1+\delta} + 1}{1+\delta}\right)$$

so $\varphi(\infty) < \infty$. To use (2.1) now, we note that $a(y) = 1$ so $m(y) = 1/\varphi'(y)$ so there is no explosion if and only if

$$\infty = \int_0^\infty dv \, \exp\left(\frac{(1+v)^{1+\delta} - 1}{1+\delta}\right) \int_v^\infty du \, \exp\left(\frac{-(1+u)^{1+\delta} + 1}{1+\delta}\right)$$

$$= \int_0^\infty dv \int_v^\infty du \, \exp\left(\frac{(1+v)^{1+\delta} - (1+u)^{1+\delta}}{1+\delta}\right)$$

$$= \int_1^\infty dy \int_y^\infty dx \, \exp\left(\frac{y^{1+\delta} - x^{1+\delta}}{1+\delta}\right)$$

where in the last step we have changed variables $x = u+1$, $y = v+1$ to get rid of the 1's. The next lemma estimates the last expression and shows that it is

$$< \infty \quad \text{if} \quad \delta > 1$$
$$= \infty \quad \text{if} \quad 0 < \delta \leq 1$$

(2.8) Lemma. If $\delta > 0$,

$$y^\delta \int_y^\infty dx \, \exp\left(\frac{y^{1+\delta} - x^{1+\delta}}{(1+\delta)}\right) \to 1 \quad \text{as } y \to \infty$$

Proof Changing variables $x = y + zy^{-\delta}$ we have

$$\int_y^\infty dx \, \exp\left(\frac{y^{1+\delta} - x^{1+\delta}}{1+\delta}\right) = y^{-\delta} \int_0^\infty dz \, \exp\left(-\int_y^{y+zy^{-\delta}} w^\delta \, dw\right)$$

Since $z \leq \int_y^{y+zy^{-\delta}} w^\delta \, dw \leq zy^{-\delta}(y + zy^{-\delta})^\delta \to z$ as $y \to \infty$, the desired result follows from the dominated convergence theorem. □

6.3. Recurrence and Transience

The natural scale, which was used in Section 6.1 to construct one dimensional diffusions, can also be used to study their recurrence and transience. Let X_t be a solution of MP(b, a) and suppose

Chapter 6 One Dimensional Diffusions

(1D) b and σ are continuous with $\sigma(x) > 0$ for all x

Let $a(x) = \sigma^2(x)$, and let φ be the natural scale defined by

$$\varphi(x) = \int_0^x \exp\left(\int_0^y -\frac{2b(z)}{a(z)} dz\right) dy$$

Let $T_y = \inf\{t > 0 : X_t = y\}$ and let $\tau = T_a \wedge T_b$. The first detail is to show

(3.1) Lemma. If $a < x < b$ then $P_x(\tau < \infty) = 1$.

Proof We saw in Section 6.1 that $Y_t = \varphi(X_t)$ is a solution of MP$(0, h)$ where h given by (1.5) is positive and continuous. Thus (6.1) in Chapter 5 implies that Y can be constructed as a time change of Brownian motion, B_t:

$$Y_s = B(\gamma_s) \text{ where } \gamma_s = \inf\{t : \sigma_t > s\} \text{ and } \sigma_t = \int_0^t ds/h(B_s)$$

Since Brownian motion will exit $(\varphi(a), \varphi(b))$ with probability one, Y_t will exit $(\varphi(a), \varphi(b))$ with probability one, and X_t will exit (a, b) with probability one. □

With (3.1) established we can study the recurrence and transience as we did for Brownian motion. $\varphi(X_{t \wedge \tau})$ is a uniformly bounded martingale, so the optional stopping theorem implies

$$\varphi(x) = E_x \varphi(X_\tau) = \varphi(a) P_x(T_a < T_b) + \varphi(b)\{1 - P_x(T_a < T_b)\}$$

and solving we have

$$(3.2) \quad P_x(T_a < T_b) = \frac{\varphi(b) - \varphi(x)}{\varphi(b) - \varphi(a)} \quad P_x(T_b < T_a) = \frac{\varphi(x) - \varphi(a)}{\varphi(b) - \varphi(a)}$$

Letting $\varphi(\infty) = \lim_{b \to \infty} \varphi(b)$ and $\varphi(-\infty) = \lim_{a \to -\infty} \varphi(a)$ (the limits exist since φ is strictly increasing) we have

(3.3) Theorem. Suppose $a < x < b$.
  $P_x(T_a < \infty) = 1$ if and only if $\varphi(\infty) = \infty$.
  $P_x(T_b < \infty) = 1$ if and only if $\varphi(-\infty) = -\infty$.

In one dimension we say that X is recurrent if $P_x(T_y < \infty) = 1$ for all y. From (3.3) it follows that

Section 6.3 Recurrence and Transience

(3.4) Corollary. X is recurrent if and only if $\varphi(\mathbf{R}) = \mathbf{R}$.

This conclusion should not be surprising. $Y_t = \varphi(X_t)$ is a local martingale so it is a time change of a Brownian motion run for a random amount of time and if $\varphi(\mathbf{R}) \neq \mathbf{R}$ that random amount of time must be finite.

To understand the dividing line between recurrence and transience we consider some examples. We begin by noting that the natural scale only depends on b and σ through the ratio $2b/a$.

Exercise 3.1. (i) Show that if $b(y) \leq 0$ for $y \geq y_0$ then $P_x(T_0 < \infty) = 1$ for all $x > 0$. (ii) Show that if $b(y)/a(y) \geq \epsilon > 0$ for $y \geq y_0$ then $P_x(T_0 < \infty) < 1$ for all $x > 0$.

Example 3.1. The exercise identifies the interesting case as being $b(y) \geq 0$ with $b(y)/a(y) \to 0$ as $y \to \infty$. Suppose that for $x \geq 0$

$$2b(x)/a(x) = C(1+x)^{-r} \quad \text{where } C, r > 0$$

If $r > 1$ then

$$\int_0^\infty -C(1+z)^{-r}\, dz = -K > -\infty$$

so $\varphi'(y) \geq e^{-K}$ for all $y > 0$ and $\varphi(\infty) = \infty$. If $r < 1$

$$\int_0^y -C(1+z)^{-r}\, dz = -\frac{C}{1-r}\{(1+y)^{1-r} - 1\}$$

and it follows that

$$\varphi(\infty) = \int_0^\infty \exp\left(-\frac{C}{1-r}\{(1+y)^{1-r} - 1\}\right) dy < \infty$$

In the borderline case $r = 1$ the outcome depends on the value of C.

$$\int_0^y -C(1+z)^{-1}\, dz = -C\ln(1+y)$$

so

$$\varphi(\infty) = \int_0^\infty (1+y)^{-C}\, dy \quad \begin{cases} < \infty & \text{if } C > 1 \\ = \infty & \text{if } C \leq 1 \end{cases} \qquad \square$$

To check the last calculation note that Example 1.2 in Chapter 5 shows that the radial part of d dimensional Brownian motion has $b(r)/a(r) = (d-1)/2r$, while Section 3.1 shows that Brownian motion is recurrent in $d \leq 2$ and transient in $d > 2$, which corresponds to $C \leq 1$ and to $C > 1$ respectively. Sticklers for detail may complain that we do not have any Brownian motions defined for dimensions $2 < d < 3$. However, this analysis suggests that if we did then they would be transient.

Exercise 3.2. Suppose a and b satisfy (1D) and are periodic with period 1, i.e., $a(y+1) = a(y)$ and $b(y+1) = b(y)$ for all y. Find a necessary and sufficient condition for X_t to be recurrent.

6.4. Green's Functions

Suppose X_t is a solution of MP(b, a), where b and a satisfy (1D). Let $a < b$ be real numbers, which you should not confuse with the coefficients, let $D = (a, b)$, and let $\tau = \inf\{t : X_t \notin (a, b)\}$. Our goals in this section, which are accomplished in (4.5) and (4.4) below, are to show that (i) if g is bounded and measurable then

$$E_x \int_0^\tau g(X_s)\,ds = \int G_D(x,y) g(y)\,dy$$

and (ii) to give a formula for the Green's function $G_D(x, y)$. As in the discussion of (8.1) of Chapter 4, we think of $G_D(x, y)$ as the expected amount of time spent at y before the process exits D when it starts at x. The rigorous meaning of the last sentence is given by (i).

The analysis here is similar to that in Section 4.5 but instead of the Laplacian, Δ, we are concerned with

$$Lf(x) = \frac{1}{2}a(x)f''(x) + b(x)f'(x)$$

The first step is to generalize (1.3) from Chapter 3.

(4.1) Lemma. $\sup_{x \in (a,b)} E_x \tau < \infty$.

Proof Let φ be the natural scale defined in (1.4) and consider $Y_t = \varphi(X_t)$. By (1.5) Y_t is a local martingale with $\langle Y \rangle_t = \int_0^t h(Y_s)\,ds$ where $h(y)$ is positive and continuous. To estimate $\tau = \inf\{t : Y_t \notin (\varphi(a), \varphi(b))\}$ we let

$\nu = (\varphi(a) + \varphi(b))/2$ be the midpoint of the interval,

$\ell = (\varphi(b) - \varphi(a))/2$ be half the length of the interval,

$\rho = \inf\{h(y) : y \in (\varphi(a), \varphi(b))\} > 0$,

$f(x) = (\ell^2 - (x - \nu)^2)/\rho$,

and note that $f \geq 0$ in $(\varphi(a), \varphi(b))$ with $f''(x) = -2/\rho$. Itô's formula implies

$$f(Y_t) - f(Y_0) = \text{local mart.} - \int_0^t \frac{h(Y_s)}{\rho}\, ds$$

Our choice of ρ implies $h(y) \geq \rho$, so $f(Y_t) + t$ is a local supermartingale on $[0, \tau)$. Using the optional stopping theorem at time $\tau_n \wedge n$ where $\tau_n = \inf\{t : Y_t \notin (\varphi(a) + 1/n, \varphi(b) - 1/n)\}$ we have

$$f(x) \geq E_x f(Y_{\tau_n \wedge n}) + E_x(\tau_n \wedge n)$$

Since $0 \leq f(x) \leq \ell^2/\rho$ for $x \in [\varphi(a), \varphi(b)]$ we have $\ell^2/\rho \geq E_x(\tau_n \wedge n)$. Now, let $n \to \infty$ and use the monotone convergence theorem. □

(4.2) Theorem. Suppose g is bounded. If there is a function v with
(i) $v \in C^2$, $Lv = -g$ in (a, b)
(ii) v is continuous at a and b with $v(a) = v(b) = 0$ then

$$v(x) = E_x \int_0^\tau g(X_s)\, ds$$

Proof Let $M_t = v(X_t) + \int_0^t g(X_s)\, ds$. (i) and (1.1) imply that for $t < \tau$

$$v(X_t) - v(X_0) = \text{local mart.} - \int_0^t g(X_s)\, ds$$

so M_t is a local martingale on $[0, \tau)$. If v and g are bounded then for $t < \tau$

$$|M_t| \leq \|v\|_\infty + \tau\|g\|_\infty$$

(4.1) implies that the right-hand side is an integrable random variable so (2.7) in Chapter 2 and (ii) imply

$$M_\tau \equiv \lim_{t \uparrow \tau} M_t = \int_0^\tau g(B_t)\, dt$$

$$v(x) = E_x M_0 = E_x M_\tau = E_x \int_0^\tau g(B_t)\, dt \qquad \square$$

Solving (4.2). The next two pages will be devoted to deriving the solution to the equation in (4.2). Readers who are content to guess and verify the answer can skip to (4.4) now. To solve the equation it is convenient to introduce

$$m(x) = \frac{1}{\varphi'(x)a(x)}$$

where $\varphi(x)$ is the natural scale and note that

$$(4.3) \quad \frac{1}{2m(x)}\frac{d}{dx}\left(\frac{1}{\varphi'(x)}\frac{df}{dx}\right) = \frac{a(x)}{2}\frac{d^2f}{dx^2} + \frac{a(x)}{2}\left(\frac{-\varphi''(x)}{\varphi'(x)}\right)\frac{df}{dx} = Lf(x)$$

since the definition of the natural scale implies $\varphi''(x)/\varphi'(x) = -2b(x)/a(x)$. m is called the (density function of the) **speed measure**, though as we will see in Example 4.1 the rate of movement of the process near x is inversely proportional to $m(x)$! To simplify notation and to facilitate comparison with our source for this material, Karlin and Taylor (1981), Vol. II, we let $s(x) = \varphi'(x)$ be the derivative of the natural scale.

To solve equation $Lf = -g$ now, we use (4.3) to write

$$\frac{d}{dx}\left(\frac{1}{s(x)}\frac{dv}{dx}\right) = -2m(x)g(x)$$

and integrate to conclude that for some constant β

$$\frac{1}{s(y)}\frac{dv}{dy} = \beta - 2\int_a^y dz\, m(z)g(z)$$

Multiplying by $s(y)$ on each side, integrating y from a to x, and recalling that $v(a) = 0$ and $s = \varphi'$ we have

(a) $$v(x) = \beta(\varphi(x) - \varphi(a)) - 2\int_a^x dy\, s(y) \int_a^y dz\, m(z)g(z)$$

In order to have $v(b) = 0$ we must have

$$\beta = \frac{2}{\varphi(b) - \varphi(a)}\int_a^b dy\, s(y) \int_a^y dz\, m(z)g(z)$$

Plugging the formula for β into (a) and writing

$$u(x) = \frac{\varphi(x) - \varphi(a)}{\varphi(b) - \varphi(a)} = P_x(T_b < T_a)$$

we have

(b) $$\begin{aligned}v(x) &= 2u(x)\int_a^b dy\, s(y) \int_a^y dz\, m(z)g(z) \\ &\quad - 2\int_a^x dy\, s(y) \int_a^y dz\, m(z)g(z)\end{aligned}$$

Breaking the first $\int_a^b dy$ into an integral over $[a, x]$ and one over $[x, b]$ we have

(c)
$$v(x) = 2(u(x) - 1) \int_a^x dy\, s(y) \int_a^y dz\, m(z)g(z)$$
$$+ 2u(x) \int_x^b dy\, s(y) \int_a^y dz\, m(z)g(z)$$

Recalling $s(y) = \varphi'(y)$ we have

$$u(x) \int_x^b dy\, s(y) = \frac{\varphi(x) - \varphi(a)}{\varphi(b) - \varphi(a)} \cdot (\varphi(b) - \varphi(x)) = (1 - u(x)) \int_a^x dy\, s(y)$$

Multiplying the last identity by $2 \int_a^x dz\, m(z)g(z)$, we have

$$2u(x) \int_x^b dy\, s(y) \int_a^x dz\, m(z)g(z) = 2(1 - u(x)) \int_a^x dy\, s(y) \int_a^x dz\, m(z)g(z)$$

Using this in (c) we can break off part of the second term and cancel with the first one to end up with

(d)
$$v(x) = 2(1 - u(x)) \int_a^x dy\, s(y) \int_y^x dz\, m(z)g(z)$$
$$+ 2u(x) \int_x^b dy\, s(y) \int_x^y dz\, m(z)g(z)$$

Using Fubini's theorem now gives

(e)
$$v(x) = 2(1 - u(x)) \int_a^x dz\, m(z)g(z) \int_a^z dy\, s(y)$$
$$+ 2u(x) \int_x^b dz\, m(z)g(z) \int_z^b dy\, s(y)$$

If we define $G(x, z)$ to be

(4.4)
$$2 \frac{\varphi(x) - \varphi(a)}{\varphi(b) - \varphi(a)} \cdot (\varphi(b) - \varphi(z))\, m(z) \quad \text{when } z \geq x$$
$$2 \frac{\varphi(b) - \varphi(x)}{\varphi(b) - \varphi(a)} \cdot (\varphi(z) - \varphi(a))\, m(z) \quad \text{when } z \leq x$$

then we have

(f)
$$v(x) = \int_a^b G(x, z)g(z)\, dz$$

Combining the calculations above with (4.2) shows

(4.5) Theorem. If g is bounded and measurable then

$$E_x \int_0^T g(X_s)\,ds = \int_a^b G(x,z) g(z)\,dz$$

Proof By the monotone class theorem it suffices to prove the result when g is continuous. To do this it suffices to show that the v defined in (f) satisfies the hypotheses of (4.2). To check (ii), we note that as $x \to a$ or $x \to b$, $G(x,z) \to 0$, and G is bounded so the bounded convergence theorem implies $v(x) \to 0$. To check that $v \in C^2$ we note that

$$\frac{\varphi(b) - \varphi(a)}{2} v(x) = (\varphi(b) - \varphi(x)) \int_a^x (\varphi(z) - \varphi(a)) m(z) g(z)\,dz$$
$$+ (\varphi(x) - \varphi(a)) \int_x^b (\varphi(b) - \varphi(z)) m(z) g(z)\,dz$$

Differentiating and noting that the terms from the limits of the integrals cancel we have

$$\frac{\varphi(b) - \varphi(a)}{2} v'(x) = -\varphi'(x) \int_a^x (\varphi(z) - \varphi(a)) m(z) g(z)\,dz$$
$$+ \varphi'(x) \int_x^b (\varphi(b) - \varphi(z)) m(z) g(z)\,dz$$

Differentiating again we have

$$\frac{\varphi(b) - \varphi(a)}{2} v''(x) = -\varphi''(x) \int_a^x (\varphi(z) - \varphi(a)) m(z) g(z)\,dz$$
$$+ \varphi''(x) \int_x^b (\varphi(b) - \varphi(z)) m(z) g(z)\,dz$$
$$- \varphi'(x)(\varphi(b) - \varphi(a)) m(x) g(x)$$

This shows $v \in C^2$. Multiplying the last two equations by $b(x)$ and $a(x)/2$, adding, then using $L\varphi = 0$ and the definition of $m(x)$ we have

$$\frac{\varphi(b) - \varphi(a)}{2} Lv(x) = -\frac{a(x)}{2} \cdot \varphi'(x)(\varphi(b) - \varphi(a)) \frac{1}{\varphi'(x) a(x)} g(x)$$

so $Lv = -g$. This shows that v satisfies (i) in (4.2) and the result follows. □

Example 4.1. Speed measure? If X has no drift (e.g., Brownian motion or any diffusion on its natural scale) then $\varphi(x) = x$ and (4.4) becomes

$$G(x,z) = \begin{cases} 2m(z)(b-z)(x-a)/(b-a) & \text{when } z \geq x \\ 2m(z)(b-x)(z-a)/(b-a) & \text{when } z \leq x \end{cases}$$

Taking $g = 1$, $a = x - h$, and $b = x + h$ the last expression becomes

$$G(x,z) = \begin{cases} m(z)(x+h-z) & \text{when } z \geq x \\ m(z)(z-x+h) & \text{when } z \leq x \end{cases}$$

Letting $\tau_{(a,b)} = \inf\{t : X_t \notin (a,b)\}$, and using (4.5) with $g = 1$ we have

(4.6)
$$E_x \tau_{(x-h,x+h)} = \int_x^{x+h} (x+h-z)\, m(z)\, dz$$
$$+ \int_{x-h}^x (z-x+h)\, m(z)\, dz \sim m(x)h^2$$

as $h \to 0$. Thus $m(x)$ gives us the *time* that X_t takes to exit a small interval centered at x, or to be precise, the ratio of the time for X_t to that for Brownian motion.

For another interpretation of m note that (11.7) in Chapter 2 implies that if L_t^x is the local time at x up to time t then for a bounded measurable g

$$\int L_t^x g(x)\, dx = \int_0^t g(X_s)\, d\langle X \rangle_t = \int_0^t g(X_s) a(X_s)\, ds$$

When X_t has no drift $\varphi'(x) = 1$ so $m(x) = 1/a(x)$ and taking $g(x) = f(x)m(x)$ we have

$$\int m(x) L_t^x f(x)\, dx = \int_0^t f(X_s)\, ds$$

Thus multiplying the local times by $m(x)$ converts them into occupation times. This and the previous interpretation suggest that it would be natural to call $m(x)\, dx$ the **occupation measure**. However, it is too late to try to change the original name, speed measure. □

Example 4.2. Second Proof of Feller's test. Let $T_x = \inf\{t : X_t = x\}$, $T_\alpha = \lim_{x \downarrow \alpha} T_x$ and $T_{(a,b)} = T_a \wedge T_b$. We content ourselves to establish a variant of (b) in (2.1). The first of two steps is

(4.7) Lemma. Let $0 < b < \beta$. Then

$$\int_\alpha^0 (\varphi(z) - \varphi(\alpha))\, m(z)\, dz < \infty$$

if and only if $\inf_{\alpha < a < 0} P_0(T_a < T_b) > 0$ and $\sup_{\alpha < a < 0} E_0(T_a \wedge T_b) < \infty$.

Proof (3.2) implies

$$P_0(T_a < T_b) = \frac{\varphi(b) - \varphi(0)}{\varphi(b) - \varphi(a)}$$

so $\inf_{\alpha < a < 0} P_0(T_a < T_b) > 0$ if and only if $\varphi(\alpha) > -\infty$. Taking $g = 1$ in (4.5) and using (4.4) we have

(4.8)
$$\begin{aligned} E_x \tau_{(a,b)} &= 2 \frac{\varphi(x) - \varphi(a)}{\varphi(b) - \varphi(a)} \int_x^b (\varphi(b) - \varphi(z))\, m(z)\, dz \\ &+ 2 \frac{\varphi(b) - \varphi(x)}{\varphi(b) - \varphi(a)} \int_a^x (\varphi(z) - \varphi(a))\, m(z)\, dz \end{aligned}$$

The first integral always stays bounded as $a \downarrow \alpha$. So $E_0 \tau_{(a,b)}$ stays bounded as $a \to \alpha$ if and only if $\varphi(\alpha) > -\infty$ and

$$\int_\alpha^0 (\varphi(z) - \varphi(\alpha)) m(z)\, dz < \infty \qquad \square$$

The two conditions in (4.7) are equivalent to $P_0(T_\alpha < T_b) > 0$ and $E_0(T_\alpha \wedge T_b) < \infty$. To complete the proof now we will show

(4.9) Lemma. If $P_0(T_\alpha < T_b) > 0$ then $E_0(T_\alpha \wedge T_b) < \infty$.

Proof $P_0(T_\alpha < T_b) > 0$ implies $P_0(T_\alpha < \infty) > 0$. Pick M large enough so that $P_0(T_\alpha \leq M) \geq \epsilon > 0$ and that $P_0(T_b \leq M) \geq \epsilon > 0$. By considering the first time the process starting from 0 hits x and using the strong Markov property, it follows that if $0 < x < b$ then

$$P_0(T_b \leq t) = E_0(P_x(T_b \leq t - T_x); T_x \leq t) \\ \leq P_x(T_b \leq t) P_0(T_x \leq t) \leq P_x(T_b \leq t)$$

A similar argument shows

$$P_x(T_\alpha \leq t) \geq P_0(T_\alpha \leq t) \qquad \text{for } \alpha < x < 0$$

Combining the last two results shows that

$$P_x(T_\alpha \wedge T_b \leq M) \geq \epsilon > 0 \qquad \text{for all } \alpha < x < b$$

Letting $\tau = T_a \wedge T_b$ and repeating the proof of (1.2) in Chapter 3 now, we have

$$P_x(\tau > kM) = E_x(P_{X_M}(\tau > (k-1)M); \tau > M)$$
$$\leq (1-\epsilon) \sup_{a < y < b} P_y(\tau > (k-1)M)$$

It follows by induction that for all integers $k \geq 1$

$$\sup_{x \in (a,b)} P(\tau > kM) \leq (1-\epsilon)^k$$

Using (5.7) in Chapter 1 of Durrett (1995) now, we get an upper bound on all the moments of the exit time τ. □

6.5. Boundary Behavior

In Section 6.1 when we constructed a diffusion on a half line or a bounded interval, we gave up when the process reached the boundary and we said that it had exploded. In this section, we will identify situations in which we can extend the life of the process. To explain how we might continue we begin with a trivial example.

Example 5.1. Reflecting boundary at 0. Suppose $b(x) \equiv 0$, and $\sigma(x)$ is positive and continuous on $[0, \infty)$. To define a solution to $dX_t = \sigma(X_t) dB_t$ with a "reflecting boundary at 0," we extend σ to $\mathbf{R}$ by setting $\sigma(-x) = \sigma(x)$, let Y_t be a solution of $dY_t = \sigma(Y_t) dB_t$ on $\mathbf{R}$ and then let $X_t = |Y_t|$.

To use the last recipe in general we start with X_t a solution of MP(b, a) and let φ be its natural scale. $Y_t = \varphi(X_t)$ solves MP($0, h$) where

$$h(y) = \{\varphi'(\varphi^{-1}(y))\}^2 a(\varphi^{-1}(y))$$

To see if we can start the process Y_t at 0 we extend h to the negative half-line by setting $h(-y) = h(y)$ and let Z_t be a solution of MP($0, h$) on $\mathbf{R}$. If we let $\bar{m}(y) = 1/h(|y|)$ be the speed measure for Z_t, which is on its natural scale, we can use (4.6) and the symmetry $\bar{m}(-y) = \bar{m}(y)$ to conclude

$$\frac{1}{2} E_0 \tau_{(-\epsilon, \epsilon)} = \int_0^\epsilon (\epsilon - y) \bar{m}(y) \, dy$$

Changing variables $y = \varphi(x)$, $dy = \varphi'(x) dx$, $\epsilon = \varphi(\delta)$ the above

$$= \int_0^\delta (\varphi(\delta) - \varphi(x)) \frac{1}{\varphi'(x) a(x)} \, dx$$

Now, (i) recalling that the speed measure for X_t is $m(x) = 1/\varphi'(x)a(x)$, and changing notation $\varphi'(z) = s(z)$, and (ii) using Fubini's theorem, the above

$$= \int_0^\delta \left(\int_x^\delta s(z)\, dz \right) m(x)\, dx = \int_0^\delta \int_0^z m(x)\, dx\, s(z)\, dz$$

Introducing M as the antiderivative of m to bring out the analogy with the condition in Feller's test (2.1), we have

(5.1) $$E_0 \tau_{(-\epsilon,\epsilon)} = 2 \int_0^\delta (M(z) - M(0))\, s(z)\, dz$$

To see that this means that the process cannot escape from 0, we note that repeating the proof of (4.9) shows

(5.2) **Lemma.** If $P_0(\tau_{(-\epsilon,\epsilon)} < \infty) > 0$ then $E_0 \tau_{(-\epsilon,\epsilon)} < \infty$.

Proof Pick M so that $P_0(\tau_{(-\epsilon,\epsilon)} \leq M) = \delta > 0$. Using the strong Markov property as in the proof of (4.9) we first get

$$\sup_{x \in (-\epsilon,\epsilon)} P_x(\tau_{(-\epsilon,\epsilon)} > M) \leq 1 - \delta$$

and then conclude that for each integer $k \geq 1$

$$\sup_{x \in (-\epsilon,\epsilon)} P_x(\tau_{(-\epsilon,\epsilon)} > kM) \leq (1-\delta)^k$$

from which the desired result follows. □

Consider a diffusion on $(0,r)$ where $r \leq \infty$, let $q \in (0,r)$, and let

$$I = \int_0^q (\varphi(z) - \varphi(0))\, m(z)\, dz$$

$$J = \int_0^q (M(z) - M(0))\, s(z)\, dz$$

Feller's test implies that

when $I < \infty$ we can get IN to the boundary point.

The analysis above shows that

when $J < \infty$ we can get OUT from the boundary point.

Section 6.5 Boundary Behavior   231

This leads to four possible combinations, which were named by Feller as follows

I	J	name
$< \infty$	$< \infty$	regular
$< \infty$	$= \infty$	absorbing
$= \infty$	$< \infty$	entrance
$= \infty$	$= \infty$	natural

The second case is called absorbing because we can get in to 0 but cannot get out. The third is called an entrance boundary because we cannot get to 0 but we can start the process there. Finally, in the fourth case the process can neither get to nor start at 0, so it is reasonable to exclude 0 from the state space. We will now give examples of the various possibilities.

Example 5.2. Feller's branching diffusion. Let
$$dX_t = \beta X_t \, dt + \sigma \sqrt{X_t} dB_t$$

Of course we want to suppose $\sigma > 0$ but we will also suppose $\beta > 0$ since the calculations are somewhat different in the cases $\beta = 0$ and $\beta < 0$. See Exercise 5.1 below. Using the formula in (1.4), the natural scale is

$$\varphi(x) = \int_0^x \exp\left(\int_0^y \frac{-2\beta z}{\sigma^2 z} dz\right) dy$$
$$= \int_0^x e^{-2\beta y/\sigma^2} dy = \frac{\sigma^2}{2\beta}(1 - e^{-2\beta x/\sigma^2})$$

which maps $[0, \infty)$ onto $[0, \sigma^2/2\beta)$. The speed measure is

$$m(x) = \frac{1}{\varphi'(x)a(x)} = e^{2\beta x/\sigma^2}/(\sigma^2 x)$$

To investigate the boundary at 0, we note that

$$I = \int_0^1 m(x)(\varphi(x) - \varphi(0)) \, dx = \int_0^1 \frac{1}{2\beta x}\left(e^{2\beta x/\sigma^2} - 1\right) dx < \infty$$

since the integrand converges to $1/\sigma^2$ as $x \to 0$. To calculate J we note that $m(x) \sim 1/(\sigma^2 x)$ as $x \to 0$ so $M(0) = -\infty$ and $J = \infty$. The combination $I < \infty$ and $J = \infty$ says that the process can get into 0 but not get out, so

0 is an absorbing boundary.

To investigate the boundary at ∞, we note that

$$\int_1^\infty m(x)(\varphi(\infty) - \varphi(x)) \, dx = \int_1^\infty \frac{1}{2\beta x} dx = \infty$$

To calculate J we note that when $\beta \geq 0$, $m(x) \geq 1/(\sigma^2 x)$ so $M(\infty) = \infty$ and $J = \infty$. The combination $I = \infty$ and $J = \infty$ says that the process cannot get into or out of ∞, so

$$\infty \text{ is a natural boundary.}$$

Exercise 5.1. Show that 0 is an absorbing boundary and ∞ is a natural boundary when (a) $\beta = 0$, (b) $\beta < 0$.

Example 5.3. Bessel process. Consider

$$dX_t = \frac{\gamma}{2X_t} dt + dB_t$$

Here $\gamma > -1$ is the **index** of the Bessel process. To explain the restriction on γ and to prepare for the results we will derive, note that Example 1.2 in Chapter 5 shows that the radial part of d dimensional Brownian motion is a Bessel process with $\gamma = (d-1)$. The natural scale is

$$\varphi(x) = \int_1^x \exp\left(-\int_1^y \gamma/z \, dz\right) dy$$

$$= \int_1^x y^{-\gamma} dy = \begin{cases} \ln x & \text{if } \gamma = 1 \\ (x^{1-\gamma} - 1)/(1-\gamma) & \text{if } \gamma \neq 1 \end{cases}$$

From the last computation we see that if $\gamma \geq 1$ then $\varphi(0) = -\infty$ and $I = \infty$. To handle $-1 < \gamma < 1$ we observe that the speed measure

$$m(z) = \frac{1}{\varphi'(z) a(z)} = z^\gamma$$

So taking $q = 1$ in the definition of I

$$I = \int_0^1 \frac{z^{1-\gamma}}{1-\gamma} z^\gamma \, dz < \infty$$

To compute J we observe that for any $\gamma > -1$, $M(z) = z^{\gamma+1}/(\gamma+1)$ and

$$J = \int_0^1 \frac{z^{\gamma+1}}{\gamma+1} z^{-\gamma} \, dz < \infty$$

Combining the last two conclusions we see that 0 is an

$$\begin{array}{ll} \text{entrance boundary} & \text{if } \gamma \in [1, \infty) \\ \text{regular boundary} & \text{if } \gamma \in (-1, 1) \end{array}$$

The first conclusion is reasonable since in $d \geq 2$ we can start Brownian motion at 0 but then it does not hit 0 at positive times. The second conclusion can be thought of as saying that in dimensions $d < 2$ Brownian motion will hit 0.

Exercise 5.2. Show that 0 is an absorbing boundary if $\gamma \leq -1$. We leave it to the reader to ponder the meaning of Brownian motion in dimension $d \leq 0$.

Example 5.4. Power noise. Consider
$$dX_t = X_t^\delta \, dB_t$$
on $(0, \infty)$. The natural scale is $\varphi(x) = x$ and the speed measure is $m(x) = 1/(\varphi'(x)a(x)) = x^{-2\delta}$ so
$$I = \int_0^1 x^{1-2\delta} \, dx = \begin{cases} < \infty & \text{if } \delta < 1 \\ = \infty & \text{if } \delta \geq 1 \end{cases}$$

When $\delta \geq 1/2$, $M(0) = -\infty$ and hence $J = \infty$. When $\delta < 1/2$
$$J = \int_0^1 \frac{z^{1-2\delta}}{1-2\delta} \, dz < \infty$$

Combining the last two conclusions we see that the boundary point 0 is

natural	if	$\delta \in [1, \infty)$
absorbing	if	$\delta \in [1/2, 1)$
regular	if	$\delta \in (0, 1/2)$

The fact that we can start at 0 if and only if $\delta < 1/2$ is suggested by (3.3) and Example 6.1 in Chapter 5. The new information here is that the solution can reach 0 if and only if $\delta < 1$. For another proof of that result, recall the time change recipe in Example 6.1 of Chapter 5 for constructing solutions of the equation above and do

Exercise 5.3. Consider
$$H_\delta = \int_0^{T_0} B_s^{-2\delta} 1_{(B_s \leq 1)} \, ds \quad \text{and} \quad H_\delta' = \int_0^{T_0} B_s^{-2\delta} 1_{(B_s > 1)} \, ds$$

Show that (a) for any $\delta > 0$, $P_1(H_\delta' < \infty) = 1$. (b) $E_1 H_\delta < \infty$ when $\delta < 1$. (c) $P_1(H_\delta = \infty) = 1$ when $\delta \geq 1$.

Exercise 5.4. Show that the boundary at ∞ in Example 5.4 is natural if $\delta \leq 1$ and entrance if $\delta > 1$.

Remark. The fact that we cannot reach ∞ is expected. (5.3) in Chapter 5 implies that these processes do not explode for any δ. The second conclusion is surprising at first glance since the process starting at ∞ is a time change of a Brownian motion starting from ∞. However the function we use in (5.1) of Chapter 5 is $g(x) = x^{-2\delta}$ so (2.9) in Chapter 3 implies

$$E_M \int_0^{T_1} g(B_s) 1_{(B_s \leq M)} \, ds = \int_1^M x^{-2\delta} \cdot 2x \, dx$$

which stays bounded as $M \to \infty$ if and only if $\delta > 1$.

Exercise 5.5. Wright-Fisher diffusion. Recall the definition given in Example 1.7 in Chapter 5. Show that the boundary point 0 is

$$\begin{array}{lll} \text{absorbing} & \text{if} & \beta = 0 \\ \text{regular} & \text{if} & \beta \in (0, 1/2) \\ \text{entrance} & \text{if} & \beta \geq 1/2 \end{array}$$

Hint: simplify calculations by first considering the case $\alpha = 0$ and then arguing that the value of α is not important.

6.6. Applications to Higher Dimensions

In this section we will use the technique of comparing a multidimensional diffusion with a one dimensional one to obtain sufficient conditions (a) for no explosion, (b) for recurrence and transience, and (c) for diffusions to hit points or not. Let $S_r = \inf\{t : |X_t| = r\}$ and $S_\infty = \lim_{r \to \infty} S_r$. Throughout this section we will suppose that X_t is a solution to MP(b, a) on $[0, S_\infty)$, and that the coefficients satisfy

(i) b is measurable and locally bounded

(ii) a is continuous and **nondegenerate** for each x, i.e.,

$$\sum_{i,j} y^i a_{ij}(x) y^j > 0 \quad \text{when } y \neq 0$$

The first detail is to show that X_t will eventually exit from any ball.

(6.1) Theorem. Let $S_r = \inf\{t : |X_t| = r\}$. Then there is a constant $C < \infty$ so that $E_x S_r \leq C$ for all x with $|x| \leq r$.

Proof Assumptions (i) and (ii) imply that we can pick an α large so that if $|x| \leq r$

$$\frac{\alpha}{2} a_{ii}(x) \geq 1 + |b_i(x)|$$

Section 6.6 Applications to Higher Dimensions   235

for all x with $|x| \leq r$. Let $h(x) = \sum_{i=1}^{d} \cosh(\alpha x_i)$. Using Itô's formula we have

$$h(X_t) - h(X_0) = \int_0^t \sum_{i=1}^{d} \alpha \sinh(\alpha X_s^i) b_i(X_s)\, ds + \text{local mart.}$$

$$+ \frac{1}{2} \int_0^t \sum_{i=1}^{d} \alpha^2 \cosh(\alpha X_s^i) a_{ii}(X_s)\, ds$$

Noting that $\cosh(y) \geq 1$ and

$$\cosh(y) \geq \max\left(\frac{e^y}{2}, \frac{e^{-y}}{2}\right) \geq |\sinh(y)|$$

our choice of α implies

$$\frac{\alpha^2}{2} \cosh(\alpha X_s^i) a_{ii}(X_s) + \alpha \sinh(\alpha X_s^i) b_i(X_s)$$
$$\geq \alpha \cosh(\alpha X_s^i) \left(\frac{\alpha}{2} a_{ii}(X_s) - |b_i(X_s)|\right) \geq \alpha \cosh(\alpha X_s^i) \geq \alpha$$

so $h(X_t) - t\alpha d$ is a local submartingale on $[0, S_r)$. Using the optional stopping theorem at time $S_r \wedge t$ we have

$$0 \leq h(x) \leq E_x\{h(X_{S_r \wedge t}) - \alpha d(S_r \wedge t)\}$$

Since $h(X_{S_r \wedge t}) \leq d \cosh(\alpha r)$ it follows that

$$E_x(S_r \wedge t) \leq \cosh(\alpha r)/\alpha$$

Letting $t \to \infty$ gives the desired result.   □

Remark. Tracing back through the proof shows $C = \cosh(\alpha r)/\alpha$ where α is chosen so that

$$\frac{\alpha}{2} a_{ii}(x) \geq 1 + |b_i(x)|$$

To see that the last estimate is fairly sharp consider a special case

Example 6.1. Suppose $d = 1$, $a(x) = 1$, and $b(x) = -\beta \operatorname{sgn}(x)$ with $\beta > 0$. The natural scale has

$$\varphi(x) = \int_0^x e^{2\beta y}\, dy = \frac{1}{2\beta}\left(e^{2\beta x} - 1\right)$$

for $x \geq 0$ and $\varphi(-x) = -\varphi(x)$, so the speed measure

$$m(x) = \frac{1}{\varphi'(x)} = e^{-2\beta|x|}$$

Using (4.8) now with $b = r$, $a = -r$, $x = 0$ and noting that

$$2\frac{\varphi(r) - \varphi(0)}{\varphi(r) - \varphi(-r)} = 2\frac{\varphi(0) - \varphi(-r)}{\varphi(r) - \varphi(-r)} = 1$$

we have

$$E_0 \tau_{(-r,r)} = \int_0^r \frac{1}{2\beta} \left(e^{2\beta r} - e^{2\beta z}\right) e^{-2\beta z}\, dz$$

$$+ \int_{-r}^0 \frac{1}{2\beta} \left(-e^{2\beta|z|} + e^{2\beta r}\right) e^{-2\beta|z|}\, dz$$

The two integrals are equal so

$$E_0 \tau_{(-r,r)} = \frac{1}{\beta} \int_0^r \left(e^{2\beta(r-z)} - 1\right) dz$$

$$= \left. \left(-\frac{1}{2\beta^2} e^{2\beta(r-z)} - \frac{z}{\beta}\right)\right|_0^r = \frac{1}{2\beta^2}\left(e^{2\beta r} - 1\right) - \frac{r}{\beta}$$

To compare with (6.1) now, note that when $a(x) = 1$ and $b(x) = -\beta \text{sgn}(x)$, the remark after the proof says we can pick $\alpha = 2\beta + 2$ so the inequality in (6.1) is

$$E_0 \tau_{(-r,r)} \leq \frac{\cosh((2\beta + 2)r)}{2\beta + 2} \qquad \square$$

a. Explosions

With the little detail of the finiteness of $E_x S_r$ out of the way we now introduce a comparison between a multidimensional diffusion and a one dimensional one, which is due to Khasminskii (1960). We begin by introducing three pairs of definitions that may look strange but are tailor made for computations in (d), (e), and (a) of the proof of (6.2).

$$\alpha(x) = 2x^T a(x) x \quad \text{and} \quad \beta(x) = \sum_{i=1}^d 2x_i b_i(x) + a_{ii}(x)$$

$$\nu(r) = \sup\left\{\frac{\beta(x)}{\alpha(x)} : |x|^2 = r\right\} \quad \text{and} \quad \rho(r) = \sup\{\alpha(x) : |x|^2 = r\}$$

$$\mu(r) = \rho(r)\nu(r) \quad \text{and} \quad \theta(r) = \sqrt{2\rho(r)}$$

Section 6.6 Applications to Higher Dimensions

Our plan is to compare $R_t = |X_t|^2$ with

$$dZ_t = \mu(Z_t)\,dt + \theta(Z_t)\,dB_t \quad \text{on } (0, \infty)$$

So we introduce the natural scale $\varphi(x)$ and speed measure $m(x)$ for Z_t:

$$s(x) = \exp\left(-\int_1^x \frac{2\mu(y)}{\theta^2(y)}\,dy\right)$$

$$\varphi(x) = \int_1^x s(y)\,dy$$

$$m(x) = 1/(s(x)\theta^2(x))$$

In view of Feller's test, (2.1), the next result says that if Z_t can't reach ∞ in finite time then X_t does not explode.

(6.2) Theorem. There is no explosion if

$$\int_1^\infty dx\, m(x)(\varphi(\infty) - \varphi(x)) = \infty$$

Proof To prove the result we will let $R_t = |X_t|^2$ and find a function g with $g(r) \to \infty$ as $r \to \infty$ so that $e^{-t}g(R_t)$ is a supermartingale while $R_t \geq 1$. To see that this is enough, let $S_r = \inf\{t : R_t = r\}$ and use the optional stopping theorem at time $S_1 \wedge S_n \wedge t$ to conclude that for $|x| > 1$

$$g(|x|^2) \geq e^{-t}g(n^2)P_x(S_n < S_1 \wedge t)$$

Letting $S_\infty = \lim_{r \uparrow \infty} S_r$, then $n \to \infty$ and $t \to \infty$ we have $P_x(S_\infty < S_1) = 0$.

To define the function g, we follow the proof of (2.1). Let $Y_t = \varphi(Z_t)$, use the construction there to produce a function f so that $e^{-t}f(Y_t)$ is a local martingale, then take $g = f \circ \varphi$ where φ is the natural scale of Z_t. It follows from our assumption and (2.2a) that $g(r) \to \infty$ as $r \to \infty$.

Since $g \in C^2$ it follows from Itô's formula that

$$e^{-t}g(Z_t) - g(Z_0) = -\int_0^t e^{-s}g(Z_s)\,ds$$

$$+ \int_0^t e^{-s}g'(Z_s)\mu(Z_s)\,ds + \text{local mart.}$$

$$+ \frac{1}{2}\int_0^t e^{-s}g''(Z_s)\theta^2(Z_s)\,ds$$

Since $e^{-t}g(Z_t)$ is a local martingale it follows that

$$\frac{1}{2}\theta^2(z)g''(z) + \mu(z)g'(z) = g(z)$$

Using the definition of θ and μ now we can rewrite this as

(a) $$\rho(z)g''(z) + \nu(z)\rho(z)g'(z) = g(z)$$

Before plunging into the proof we need one more property which follows from (2.6c) and the fact that our natural scale is an increasing function with $\varphi(1) = 0$:

(b) $$g'(r) \geq 0 \quad \text{for } r \geq 1$$

A little calculus gives

(c) $$\begin{aligned} D_i g(|x|^2) &= g'(|x|^2)2x_i \\ D_{ij} g(|x|^2) &= g''(|x|^2)4x_i x_j \quad i \neq j \\ D_{ii} g(|x|^2) &= g''(|x|^2)4x_i^2 + g'(|x|^2)2 \end{aligned}$$

Using Itô's formula now we have

$$e^{-t}g(R_t) - g(R_0) = \int_0^t -e^{-s}g(R_s)\,ds$$
$$+ \sum_i \int_0^t e^{-s}g'(R_s)2X_s^i b_i(X_s)\,ds + \text{local mart.}$$
$$+ \frac{1}{2}\sum_{i,j} \int_0^t e^{-s}g''(R_s)4X_s^i X_s^j a_{ij}(X_s)\,ds$$
$$+ \frac{1}{2}\sum_i \int_0^t e^{-s}g'(R_s)2a_{ii}(X_s)\,ds$$

Using the functions $\alpha(r)$ and $\beta(r)$ introduced above we can write the last equation compactly as

(d) $$\begin{aligned} e^{-t}g(R_t) - g(R_0) &= \text{local mart.} \\ &+ \int_0^t e^{-s}\{-g(R_s) + \beta(X_t)g'(R_s) + \alpha(X_s)g''(R_s)\}\,ds \end{aligned}$$

To bound the integral when $R_s \geq 1$ we use (i) the definition of ν and (b), then (ii) the equation in (a), the fact that $g \geq 0$, and the definition of ρ

(e) $$\begin{aligned} -g(R_s) &+ \left\{\frac{\beta(X_s)}{\alpha(X_s)}g'(R_s) + g''(R_s)\right\}\alpha(X_s) \\ &\leq -g(R_s) + \{\nu(R_s)g'(R_s) + g''(R_s)\}\alpha(X_s) \\ &\leq -g(R_s) + \left\{\frac{g(R_s)}{\rho(R_s)}\right\}\rho(R_s) \leq 0 \end{aligned}$$

Combining (d) and (e) shows that $e^{-t}g(R_t)$ is a local supermartingale while $R_t \geq 1$ and completes the proof. □

b. Recurrence and Transience

The next two results give sufficient conditions for transience and recurrence in $d > 1$. In this part we use the formulation of Meyers and Serrin (1960). Let

$$\bar{a}(x) = \frac{x}{|x|} \cdot a(x) \frac{x}{|x|} \qquad d_e(x) = \frac{2x \cdot b(x) + \operatorname{tr}(a(x))}{\bar{a}(x)}$$

where $\operatorname{tr}(a(x)) = \sum_i a_{ii}(x)$ is the trace of a. We call $d_e(x)$ the **effective dimension** at x because (6.3) and (6.4) will show that there will be recurrence or transience if the dimension is < 2 or > 2 in a neighborhood of infinity. To formulate a result that allows $d_e(x)$ to approach 2 as $|x| \to \infty$ we need a definition: $\delta(t)$ is said to be a **Dini function** if and only if

$$\int_1^\infty \frac{\delta(t)}{t} dt < \infty$$

In the next two results we will apply the definition to

$$\delta(t) = \exp\left(-\int_1^t \frac{\epsilon(s)}{s} ds\right)$$

To see what the conditions say note that $\delta(t) = (\log t)^{-p}$ is a Dini function if $p > 1$ but not if $p \leq 1$ and this corresponds to $\epsilon(s) = p/\log s$.

(6.3) Theorem. Suppose $d_e(x) \geq 2(1 + \epsilon(|x|))$ for $|x| \geq R$ and $\delta(t)$ is a Dini function, then X_t is transient. To be precise, if $|x| > R$ then $P_x(S_R < \infty) < 1$.

Here $S_r = \inf\{t : |X_t| = r\}$, and being transient includes the possibility of explosion.

(6.4) Theorem. Suppose $d_e(x) \leq 2(1+\epsilon(|x|))$ for $|x| \geq R$ and $\delta(t)$ is not a Dini function, then X_t is recurrent. To be precise, if $|x| > R$ then $P_x(S_R < \infty) = 1$.

Proof of (6.3) The first step is to let

$$\varphi(r) = \int_r^\infty \exp\left(-\int_1^s \frac{\epsilon(t)}{t} dt\right) \frac{ds}{s}$$

which has

$$\varphi'(r) = \frac{-1}{r} \exp\left(-\int_1^r \frac{\epsilon(t)}{t}\, dt\right) \le 0$$

$$\varphi''(r) = \frac{1+\epsilon(r)}{r^2} \exp\left(-\int_1^r \frac{\epsilon(t)}{t}\, dt\right)$$

and hence satisfies

$$r\varphi''(r) + (1+\epsilon(r))\varphi'(r) = 0$$

Letting $R_t = |X_t|^2$ and using Itô's formula we get the following which we will use four times below:

(6.5) Lemma. If $rg''(r) + \gamma(r)g'(r) = 0$ for $r \in (u^2, v^2)$ and $g'(|x|^2) \cdot (d_e(x) - 2\gamma(|x|^2)) \le 0$ when $|x| \in (u,v)$, then $g(R_t)$ is a local supermartingale while $u < |X_t| < v$.

Proof Using Itô's formula with (c) from the proof of (6.2)

$$g(R_t) - g(R_0) = \sum_i \int_0^t g'(R_s) 2 X_s^i b_i(X_s)\, ds + \text{local mart.}$$

$$+ \frac{1}{2}\sum_{ij} g''(R_s) 4 X_s^i X_s^j a_{ij}(X_s)\, ds$$

$$+ \frac{1}{2}\sum_i g'(R_s) 2 a_{ii}(X_s)\, ds$$

$$= \text{local mart.} + \int_0^t 2\bar{a}(X_s)\left(R_s g''(R_s) + g'(R_s)\frac{d_e(X_s)}{2}\right) ds$$

Using our two assumptions on g, we get for $u < |X_t| < v$

$$R_s g''(R_s) + g'(R_s)\frac{d_e(X_s)}{2} = g'(R_s)\left(\frac{d_e(X_s)}{2} - \gamma(R_s)\right) \le 0$$

which proves the result. □

(6.5) implies that $\varphi(R_t)$ is a local supermartingale while $R_s \ge R^2$. If $R \le r \le s < \infty$ then using the optional stopping theorem at time $\tau = S_r \wedge S_s$ ((6.1) implies that $P_x(\tau < \infty) = 1$) we see that

$$\varphi(|x|^2) \ge E_x \varphi(R_\tau) \ge \varphi(r^2) P_x(S_r < S_s)$$

Rearranging gives

$$P_x(S_r < S_s) \le \frac{\varphi(|x|^2)}{\varphi(r^2)} < 1$$

The upper bound is independent of s and (6.3) follows. □

Proof of (6.4) This time we let

$$\psi(r) = \int_1^r \exp\left(-\int_1^s \frac{\epsilon(t)}{t} dt\right) \frac{ds}{s}$$

Since $\psi'(r) = -\varphi'(r)$ we again have

$$r\psi''(r) + (1 + \epsilon(r))\psi'(r) = 0$$

but this time $\psi'(r) \geq 0$. Since we have assumed $d_e(x) \leq 2(1+\epsilon(|x|))$ for $|x| \geq R$ it follows from (6.5) that $\psi(R_t)$ is a local supermartingale while $R_t \geq R^2$. If $R \leq r < s < \infty$ using the optional stopping theorem at time $\tau = S_r \wedge S_s$ ((6.1) implies that $P_x(\tau < \infty) = 1$) we see that

$$\psi(|x|^2) \geq E_x\psi(R_\tau) = \psi(r^2)P_x(S_r < S_s) + \psi(s^2)\{1 - P(S_r < S_s)\}$$

Rearranging gives

$$P_x(S_r < S_s) \geq \frac{\psi(s^2) - \psi(|x|^2)}{\psi(s^2) - \psi(r^2)}$$

Letting $s \to \infty$ and noting that $\psi(s^2) \to \infty$ since $\delta(t)$ is not a Dini function gives the desired result. □

c. Hitting Points

The proofs of (6.3) and (6.4) generalize easily to investigate whether multidimensional diffusions hit 0. Let $T_0 = \inf\{t > 0 : X_t = 0\}$.

(6.6) Theorem. If $d_e(x) \geq 2(1 - \epsilon(|x|))$ for $|x| \leq \eta$ and

$$\int_0^1 \exp\left(-\int_y^1 \frac{\epsilon(z)}{z} dz\right) \frac{dy}{y} = \infty$$

then $P_x(T_0 < \infty) = 0$.

(6.7) Theorem. If $d_e(x) \leq 2(1 - \epsilon(|x|))$ for $|x| \leq \eta$ and

$$\int_0^1 \exp\left(-\int_y^1 \frac{\epsilon(z)}{z} dz\right) \frac{dy}{y} < \infty$$

then $P_x(T_0 < \infty) > 0$.

242 Chapter 6 One Dimensional Diffusions

We have changed the sign of ϵ to make the proofs more closely parallel the previous pair. Another explanation of the change is that Brownian motion

is recurrent in $d \le 2$ hits points in $d < 2$
is transient in $d > 2$ doesn't hit points in $d \ge 2$

so the boundary is now a little below 2 dimensions rather than a little above.

Proof of (6.6) Let

$$\varphi(r) = \int_r^1 \exp\left(-\int_s^1 \frac{\epsilon(t)}{t} dt\right) \frac{ds}{s}$$

which has

$$\varphi'(r) = \frac{-1}{r} \exp\left(-\int_r^1 \frac{\epsilon(t)}{t} dt\right) \le 0$$

$$\varphi''(r) = \frac{1 - \epsilon(r)}{r^2} \exp\left(-\int_r^1 \frac{\epsilon(t)}{t} dt\right)$$

and hence satisfies

$$r\varphi''(r) + (1 - \epsilon(r))\varphi'(r) = 0$$

Letting $R_t = |X_t|^2$ and using (6.5) shows $\varphi(R_t)$ is a local supermartingale while $R_t \in (0, \eta^2)$. If $0 < r < s \le \eta$ using the optional stopping theorem at time $\tau = S_r \wedge S_s$ we see that

$$\varphi(|x|^2) \ge E_x \varphi(R_\tau) \ge \varphi(r^2) P_x(S_r < S_s)$$

Rearranging gives

$$P_x(S_r < S_s) \le \varphi(|x|^2)/\varphi(r^2)$$

Letting $r \to 0$ and noting $\varphi(r^2) \to \infty$ gives

$$P_x(T_0 < S_\eta) = 0 \quad \text{for } 0 < |x| < \eta$$

To replace S_η by ∞ in the last equation let $\sigma_0 = 0$ and for $n \ge 1$ let

$$\tau_n = \inf\{t > \sigma_{n-1} : |X_t| = \eta\}$$
$$\sigma_n = \inf\{t > \tau_n : |X_t| = \eta/2\}$$

The strong Markov property implies that the $\sigma_n - \tau_n$ are i.i.d. so $\sigma_n \to \infty$ as $n \to \infty$. It is impossible for the process to hit 0 in $[\tau_n, \sigma_n]$ and hitting 0 in $[\sigma_n, \tau_{n+1}]$ has probability 0 so we have

$$P_x(T_0 < \infty) = 0 \quad \text{for } 0 < |x| < \eta$$

Using (3.2) and the Markov property now extends the conclusion to $|x| \geq \eta$. □

Proof of (6.7) This time we let

$$\psi(r) = \int_0^r \exp\left(-\int_s^1 \frac{\epsilon(t)}{t} dt\right) \frac{ds}{s}$$

Since $\psi'(r) = -\varphi'(r)$ we have $\psi'(r) \geq 0$ and

$$r\psi''(r) + (1 - \epsilon(r))\psi'(r) = 0$$

Letting $R_t = |X_t|^2$ and using (6.5) shows $\psi(R_t)$ is a local supermartingale while $R_t \in (0, \eta^2)$. If $0 < r < s \leq \eta$ using the optional stopping theorem at time $\tau = S_r \wedge S_s$ we see that

$$\psi(|x|^2) \geq E_x \psi(R_\tau) = \psi(r^2) P_x(S_r < S_s) + \psi(s^2)\{1 - P_x(S_r < S_s)\}$$

Rearranging gives

$$P_x(S_r < S_s) \geq \frac{\psi(s^2) - \psi(|x|^2)}{\psi(s^2) - \psi(r^2)}$$

Letting $r \to 0$ and noting $\psi(r^2) \to 0$ gives

$$P_x(T_0 < \infty) > 0 \quad \text{for } |x| < \eta$$

Using (3.2) and the Markov property now extends the conclusion to $|x| \geq \eta$. □

(6.8) Corollary. If (i) $d \geq 3$ or (ii) $d = 2$ and a is Hölder continuous then $P_x(T_0 < \infty) = 0$ for all x.

Proof Referring to (4.2) in Chapter 5, we can write $a(0) = U^T \Lambda U$ where Λ is a diagonal matrix with entries λ_i. Let Γ be the diagonal matrix with entries $1/\sqrt{\lambda_i}$ and let $V = U^T \Gamma$. The formula for the covariance of stochastic integrals implies that $\hat{X}_t = V X_t$ has

$$\hat{a}(0) = V^T a(0) V = I$$

So we can without loss of generality suppose that $a(0) = I$. In $d \geq 3$, the continuity of a implies $d_e(x) \to d$ as $x \to 0$ so we can take $\epsilon(r) \equiv 0$ and the desired result follows from (6.6). If $d = 2$ and a is Hölder continuous with exponent δ we can take $\epsilon(r) = Cr^\delta$. To extract the desired result from (6.6) we note that

$$\int_0^1 \exp\left(-\int_y^1 Cz^{\delta-1} dz\right) \frac{dy}{y} = \int_0^1 \exp\left(-\frac{C}{\delta}(1 - y^\delta)\right) \frac{dy}{y} = \infty$$

since the exponential in the integrand converges to a positive limit as $y \to 0$. □

It follows from (6.7) that a two dimensional diffusion can hit 0.

Example 6.1. Suppose $b(x) \equiv 0$, $a(0) = I$, and for $0 < |x| < 1/e$ let $a(x)$ be the matrix with eigenvectors

$$v_1 = \frac{1}{|x|}(x_1, x_2) \qquad v_2 = \frac{1}{|x|}(-x_2, x_1)$$

and associated eigenvalues

$$\lambda_1 = 1 \qquad \lambda_2 = 1 - 2p/\log|x|$$

These definitions are chosen so that $d_e(x) = 2 - 2p/\log(|x|)$ and hence we can take $\epsilon(r) = p/\log(r)$. Plugging into the integral

$$\int_0^{1/e} \exp\left(-\int_y^{1/e} \frac{p\,dz}{z \log z}\right) \frac{dy}{y} = \int_0^{1/e} \exp\left(p \log |\log y|\right) \frac{dy}{y}$$

$$= \int_0^{1/e} \frac{dy}{y |\log y|^p} < \infty \quad \text{if } p > 1$$

Remark. The problem of whether or not diffusions can hit points has been investigated in a different guise by Gilbarg and Serrin (1956), see page 315.

7 Diffusions as Markov Processes

In this chapter we will assume that the martingale problem $MP(b, a)$ has a unique solution and hence gives rise to a strong Markov process. As the title says, we will be interested in diffusions as Markov processes, in particular, in their asymptotic behavior as $t \to \infty$.

7.1. Semigroups and Generators

The results in this section hold for any Markov process, X_t. For any bounded measurable f, and $t \geq 0$, let

$$T_t f(x) = E_x f(X_t)$$

The Markov property implies that

$$E_x(f(X_{s+t})|\mathcal{F}_s) = E_{X_s} f(X_t) = T_t f(X_s)$$

So taking expected values gives the **semigroup property**

(1.1) $$T_{s+t} f(x) = T_s(T_t f)(x)$$

for $s, t \geq 0$. Introducing the norm $\|f\| = \sup |f(x)|$ and $L^\infty = \{f : \|f\| < \infty\}$ we see that T_t is a **contraction semigroup** on L^∞, i.e.,

(1.2) $$\|T_t f\| \leq \|f\|$$

Following Dynkin (1965), but using slightly different notation, we will define the domain of T to be

$$\mathcal{D}(T) = \{f \in L^\infty : \|T_t f - f\| \to 0 \text{ as } t \to 0\}$$

As the next result shows, this is a reasonable choice.

(1.3) Theorem. $\mathcal{D}(T)$ is a vector space, i.e., if $f_1, f_2 \in \mathcal{D}(T)$ and $c_1, c_2 \in \mathbf{R}$, $c_1 f_1 + c_2 f_2 \in \mathcal{D}(T)$. $\mathcal{D}(T)$ is closed. If $f \in \mathcal{D}(T)$ then $T_s f \in \mathcal{D}(T)$ and $s \to T_s f$ is continuous.

Remark. Here and in what follows, limits are always taken in, and hence continuity is defined for the uniform norm $\|\cdot\|$.

Proof Since $T_t(c_1 f_1 + c_2 f_2) = c_1 T_t f_1 + c_2 T_t f_2$, the first conclusion follows from the triangle inequality. To prove $\mathcal{D}(T)$ is closed, let $f_n \in \mathcal{D}(T)$ with $\|f_n - f\| \to 0$. Let $\epsilon > 0$. If we pick n large $\|f_n - f\| < \epsilon/3$. $f_n \in \mathcal{D}(T)$ so $\|T_t f_n - f_n\| < \epsilon/3$ for $t < t_0$. Using the triangle inequality and the contraction property (1.2) now, it follows that for $t < t_0$

$$\|T_t f - f\| \leq \|T_t f - T_t f_n\| + \|T_t f_n - f_n\| + \|f_n - f\|$$
$$\leq \|f - f_n\| + \epsilon/3 + \epsilon/3 < \epsilon$$

To prove the last claim in the theorem we note that the semigroup and contraction properties, (1.1) and (1.2), imply that if $s < t$

$$\|T_t f - T_s f\| = \|T_s(T_{t-s} f - f)\| \leq \|T_{t-s} f - f\| \qquad \square$$

Remark. While $\mathcal{D}(T)$ is a reasonable choice it is certainly not the only one. A common alternative is define the domain to be C_0, the continous functions which converge to 0 at ∞ equipped with the sup norm. When the semi-group T_t maps C_0 into itself, it is called a **Feller semi-group**. We will not follow this approach since this assumption does not hold, for example, in $d = 1$ when ∞ is an entrance boundary.

The **infinitesimal generator** of a semigroup is defined by

$$Af = \lim_{h \downarrow 0} \frac{T_h f - f}{h}$$

Its domain $\mathcal{D}(A)$ is the set of f for which the limit exists, that is, the f for which there is a g so that

$$\left\| \frac{T_h f - f}{h} - g \right\| \to 0 \quad \text{as } h \to 0$$

Before delving into properties of generators, we pause to identify some operators that cannot be generators.

(1.4) Theorem. If $f \in \mathcal{D}(A)$ and $f(x_0) \geq f(y)$ for all y then $Af(x_0) \leq 0$.

Proof Since x_0 is a maximum, $T_t f(x_0) - f(x_0) \leq 0$ so $Af(x_0) \leq 0$. $\square$

Example 1.1. The property in (1.4) looks innocent but it eliminates a number of operators. For example, when $k \geq 3$ is an integer we cannot have $Af = f^{(k)}$, the kth derivative. To prove this, we begin by noting that $g_k(x) = 1 - x^2 + x^k$ has a local maximum at 0. From this it follows that we can pick a small $\delta > 0$ and define a C^∞ function $f_k \geq 0$ that agrees with g_k on $(-\delta, \delta)$ and has a strict global maximum there. Since $Af_k(0) = k! > 0$, this contradicts (1.4). □

Returning to properties of actual generators, we have

(1.5) Theorem. $\mathcal{D}(A)$ is dense in $\mathcal{D}(T)$. If $f \in \mathcal{D}(A)$ then $Af \in \mathcal{D}(T)$, $T_t f \in \mathcal{D}(A)$, and

$$\frac{d}{dt} T_t f = A T_t f = T_t A f$$

$$T_t f - f = \int_0^t T_s A f \, ds$$

Remark. Here we are differentiating and integrating functions that take values in a Banach space, i.e., that map $s \in [0, \infty)$ into L^∞. Most of the proofs from calculus apply if you replace the absolute value by the norm. Readers who want help justifying our computations can consult Dynkin (1965), Volume 1, pages 19–22, or Ethier and Kurtz (1986), pages 8–10.

Proof Let $f \in \mathcal{D}(T)$ and $g_a = \int_0^a T_s f \, ds$. To explain the motivation for the last definition, we note that

$$\left\| \frac{g_a}{a} - f \right\| \leq \frac{1}{a} \int_0^a \| T_s f - f \| \, ds \to 0$$

as $a \to 0$ since $f \in \mathcal{D}(T)$. Now

$$T_h g_a = \int_0^a T_{s+h} f \, ds = g_a + \int_a^{a+h} T_s f \, ds - \int_0^h T_s f \, ds$$

so we have

$$\left| \frac{T_h g_a - g_a}{h} - (T_a f - f) \right| \leq \frac{1}{h} \int_a^{a+h} \| T_s f - T_a f \| \, ds$$

$$+ \frac{1}{h} \int_0^h \| T_s f - f \| \, ds \to 0$$

This shows that $g_a \in \mathcal{D}(A)$ and $A g_a = T_a f - f$. Since the first calculation shows $g_a/a \in \mathcal{D}(T)$ and converges to f as $a \to 0$ the first claim follows.

The fact that $Af \in \mathcal{D}(T)$ follows from (1.3) which implies $(T_h f - f)/h \in \mathcal{D}(T)$ and $\mathcal{D}(T)$ is closed. To prove that $T_t f \in \mathcal{D}(A)$ we note that the contraction property and the fact that $f \in \mathcal{D}(A)$ imply

$$\left\| \frac{T_h T_t f - T_t f}{h} - T_t Af \right\| = \left\| T_t \left(\frac{T_h f - f}{h} - Af \right) \right\|$$
$$\leq \left\| \frac{T_h f - f}{h} - Af \right\| \to 0$$

The last equality shows that $AT_t f = T_t Af$ and that the right derivative of $T_t f$ is $T_t Af$. To see that the left derivative exists and has the same value we note that

$$\left\| \frac{T_t f - T_{t-h} f}{h} - T_t Af \right\| \leq \left\| T_{t-h} \left(\frac{T_h f - f}{h} - Af \right) \right\| + \|T_{t-h}(Af - T_h Af)\|$$
$$\leq \left\| \frac{T_h f - f}{h} - Af \right\| + \|Af - T_h Af\| \to 0$$

as $h \to 0$ since $f \in \mathcal{D}(A)$ and $Af \in \mathcal{D}(T)$. The final equation in (1.5) is obtained by integrating the one above it. □

To make the connection between generators and martingale problems we will now prove

(1.6) Theorem. If $f \in \mathcal{D}(A)$ then for any probability measure μ

$$M_t^f \equiv f(X_t) - f(X_0) - \int_0^t Af(X_s)\,ds$$

is a P_μ martingale with respect to the filtration $\mathcal{F}_t$ generated by X_t.

Proof Since f and Af are bounded, M_t^f is integrable for each t. Since M_s^f is $\mathcal{F}_s$ measurable

$$E_\mu(M_t^f | \mathcal{F}_s) = M_s^f + E_\mu \left(f(X_t) - f(X_s) - \int_s^t Af(X_r)\,dr \,\Big|\, \mathcal{F}_s \right)$$

By the Markov property the second term on the right-hand side is

$$= E_{X_s} \left(f(X_{t-s}) - f(X_0) - \int_0^{t-s} Af(X_r)\,dr \right)$$

Section 7.1 Semigroups and Generators   249

But for any y, it follows from the definition of T_r and Fubini's theorem that

$$E_y \left(f(X_{t-s}) - f(X_0) - \int_0^{t-s} Af(X_r)\, dr \right)$$
$$= T_{t-s}f(y) - f(y) - \int_0^{t-s} T_r Af(y)\, dr$$

which is 0 by (1.5)   □

Our final topic is an important concept for some developments but will play a minor role here. For each $\lambda > 0$ define the **resolvent operator** for bounded measurable f by

$$U_\lambda f(x) = \int_0^\infty e^{-\lambda t}\, T_t f(x)\, dt = E_x \int_0^\infty e^{-\lambda t} f(X_t)\, dt$$

Clearly $\|U_\lambda f\| \leq \|f\|/\lambda$. A more subtle fact is that

(1.7) Theorem. U_λ maps $\mathcal{D}(T)$ 1-1 onto $\mathcal{D}(A)$. Its inverse is $(\lambda - A)$.

Proof Let $A_h f = (T_h f - f)/h$. Plugging in the definition of U_λ and changing variables $s = t + h$ and $s = t$ we have

$$A_h U_\lambda f = \frac{1}{h} \int_0^\infty e^{-\lambda t}(T_{t+h}f - T_t f)\, dt$$
$$= \frac{e^{\lambda h} - 1}{h} \int_h^\infty e^{-\lambda s} T_s f\, ds - \frac{1}{h} \int_0^h e^{-\lambda s} T_s f\, ds$$

To conclude that $U_\lambda f \in \mathcal{D}(A)$ and $AU_\lambda f = \lambda U_\lambda f - f$ we note that

$$\lambda U_\lambda f - f = \lambda \int_0^\infty e^{-\lambda s} T_s f\, ds - \frac{1}{h} \int_0^h f\, ds$$

and compare the last two displays to conclude (using $\|T_t f\| \leq \|f\|$) that

$$\|A_h U_\lambda f - (\lambda U_\lambda f - f)\| \leq \left| \frac{e^{\lambda h} - 1}{h} - \lambda \right| \int_h^\infty e^{-\lambda s} \|f\|\, ds + \lambda \int_0^h e^{-\lambda s} \|f\|\, ds$$
$$+ \frac{1}{h} \int_0^h (1 - e^{-\lambda s}) \|f\|\, ds + \frac{1}{h} \int_0^h \|T_s f - f\|\, ds \to 0$$

as $h \to 0$. The formula for $AU_\lambda f$ implies that $(\lambda - A)U_\lambda f = f$. Thus U_λ is 1-1 and its inverse is $\lambda - A$.

To prove that U_λ is onto, let $f \in \mathcal{D}(A)$ and note (1.5) implies

$$U_\lambda(\lambda f - Af) = \int_0^\infty e^{-\lambda t} T_t(\lambda f - Af)\, dt$$
$$= \int_0^\infty \lambda e^{-\lambda t} T_t f\, dt - \int_0^\infty e^{-\lambda t} \frac{d}{dt} T_t f\, dt$$

Integrating by parts

$$\int_0^\infty e^{-\lambda t} \frac{d}{dt} T_t f\, dt = \left. (e^{-\lambda t} T_t f) \right|_0^\infty + \int_0^\infty \lambda e^{-\lambda t} T_t f\, dt$$

Combining the last two displays shows $U_\lambda(\lambda f - Af) = f$ and the proof is complete. □

7.2. Examples

In this section we will consider two families of Markov processes and compute their generators.

Example 2.1. Pure jump Markov processes. Let $(S, \mathcal{S})$ be a measurable space and let $Q(x, A) : S \times \mathcal{S} \to [0, \infty)$ be a **transition kernel**. That is,

(i) for each $A \in \mathcal{S}$, $x \to Q(x, A)$ is measurable

(ii) for each $x \in S$, $A \to Q(x, A)$ is a finite measure

(iii) $Q(x, \{x\}) = 0$

Intuitively $Q(x, A)$ gives the rate at which our Markov chains makes jumps from x into A and we exclude jumps from x to x since they are invisible.

To be able to construct a process X_t with a minimum of fuss, we will suppose that $\lambda = \sup_{x \in S} Q(x, S) < \infty$. Define a transition probability P by

$$P(x, A) = \frac{1}{\lambda} Q(x, A) \quad \text{if } x \notin A$$
$$P(x, \{x\}) = \frac{1}{\lambda}(\lambda - Q(x, S))$$

Let $t_1, t_2, \ldots$ be i.i.d. exponential with parameter λ, that is, $P(t_i > t) = e^{-\lambda t}$. Let $T_n = t_1 + \cdots + t_n$ for $n \geq 1$ and let $T_0 = 0$. Let Y_n be a Markov chain with transition probability $P(x, dy)$ (see e.g., Section 5.1 of Durrett (1995) for a defintion), and let

$$X_t = Y_n \quad \text{for } t \in [T_n, T_{n+1})$$

X_t defines our pure jump Markov process. To compute its generator we note that $N_t = \sup\{n : T_n \le t\}$ has a Poisson distribution with mean λt so if f is bounded and measurable

$$T_t f(x) = E_x f(X_t) = e^{-\lambda t} f(x) + \lambda t e^{-\lambda t} \int P(x, dy) f(y) + O(t^2)$$

Doing a little arithmetic it follows that

$$\frac{T_t f(x) - f(x)}{t} - \lambda \int P(x, dy)(f(y) - f(x))$$
$$= t^{-1}(e^{-\lambda t} - 1 + \lambda t) f(x)$$
$$+ \lambda(e^{-\lambda t} - 1) \int P(x, dy) f(y) + O(t)$$

Since $t^{-1}(e^{-\lambda t} - 1 + \lambda t) \to 0$, $e^{-\lambda t} \to 1$, and $P(x, dy)$ is a transition probability it follows that

$$\left\| \frac{T_t f(x) - f(x)}{t} - \lambda \int P(x, dy)(f(y) - f(x)) \right\| \to 0$$

as $t \to 0$. Recalling the definition of $P(x, dy)$ it follows that

(2.1) Theorem. $\mathcal{D}(A) = \mathcal{D}(T) = L^\infty$ and

$$Af(x) = \int Q(x, dy)(f(y) - f(x))$$

We are now ready for the main event.

Example 2.2. Diffusion processes. Suppose we have a family of measures P_x on $(C, \mathcal{C})$ so that under P_x the coordinate maps $X_t(\omega) = \omega(t)$ give the unique solution to the MP(b, a) starting from x, and $\mathcal{F}_t$ is the filtration generated by the X_t. Let C_K^2 be the C^2 functions with compact support.

(2.2) Theorem. Suppose a and b are continuous and MP(b, a) is well posed. Then $C_K^2 \subset \mathcal{D}(A)$ and for all $f \in C_K^2$

$$Af(x) = \frac{1}{2} \sum_{ij} a_{ij}(x) D_{ij} f(x) + \sum_i b_i(x) D_i f(x)$$

Proof If $f \in C^2$ then Itô's formula implies

$$f(X_t) - f(X_0) = \sum_i \int_0^t D_i f(X_s) b_i(X_s)\, ds + \text{local mart.}$$
$$+ \frac{1}{2} \sum_{i,j} \int_0^t D_{ij} f(X_s) a_{ij}(X_s)\, ds$$

If we let T_n be a sequence of times that reduce the local martingale, stopping at time $t \wedge T_n$ and taking expected value gives

$$E_x f(X_{t \wedge T_n}) - f(x) = \sum_i E_x \int_0^{t \wedge T_n} D_i f(X_s) b_i(X_s)\, ds$$
$$+ \frac{1}{2} \sum_{i,j} E_x \int_0^{t \wedge T_n} D_{ij} f(X_s) a_{ij}(X_s)\, ds$$

If $f \in C_K^2$ and the coefficients are continuous then the integrands are bounded and the bounded convergence theorem implies

(2.3)
$$E_x f(X_t) - f(x) = \sum_i E_x \int_0^t D_i f(X_s) b_i(X_s)\, ds$$
$$+ \frac{1}{2} \sum_{i,j} E_x \int_0^t D_{ij} f(X_s) a_{ij}(X_s)\, ds$$
$$= E_x \int_0^t Af(X_s)\, ds$$

To prove the conclusion now we have to show that

$$\left\| \frac{E_x f(X_t) - f(x)}{t} - Af \right\| \to 0$$

The first step is to bound the movement of the diffusion process.

(2.4) Lemma. Let $r > 0$, $K \subset \mathbf{R}^d$ be compact, and $K_r = \{y : |x - y| \leq r \text{ for some } x \in K\}$. Suppose $|b(x)| \leq B$ and $\sum_i a_{ii}(x) \leq A$ for all $x \in K_r$. If we let $S_r = \inf\{t : |X_t - X_0| \geq r\}$ then

$$\sup_{x \in K} P_x(S_r \leq t) \leq t \cdot 2 \left(\frac{B}{r} + \frac{A}{r^2} \right)$$

Proof Let $h(y) = \sum_i (y_i - x_i)^2$. Using Itô's formula we have

$$h(X_t) - h(X_0) = \sum_i \int_0^t 2(X_s^i - x_i) b_i(X_s)\, ds + \text{local mart.}$$
$$+ \frac{1}{2} \sum_i \int_0^t 2a_{ii}(X_s)\, ds$$

Letting T_n be a sequence of times that reduce the local martingale, stopping at $t \wedge S_r \wedge T_n$, taking expected value, then letting $n \to \infty$ and invoking the bounded convergence theorem we have

$$E_x h(X_{t \wedge S_r}) = E_x \int_0^{t \wedge S_r} \sum_i \{2(X_s^i - x_i)b_i(X_s) + a_{ii}(X_s)\}\, ds \le 2t(rB + A)$$

Since $h \ge 0$ and $h(X_{S_r}) = r^2$ it follows that

$$r^2 P_x(S_r \le t) \le 2t(rB + A)$$

which is the desired result. $\square$

Proof of (2.2) Pick R so that $H = \{x : |x| \le R - 1\}$ contains the support of f and let $K = \{x : |x| \le R\}$. Since $f \in C_K^2$, and the coefficients are continuous, Af is continuous and has compact support. This implies Af is uniformly continuous, that is, given $\epsilon > 0$ we can pick $\delta \in (0, 1]$ so that if $|x - y| < \delta$ then $|Af(x) - Af(y)| < \epsilon$. If we let $C = \sup_z |Af(z)|$ and use (2.3) it follows that for $x \in K$

$$\left| \frac{E_x f(X_t) - f(x)}{t} - Af(x) \right| \le \frac{1}{t} E_x \int_0^t |Af(X_s) - Af(x)|\, ds$$
$$\le \epsilon + 2C \sup_{x \in K} P_x(S_\delta < t)$$

For $x \notin K$, $f(x) = 0$ and $Af = 0$. So using (2.3) again gives that for $x \notin K$

$$\left| \frac{E_x f(X_t) - f(x)}{t} - Af(x) \right| = \left| \frac{E_x f(X_t)}{t} \right| \le C P_x(T_H < t)$$
$$\le C \sup_{y \in \partial K} P_y(S_1 < t)$$

where the second equation comes from using the strong Markov property at time T_K. (2.4) shows that

$$\sup_{x \in \partial K} P_x(S_1 < t) \le \sup_{x \in K} P_x(S_\delta < t) \to 0 \quad \text{as } t \to 0$$

254 Chapter 7 Diffusions as Markov Processes

Since $\epsilon > 0$ is arbitrary, the desired result follows. □

For general b and a it is a hopelessly difficult problem to compute $\mathcal{D}(A)$. To make this point we will now use (1.7) to compute the exact domains of the semi-group and of the generator for one dimensional Brownian motion. Let C_u be the functions f on $\mathbf{R}$ that are bounded and uniformly continuous. Let C_u^2 be the functions f on $\mathbf{R}$ so that $f, f', f'' \in C_u$.

(2.5) Theorem. For Brownian motion in $d = 1$, $\mathcal{D}(T) = C_u$, $\mathcal{D}(A) = C_u^2$, and $Af(x) = f''(x)/2$ for all $f \in \mathcal{D}(A)$.

Remark. For Brownian motion in $d > 1$, $\mathcal{D}(A)$ is larger than C_u^2 and consists of the functions so that Δf exists in the sense of distribution and lies in C_u (see Revuz and Yor (1991), p. 226).

Proof If $t > 0$ and f is bounded then writing $p_t(x, z)$ for the Brownian transition probability and using the triangle inequality we have

$$|T_t f(x) - T_t f(y)| \leq \|f\| \int |p_t(x, z) - p_t(y, z)|\, dz$$

The right-hand side only depends on $|x - y|$ and converges to 0 as $|x - y| \to 0$ so $T_t f \in C_u$. If $f \in \mathcal{D}(T)$ then $\|T_t f - f\| \to 0$. We claim that this implies $f \in C_u$. To check this, pick $\epsilon > 0$, let $t > 0$ so that $\|T_t f - f\| < \epsilon/3$ then pick δ so that if $|x - y| < \delta$ then $|T_t f(x) - T_t f(y)| < \epsilon/3$. Using the triangle inequality now it follows that if $|x - y| < \delta$ then

$$|f(x) - f(y)| \leq |f(x) - T_t f(x)| + |T_t f(x) - T_t f(y)| + |T_t f(y) - f(y)| < \epsilon$$

To prove that $\mathcal{D}(A) = C_u^2$ we begin by observing that (3.10) in Chapter 3 implies

$$U_\lambda f(x) = (2\lambda)^{-1/2} \int e^{-|x-y|\sqrt{2\lambda}} f(y)\, dy$$

Our first task is to show that if $f \in C_u$ then $U_\lambda f \in C_u^2$. Letting $g(x) = U_\lambda f(x)$ we have

$$g(x) = (2\lambda)^{-1/2} \int_x^\infty e^{-(y-x)\sqrt{2\lambda}} f(y)\, dy + (2\lambda)^{-1/2} \int_{-\infty}^x e^{-(x-y)\sqrt{2\lambda}} f(y)\, dy$$

Differentiating once with respect to x and noting that the terms from differentiating the limits cancel, we have

$$g'(x) = \int_x^\infty e^{-(y-x)\sqrt{2\lambda}} f(y)\, dy - \int_{-\infty}^x e^{-(x-y)\sqrt{2\lambda}} f(y)\, dy$$

Differentiating again gives

$$g''(x) = \sqrt{2\lambda}\int_x^\infty e^{-(y-x)\sqrt{2\lambda}}f(y)\,dy + \sqrt{2\lambda}\int_{-\infty}^x e^{-(x-y)\sqrt{2\lambda}}f(y)\,dy - 2f(x)$$

It is easy to use the dominated convergence theorem to justify differentiating the integrals. From the formulas it is easy to see that $g, g', g'' \in C_u$ so $g \in C_u^2$ and

$$g''(x) = 2\lambda g(x) - 2f(x)$$

Since $Ag(x) = \lambda g(x) - f(x)$ by the proof of (1.7), it follows that $Ag(x) = g''(x)/2$.

To complete the proof now we need to show $\mathcal{D}(A) \supset C_u^2$. Let $h \in C_u^2$ and define a function $f \in C_u$ by

$$f(x) = \lambda h(x) - \frac{1}{2}h''(x)$$

The function $y = h - U_\lambda f$ satisfies

$$\frac{1}{2}y''(x) - \lambda y(x) = 0$$

All solutions of this differential equation have the form $Ae^{x\sqrt{2\lambda}} + Be^{-x\sqrt{2\lambda}}$. Since $h - U_\lambda f$ is bounded it follows that $h = U_\lambda f$ and the proof is complete. □

7.3. Transition Probabilities

In the last section we saw that if a and b are continuous and MP(b, a) is well posed then

$$\frac{d}{dt}T_t f(x) = AT_t f(x)$$

for $f \in C_K^2$, or changing notation $v(t, x) = T_t f(x)$, we have

(3.1) $$\frac{dv}{dt} = Av(t, x)$$

where A acts in the x variable. In the previous discussion we have gone from the process to the p.d.e. We will now turn around and go the other way. Consider

(3.2) (a) $u \in C^{1,2}$ and $u_t = Lu$ in $(0, \infty) \times \mathbf{R}^d$
  (b) u is continuous on $[0, \infty) \times \mathbf{R}^d$ and $u(0, x) = f(x)$

256 Chapter 7 Diffusions as Markov Processes

Here we have returned to our old notation
$$L = \frac{1}{2} \sum_{ij} a_{ij}(x) D_{ij} f(x) + \sum_i b_i(x) D_i f(x)$$

To explain the dual notation note that (3.1) says that $v(t, \cdot) \in \mathcal{D}(A)$ and satisfies the equality, while (a) of (3.2) asks for $u \in C^{1,2}$. As for the boundary condition (b) note that $f \in \mathcal{D}(A) \subset \mathcal{D}(T)$ implies that $\|T_t f - f\| \to 0$ as $t \to 0$.

To prove the existence of solutions to (3.2) we turn to Friedman (1975), Section 6.4.

(3.3) Theorem. Suppose a and b are bounded and **Hölder continuous**. That is, there are constants $0 < \delta, C < \infty$ so that

(HC) $\qquad |a_{ij}(x) - a_{ij}(y)| \leq C|x-y|^\delta \qquad |b_i(x) - b_i(y)| \leq C|x-y|^\delta$

Suppose also that a is **uniformly elliptic**, that is, there is a constant λ so that for all x, y

(UE) $\qquad \sum_{i,j} y_i a_{ij}(x) y_j \geq \lambda |y|^2$

Then there is a function $p_t(x, y) > 0$ jointly continuous in $t > 0$, x, y, and C^2 as a function of x which satisfies $dp/dt = Lp$ (with L acting on the x variable) and so that if f is bounded and continuous

$$u(t, x) = \int p_t(x, y) f(y) \, dy$$

satisfies (3.2). The maximum principle implies $|u(t, x)| \leq \|f\|$.

To make the connection with the SDE we prove

(3.4) Theorem. Suppose X_t is a nonexplosive solution of MP(b, a). If u is a bounded solution of (3.2) then

$$u(t, x) = E_x f(X_t)$$

Proof Itô's formula implies

$$u(t-s, X_s) - u(0, X_0) = \int_0^s -\frac{du}{dt}(t-r, X_r) \, dr$$
$$+ \sum_i \int_0^s D_i u(X_r) b_i(X_r) \, dr + \text{local mart.}$$
$$+ \frac{1}{2} \sum_{ij} \int_0^s D_{ij} u(X_r) a_{ij}(X_r) \, dr$$

so (a) tells us that $u(t-s, X_s)$ is a bounded martingale on $[0,t)$. (b) implies that $u(t-s, X_s) \to f(X_t)$ as $s \to t$, so the martingale convergence theorem implies

$$u(t,x) = E_x f(X_t) \qquad \square$$

$p_t(x,y)$ is called a **fundamental solution** of the parabolic equation $u_t = Lu$ since it can be used to produce solutions for any bounded continuous initial data f. Its importance for us is that (3.4) implies that

$$P_x(X_t \in B) = \int_B p_t(x,y)\,dy$$

i.e., $p_t(x,y)$ is the **transition density** for our diffusion process. Let $D = \{x : |x| < R\}$. To obtain result for unbounded coefficients, we will consider

(3.5)
    (a) $u \in C^{1,2}$ and $du/dt = Lu$ in $(0,\infty) \times D$
    (b) u is continuous on $[0,\infty) \times \bar{D}$ with $u(0,x) = f(x)$,
    and $u(t,y) = 0$ when $t > 0, y \in \partial D$

To prove existence of solutions to (3.5) we turn to Dynkin (1965), Vol. II, pages 230–231.

(3.6) Theorem. Let $D = \{x : |x| \leq R\}$ and suppose that (HC) and (UE) hold in D. Then there is a function $p_t^R(x,y) > 0$ jointly continuous in $t > 0$, $x, y \in D$, C^2 as a function of x which satisfies $dp^R/dt = Lp$ (with L acting on the x variable) and so that if f is bounded and continuous

$$u(t,x) = \int p_t^R(x,y) f(y)\,dy$$

satisfies (3.5).

To make the connection with the SDE we prove

(3.7) Theorem. Suppose X_t is any solution of MP(b,a) and let $\tau = \inf\{t : X_t \notin D\}$. If u is any solution of (3.5) then

$$u(t,x) = E_x(f(X_t); \tau > t)$$

Proof The fact that u is continuous in $[0,t] \times \bar{D}$ implies it is bounded there. Itô's formula implies that for $0 \le s < t \wedge \tau$

$$u(t-s, X_s) - u(0, X_0) = \int_0^s -\frac{du}{dt}(t-r, X_r)\,dr$$
$$+ \sum_i \int_0^s D_i u(X_r) b_i(X_r)\,dr + \text{local mart.}$$
$$+ \frac{1}{2} \sum_{ij} \int_0^s D_{ij} u(X_r) a_{ij}(X_r)\,dr$$

so (a) tells us that $u(t-s, X_s)$ is a bounded local martingale on $[0, t \wedge \tau)$. (b) implies that $u(t-s, X_s) \to 0$ as $s \uparrow \tau$ and $u(t-s, X_s) \to f(X_t)$ as $s \to t$ on $\{\tau > t\}$. Using (2.7) from Chapter 2 now, we have

$$u(t,x) = E_x(f(X_t); \tau > t) \qquad \square$$

Combining (3.6) and (3.7) we see that

$$E_x(f(X_t); \tau > t) = \int_D p_t^R(x,y) f(y)\,dy$$

Letting $R \uparrow \infty$ and $p_t(x,y) = \lim_{R \to \infty} p_t^R(x,y)$, which exists since (3.7) implies $R \to p_t^R(x,y)$ is increasing, we have

(3.8) Theorem. Suppose that the martingale problem for MP(b,a) is well posed and that a and b satisfy (HC) and (UE) hold locally. That is, they hold in $\{x : |x| \le R\}$ for any $R < \infty$. Then for each $t > 0$ there is a lower semicontinuous function $p(t,x,y) > 0$ so that if f is bounded and continuous then

$$E_x f(X_t) = \int p_t(x,y)\,dy$$

Remark. The energetic reader can probably show that $p_t(x,y)$ is continuous. However, lower semicontinuity implies that $p_t(x,y)$ is bounded away from 0 on compact sets and this will be enough for our results in Section 7.5.

7.4. Harris Chains

In this section we will give a quick treatment of the theory of Harris chains to prepare for applications to diffusions in the next section. In this section and the next, we will assume that the reader is familiar with the basic theory of Markov chains on a countable state space as explained for example in the first five sections of Chapter 5 of Durrett (1995). This section is a close relative of Section 5.6 there.

We will formulate the results here for a **transition probability** defined on a measurable space $(S, \mathcal{S})$. Intuitively, $P(X_{n+1} \in A | X_n = x) = p(x, A)$. Formally, it is a function $p : S \times \mathcal{S} \to [0, 1]$ that satisfies

for fixed $A \in \mathcal{S}$, $x \to p(x, A)$ is measurable

for fixed $x \in S$, $A \to p(x, A)$ is a probability measure

In our applications to diffusions we will take $S = \mathbf{R}^d$, and let

$$p(x, A) = \int_A p_1(x, y) \, dy$$

where $p_t(x, y)$ is the transition probability introduced in the previous section. By taking $t = 1$ we will be able to investigate the asymptotic behavior of X_n as $n \to \infty$ through the integers but this and the Markov property will allow us to get results for $t \to \infty$ through the real numbers.

We say that a Markov chain X_n with transition probability p is a **Harris chain** if we can find sets $A, B \in \mathcal{S}$, a function q with $q(x, y) \geq \epsilon > 0$ for $x \in A$, $y \in B$, and a probability measure ρ concentrated on B so that:

(i) If $\tau_A = \inf\{n \geq 0 : X_n \in A\}$, then $P_z(\tau_A < \infty) > 0$ for all $z \in S$

(ii) If $x \in A$ and $C \subset B$ then $p(x, C) \geq \int_C q(x, y) \rho(dy)$

In the diffusions we consider we can take $A = B$ to be a ball with radius r, ρ to be a constant c times Lebesgue measure restricted to B, and $q(x, y) = p_1(x, y)/c$. See (5.2) below. It is interesting to note that the new theory still contains most of the old one as a special case.

Example 4.1. Countable State Space. Suppose X_n is a Markov chain on a countable state space S. In order for X_n to be a Harris chain it is necessary and sufficient that there be a state u with $P_x(X_n = u \text{ for some } n \geq 0) > 0$ for all $x \in S$.

Proof To prove sufficiency, pick v so that $p(u, v) > 0$. If we let $A = \{u\}$ and $B = \{v\}$ then (i) and (ii) hold. To prove necessity, let $\{u\}$ be a point with $\rho(\{u\}) > 0$ and note that for all x

$$P_x(X_n = u \text{ for some } n \geq 0) \geq E_x\{q(X_{\tau_A})\rho(\{u\})\} > 0 \qquad \square$$

The developments in this section are based on two simple ideas. (i) To make the theory of Markov chains on a countable state space work, all we need is to have one point in the state space which is hit. (ii) In a Harris chain we can manufacture such a point (called α below) which corresponds to "being in B with distribution ρ." The notation needed to carry out this plan is occasionally obnoxious but we hope these words of wisdom will help the reader stay focused on the simple ideas that underlie these developments.

Given a Harris chain on $(S, \mathcal{S})$, we will construct a Markov chain $\bar{X}_n$ with transition probability $\bar{p}$ on $(\bar{S}, \bar{\mathcal{S}})$ where $\bar{S} = S \cup \{\alpha\}$ and $\bar{\mathcal{S}} = \{B, B \cup \{\alpha\} : B \in \mathcal{S}\}$. Thinking of α as corresponding to being on B with distribution ρ, we define the modified transition probability as follows:

If $x \in S - A$, $\bar{p}(x, C) = p(x, C)$ for $C \in \mathcal{S}$
If $x \in A$, $\bar{p}(x, \{\alpha\}) = \epsilon$
$\bar{p}(x, C) = p(x, C) - \epsilon \rho(C)$ for $C \in \mathcal{S}$
If $x = \alpha$, $\bar{p}(\alpha, D) = \int \rho(dx) \bar{p}(x, D)$ for $D \in \bar{\mathcal{S}}$

Here and in what follows, we will reserve A and B for the special sets that occur in the definition and use C and D for generic elements of $\mathcal{S}$. We will often simplify notation by writing $\bar{p}(x, \alpha)$ instead of $\bar{p}(x, \{\alpha\})$, $\mu(\alpha)$ instead of $\mu(\{\alpha\})$, etc.

Our first step is to prove three technical lemmas that will help us carry out the proofs below. Define a transition probability v by

$$v(x, \{x\}) = 1 \text{ if } x \in S$$
$$v(\alpha, C) = \rho(C)$$

In words, v leaves mass in S alone but returns the mass at α to S and distributes it according to ρ.

(4.1) Lemma. (a) $v\bar{p} = \bar{p}$ and (b) $\bar{p}v = p$.

Proof Before giving the proof we would like to remind the reader that measures multiply the transition probability on the left, i.e., in the first case we want to show $\mu v \bar{p} = \mu \bar{p}$. If we first make a transition according to v and then one according to $\bar{p}$, this amounts to one transition according to $\bar{p}$, since only mass at α is affected by v and

$$\bar{p}(\alpha, D) = \int \rho(dx) \bar{p}(x, D).$$

The second equality also follows easily from the definition. In words, if $\bar{p}$ acts first and then v, then this is the same as one transition according to p since v returns the mass at α to where it came from. □

From (4.1) it follows easily that we have:

(4.2) **Lemma.** Let Y_n be an inhomogeneous Markov chain with $p_{2k} = v$ and $p_{2k+1} = \bar{p}$. Then $\bar{X}_n = Y_{2n}$ is a Markov chain with transition probability $\bar{p}$ and $X_n = Y_{2n+1}$ is a Markov chain with transition probability p.

(4.2) shows that there is an intimate relationship between the asymptotic behavior of X_n and of $\bar{X}_n$. To quantify this we need a definition. If f is a bounded measurable function on S, let $\bar{f} = vf$, i.e.,

$$\bar{f}(x) = f(x) \quad \text{for } x \in S$$
$$\bar{f}(\alpha) = \int f \, d\rho$$

(4.3) **Lemma.** If μ is a probability measure on $(S, \mathcal{S})$ then

$$E_\mu f(X_n) = E_\mu \bar{f}(\bar{X}_n)$$

Proof Observe that if X_n and $\bar{X}_n$ are constructed as in (4.2), and $P(\bar{X}_0 \in S) = 1$ then $X_0 = \bar{X}_0$ and X_n is obtained from $\bar{X}_n$ by making a transition according to v. □

Before developing the theory we give one example to explain why some of the statements to come will be messy.

Example 4.2. Perverted Brownian motion. For x that is not an integer ≥ 2, let $p(x, \cdot)$ be the transition probability of a one dimensional Brownian motion. When $x \geq 2$ is an integer let

$$p(x, \{x+1\}) = 1 - x^{-2}$$
$$p(x, C) = x^{-2}|C \cap [0,1]| \quad \text{if } x + 1 \notin C$$

p is the transition probability of a Harris chain (take $A = B = (0,1)$, $\rho =$ Lebesgue measure on B) but

$$P_2(X_n = n + 2 \text{ for all } n) > 0$$

I can sympathize with the reader who thinks that such crazy chains will not arise "in applications," but it seems easier (and better) to adapt the theory to include them than to modify the assumptions to exclude them.

a. Recurrence and transience

We begin with the dichotomy between recurrence and transience. Let $R = \inf\{n \geq 1 : \bar{X}_n = \alpha\}$. If $P_\alpha(R < \infty) = 1$ then we call the chain

recurrent, otherwise we call it **transient**. Let $R_1 = R$ and for $k \geq 2$, let $R_k = \inf\{n > R_{k-1} : \bar{X}_n = \alpha\}$ be the time of the kth return to α. The strong Markov property implies $P_\alpha(R_k < \infty) = P_\alpha(R < \infty)^k$ so $P_\alpha(\bar{X}_n = \alpha \text{ i.o.}) = 1$ in the recurrent case and is 0 in the transient case. From this the next result follows easily.

(4.4) Theorem. Let $\lambda(C) = \sum_n 2^{-n} \bar{p}^n(\alpha, C)$. In the recurrent case if $\lambda(C) > 0$ then $P_\alpha(\bar{X}_n \in C \text{ i.o.}) = 1$. For λ a.e. x, $P_x(R < \infty) = 1$.

Remark. Here and in what follows $\bar{p}^n(x, C) = P_x(\bar{X}_n \in C)$ is the nth iterate of the transition probability defined inductively by $\bar{p}^1 = \bar{p}$ and for $n \geq 1$

$$\bar{p}^{n+1}(x, C) = \int \bar{p}(x, dy) \bar{p}^n(y, C)$$

Proof The first conclusion follows from the following fact (see e.g., (2.3) in Chapter 5 of Durrett (1995)). Let X_n be a Markov chain and suppose

$$P\left(\cup_{m=n+1}^\infty \{X_m \in B_m\} \big| X_n\right) \geq \delta \quad \text{on } \{X_n \in A_n\}$$

Then $P(\{X_n \in A_n \text{ i.o.}\}) - P(\{X_n \in B_n \text{ i.o.}\}) = 0$. Taking $A_n \equiv \{\alpha\}$ and $B_n \equiv C$ gives the desired result. To prove the second conclusion, let $D = \{x : P_x(R < \infty) < 1\}$ and observe that if $p^n(\alpha, D) > 0$ for some n then

$$P_\alpha(\bar{X}_m = \alpha \text{ i.o.}) \leq \int \bar{p}^n(\alpha, dx) P_x(R < \infty) < 1 \qquad \square$$

Remark. Example 4.2 shows that we cannot expect to have $P_x(R < \infty) = 1$ for all x. To see that this can occur even when the state space is countable, consider a branching process in which the offspring distribution has $p_0 = 0$ and $\sum k p_k > 1$. If we take $A = B = \{0\}$ this is a Harris chain by Example 4.1. Since $p^n(0,0) = 1$ it is recurrent.

b. Stationary measures

(4.5) Theorem. Let $R = \inf\{n \geq 1 : \bar{X}_n = \alpha\}$. In the recurrent case,

$$\bar{\mu}(C) = E_\alpha \left(\sum_{n=0}^{R-1} 1_{\{\bar{X}_n \in C\}} \right) = \sum_{n=0}^\infty P_\alpha(\bar{X}_n \in C, R > n)$$

defines a σ-finite stationary measure for $\bar{p}$, with $\bar{\mu} \ll \lambda$, the measure defined in (4.4). $\mu = \bar{\mu}v$ is a σ-finite stationary measure for p.

Proof If $\lambda(C) = 0$ then the definition of λ implies $P_\alpha(\bar{X}_n \in C) = 0$, so $P_\alpha(\bar{X}_n \in C, R > n) = 0$ for all n, and $\bar{\mu}(C) = 0$. We next check that $\bar{\mu}$ is σ-finite. Let $G_{k,\delta} = \{x : \bar{p}^k(x,\alpha) \geq \delta\}$. Let $T_0 = 0$ and let $T_n = \inf\{m \geq T_{n-1} + k : X_m \in G_{k,\delta}\}$. The definition of $G_{k,\delta}$ implies

$$P(T_n < T_\alpha | T_{n-1} < T_\alpha) \leq (1-\delta)$$

so if we let $N = \inf\{n : T_n \geq T_\alpha\}$ then $EN < 1/\delta$. Since we can only have $\bar{X}_m \in G_{k,\delta}$ with $R > m$ when $T_n \leq m < T_n + k$ for some $0 \leq n < N$ it follows that $\bar{\mu}(G_{k,\delta}) \leq k/\delta$. Part (i) of the definition of a Harris chain implies $S \subset \cup_{k,m\geq 1} G_{k,1/m}$ and σ-finiteness follows.

Next we show that $\bar{\mu}\bar{p} = \bar{\mu}$.

CASE 1. Let C be a set that does not contain α. Using the definition $\bar{\mu}$ and Fubini's theorem.

$$\int \bar{\mu}(dy)\bar{p}(y,C) = \sum_{n=0}^{\infty} \int P_\alpha(\bar{X}_n \in dy, R > n)\bar{p}(y,C)$$

$$= \sum_{n=0}^{\infty} P_\alpha(\bar{X}_{n+1} \in C, R > n+1) = \bar{\mu}(C)$$

since $\alpha \notin C$ and $P_\alpha(X_0 = \alpha) = 1$.

CASE 2. To complete the proof now it suffices to consider $C = \{\alpha\}$.

$$\int \bar{\mu}(dy)\bar{p}(y,\alpha) = \sum_{n=0}^{\infty} \int P_\alpha(\bar{X}_n \in dy, R > n)\bar{p}(y,\alpha)$$

$$= \sum_{n=0}^{\infty} P_\alpha(R = n+1) = 1 = \bar{\mu}(\alpha)$$

where in the last two equalities we have used recurrence and the fact that when $C = \{\alpha\}$ only the $n = 0$ term contributes in the definition.

Turning to the properties of μ, we note that $\mu = \bar{\mu}v$ and $\bar{\mu}(\alpha) = 1$ so μ is σ-finite. To check $\mu p = \mu$, we note that (i) using the definition $\mu = \bar{\mu}v$ and (b) in (3.1), (ii) using (a) in (4.1), (iii) using $\bar{\mu}\bar{p} = \bar{\mu}$, and (iv) using the definition $\mu = \bar{\mu}v$:

$$\mu p = (\bar{\mu}v)(\bar{p}v) = \bar{\mu}\bar{p}v = \bar{\mu}v = \mu \qquad \square$$

To investigate uniqueness of the stationary measure we begin with:

(4.6) Lemma. If ν is a σ-finite stationary measure for p, then $\nu(A) < \infty$ and $\bar{\nu} = \nu \bar{p}$ is a σ-finite stationary measure for $\bar{p}$ with $\bar{\nu}(\alpha) < \infty$.

Proof We will first show that $\nu(A) < \infty$. If $\nu(A) = \infty$ then part (ii) of the definition of a Harris chain implies $\nu(C) = \infty$ for all sets C with $\rho(C) > 0$. If $B = \cup_i B_i$ with $\nu(B_i) < \infty$ then $\rho(B_i) = 0$ by the last observation and $\rho(B) = 0$ by countable subadditivity, a contradiction. So $\nu(A) < \infty$ and $\bar{\nu}(\alpha) = \nu \bar{p}(\alpha) = \epsilon \nu(A) < \infty$. Using the fact that $\nu p = \nu$, we find

$$\bar{\nu}(C) = \nu \bar{p}(C) = \nu(C) - \epsilon\, \nu(A) \rho(B \cap C)$$

the last subtraction being well defined since $\nu(A) < \infty$. From the last equality it is clear that $\bar{\nu}$ is σ-finite. Since $\bar{\nu}(\alpha) = \epsilon \nu(A)$, it also follows that $\bar{\nu} v = \nu$. To check $\bar{\nu} \bar{p} = \bar{\nu}$, we observe that (a) of (4.1), the last result, and the definition of $\bar{\nu}$ imply

$$\bar{\nu} \bar{p} = \bar{\nu} v \bar{p} = \nu \bar{p} = \bar{\nu} \qquad \square$$

(4.7) Theorem. Suppose p is recurrent. If ν is a σ-finite stationary measure then $\nu = \bar{\nu}(\alpha)\mu$ where μ is the measure constructed in the proof of (4.5).

Proof By (4.6) it suffices to prove that if $\bar{\nu}$ is a σ-finite stationary measure for $\bar{p}$ with $\bar{\nu}(\alpha) < \infty$ then $\bar{\nu} = \bar{\nu}(\alpha)\bar{\mu}$. Our first step is to observe that

$$\bar{\nu}(C) = \bar{\nu}(\alpha)\bar{p}(\alpha, C) + \int_{S-\{\alpha\}} \bar{\nu}(dy)\bar{p}(y, C)$$

Using the last identity to replace $\bar{\nu}(dy)$ on the right-hand side we have

$$\bar{\nu}(C) = \bar{\nu}(\alpha)\bar{p}(\alpha, C) + \int_{S-\{\alpha\}} \bar{\nu}(\alpha)\bar{p}(\alpha, dy)\bar{p}(y, C)$$
$$+ \int_{S-\{\alpha\}} \bar{\nu}(dx) \int_{S-\{\alpha\}} \bar{p}(x, dy)\bar{p}(y, C)$$
$$= \bar{\nu}(\alpha) P_\alpha(\bar{X}_1 \in C) + \bar{\nu}(\alpha) P_\alpha(\bar{X}_1 \neq \alpha, \bar{X}_2 \in C)$$
$$+ P_{\bar{\nu}}(\bar{X}_0 \neq \alpha, \bar{X}_1 \neq \alpha, \bar{X}_2 \in C)$$

Continuing in the obvious way we have

$$\bar{\nu}(C) = \bar{\nu}(\alpha) \sum_{m=1}^{n} P_\alpha(\bar{X}_k \neq \alpha \text{ for } 1 \leq k < m, \bar{X}_m \in C)$$
$$+ P_{\bar{\nu}}(\bar{X}_k \neq \alpha \text{ for } 0 \leq k < n+1, \bar{X}_{n+1} \in C)$$

The last term is nonnegative, so letting $n \to \infty$ we have

$$\bar{\nu}(C) \geq \bar{\nu}(\alpha)\bar{\mu}(C)$$

To turn the $\geq$ into an $=$ we observe that

$$\bar{\nu}(\alpha) = \int \bar{\nu}(dx)\bar{p}^n(x,\alpha) \geq \bar{\nu}(\alpha) \int \bar{\mu}(dx)\bar{p}^n(x,\alpha) = \bar{\nu}(\alpha)\bar{\mu}(\alpha) = \bar{\nu}(\alpha)$$

since $\bar{\mu}(\alpha) = 1$. Let $S_n = \{x : \bar{p}^n(x, \alpha) > 0\}$. By assumption $\cup_n S_n = S$. If $\bar{\nu}(D) > \bar{\nu}(\alpha)\bar{\mu}(D)$ for some D, then $\bar{\nu}(D \cap S_n) > \bar{\nu}(\alpha)\bar{\mu}(D \cap S_n)$ for some n and it follows that $\bar{\nu}(\alpha) > \bar{\nu}(\alpha)$, a contradiction. □

(4.5) and (4.7) show that a recurrent Harris chain has a σ-finite stationary measure that is unique up to constant multiples. The next result, which goes in the other direction, will be useful in the proof of the convergence theorem and in checking recurrence of concrete examples.

(4.8) Theorem. If there is a stationary probability distribution then the chain is recurrent.

Proof Let $\bar{\pi} = \pi\bar{p}$, which is a stationary distribution by (4.6), and note that part (i) of the definition of a Harris chain implies $\bar{\pi}(\alpha) > 0$. Suppose that the chain is transient, i.e., $P_\alpha(R < \infty) = q < 1$. If R_k is the time of the kth return then $P_\alpha(R_k < \infty) = q^k$ so if $x \neq \alpha$

$$E_x\left(\sum_{n=0}^\infty 1_{\{X_n = \alpha\}}\right) = \sum_{k=1}^\infty P_x(R_k < \infty) \leq \sum_{k=1}^\infty q^{k-1} = \frac{1}{1-q}$$

The last upper bound is also valid when $x = \alpha$ since

$$E_\alpha\left(\sum_{n=0}^\infty 1_{\{X_n = \alpha\}}\right) = 1 + \sum_{k=1}^\infty P_x(R_k < \infty) \leq \sum_{k=0}^\infty q^k = \frac{1}{1-q}$$

Integrating with respect to $\bar{\pi}$ and using the fact that $\bar{\pi}$ is a stationary distribution we have a contradiction that proves the result

$$\frac{1}{1-q} \geq E_{\bar{\pi}}\left(\sum_{n=0}^\infty 1_{\{X_n = \alpha\}}\right) = \sum_{n=0}^\infty \bar{\pi}(\alpha) = \infty \qquad □$$

266   Chapter 7   Diffusions as Markov Processes

c. Convergence theorem

If I is a set of positive integers, we let g.c.d.(I) be the greatest common divisor of the elements of I. We say that a recurrent Harris chain X_n is **aperiodic** if g.c.d.$(\{n \geq 1 : p^n(\alpha, \alpha) > 0\}) = 1$. This occurs, for example, if we can take $A = B$ in the definition for then $p(\alpha, \alpha) > 0$. A well known consequence of aperiodicity is

(4.9) Lemma. There is an $m_0 < \infty$ so that $\bar{p}^m(\alpha, \alpha) > 0$ for $m \geq m_0$.

For a proof see (5.4) of Chapter 5 in Durrett (1995). We are now ready to prove

(4.10) Theorem. Let X_n be an aperiodic recurrent Harris chain with stationary distribution π. If $P_x(R < \infty) = 1$ then as $n \to \infty$,

$$\|p^n(x, \cdot) - \pi(\cdot)\| \to 0$$

Remark. Here $\|\cdot\|$ denotes the total variation distance between the measures. (4.4) guarantees that λ a.e. x satisfies the hypothesis, while (4.5), (4.7), and (4.8) imply π is absolutely continuous with respect to λ.

Proof In view of (4.3) and (4.6) it suffices to show that if $\bar{\pi} = \pi \bar{p}$ then

$$\|\bar{p}^n(x, \cdot) - \bar{\pi}(\cdot)\| \to 0$$

Our second reduction is to note that if $x \neq \alpha$ then

$$\bar{p}^n(x, \cdot) = \sum_{m=1}^{n} P_x(R = m) \bar{p}^{n-m}(\alpha, \cdot)$$

so we have

$$\|\bar{p}^n(x, \cdot) - \bar{\pi}(\cdot)\| \leq \sum_{m=1}^{n} P_x(R = m) \|\bar{p}^{n-m}(\alpha, \cdot) - \bar{\pi}(\cdot)\|$$

and it suffices to prove the result when $x = \alpha$.

Let $S^2 = S \times S$. Define a transition probability $\hat{p}$ on $S \times S$ by

$$\hat{p}((x_1, x_2), C_1 \times C_2) = \bar{p}(x_1, C_1) \bar{p}(x_2, C_2)$$

That is, each coordinate moves independently. We will call this the "product chain" (think of product measure) and denote it by $\hat{Z}_n$.

We claim that the product chain is a Harris chain with $\hat{A} = \{(\alpha, \alpha)\}$ and $\hat{B} = B \times B$. Clearly (ii) is satisfied. To check (i) we need to show that for all (x_1, x_2) there is an N so that $\hat{p}^N((x_1, x_2), (\alpha, \alpha)) > 0$. To prove this let K and L be such that $\bar{p}^K(x_1, \alpha) > 0$ and $\bar{p}^L(x_2, \alpha) > 0$, let $M \geq m_0$, the constant in (4.9), and take $N = K + L + M$. From the definitions it follows that

$$\bar{p}^{K+L+M}(x_1, \alpha) \geq \bar{p}^K(x_1, \alpha)\bar{p}^{L+M}(\alpha, \alpha) > 0$$
$$\bar{p}^{K+L+M}(x_2, \alpha) \geq \bar{p}^L(x_1, \alpha)\bar{p}^{K+M}(\alpha, \alpha) > 0$$

and hence $\hat{p}^N((x_1, x_2), (\alpha, \alpha)) > 0$.

The next step is show that the product chain is recurrent. To do this, note that (4.6) implies $\bar{\pi} = \pi \bar{p}$ is a stationary distribution for $\bar{p}$, and the two coordinates move independently, so

$$\hat{\pi}(C_1 \times C_2) = \bar{\pi}(C_1)\bar{\pi}(C_2)$$

defines a stationary probability distribution for $\hat{p}$, and the desired conclusion follows from (4.8).

To prove the convergence theorem now we will write $\hat{Z}_n = (\bar{X}_n, \bar{Y}_n)$ and consider $\hat{Z}_n$ with initial distribution $\delta_\alpha \times \bar{\pi}$. That is $\bar{X}_0 = \alpha$ with probability 1 and $\bar{Y}_0$ has the stationary distribution $\bar{\pi}$. Let $T = \inf\{n : \hat{Z}_n = (\alpha, \alpha)\} = \hat{R}$. (4.8), (4.7), and (4.5) imply $\bar{\pi} \times \bar{\pi} \ll \hat{\lambda}$ so (4.4) implies

$$P_{\delta_\alpha \times \bar{\pi}}(T < \infty) = 1$$

Dropping the subscript $\delta_\alpha \times \bar{\pi}$ from P and considering the time T,

$$P(\bar{X}_n \in C, T \leq n) = \sum_{m=1}^{n} P(T = m)\bar{p}^{n-m}(\alpha, C) = P(\bar{Y}_n \in C, T \leq n)$$

since on $\{T = m\}$, $X_m = Y_m = \alpha$. Now

$$P(\bar{X}_n \in C) = P(\bar{X}_n \in C, T \leq n) + P(\bar{X}_n \in C, T > n)$$
$$= P(\bar{Y}_n \in C, T \leq n) + P(\bar{X}_n \in C, T > n)$$
$$\leq P(\bar{Y}_n \in C) + P(T > n)$$

Interchanging the roles of X and Y we have

$$P(\bar{Y}_n \in C) \leq P(\bar{X}_n \in C) + P(T > n)$$

and it follows that

$$\|\bar{p}^n(\alpha,\cdot) - \bar{\pi}(\cdot)\| = \|P(\bar{X}_n \in \cdot) - P(\bar{Y}_n \in \cdot)\|$$
$$= \sup_C |P(\bar{X}_n \in C) - P(\bar{Y}_n \in C)| \le P(T > n) \to 0$$

as $n \to \infty$. $\square$

7.5. Convergence Theorems

In this section we will apply the theory of Harris chains developed in the previous section to diffusion processes. To be able to use that theory we will suppose throughout this section that

(5.1) Assumption. MP(b, a) is well posed and the coefficients a and b satisfy (HC) and (UE) locally.

(5.2) Theorem. If $A = B = \{x : |x| \le r\}$ and ρ is Lebesgue measure on B normalized to be a probability measure then X_n is an aperiodic Harris chain.

Proof By (3.8) there is a lower semicontinuous function $p_t(x, y) > 0$ so that

$$P_x(X_t \in A) = \int_A p_t(x, y)\, dy$$

From this we see that $P_x(R = 1) > 0$ for each x so (i) holds. Lower semicontinuity implies $p_1(x, y) \ge \epsilon > 0$ when $x, y \in A$ so (ii) holds and we have $\bar{p}(\alpha, \alpha) > 0$. $\square$

To finish checking the hypotheses of the convergence theorem (4.10), we have to show that a stationary distribution π exists. Our approach will be to show that $E_\alpha R < \infty$, so the construction in (4.5) produces a stationary measure with finite total mass. To check $E_\alpha R < \infty$ we will use a slight generalization of Lemma 5.1 of Khasminskii (1960).

(5.3) Theorem. Suppose $u \ge 0$ is C^2 and has $Lu \le -1$ for all $x \notin K$ where K is a compact set. Let $T_K = \inf\{t > 0 : X_t \in K\}$. Then $u(x) \ge E_x T_K$.

Proof Itô's formula implies that $V_t = u(X_t) + t$ is a local supermartingale on $[0, T_K)$. Letting $T_n \uparrow T_K$ be a sequence of times that reduce V_t, we have

$$u(x) \ge E_x u(n \wedge T_n) + E_x(n \wedge T_n) \ge E_x(n \wedge T_n)$$

Section 7.5 Convergence Theorems 269

Letting $n \to \infty$ and using the monotone convergence theorem the desired result follows. □

Taking $u(x) = |x|^2/\epsilon$ gives the following

(5.4) Corollary. Suppose $\operatorname{tr} a(x) + 2x \cdot b(x) \leq -\epsilon$ for $x \in K^c = \{x : |x| > r\}$ then $E_x T_K \leq |x|^2/\epsilon$.

Though (5.4) was easy to prove, it is remarkably sharp.

Example 5.1. Consider $d = 1$. Suppose $a(x) = 1$ and $b(x) = -c/x$ for $|x| \geq 1$ where $c \geq 0$ and set $K = [-1, 1]$. If $c > 1/2$ then (5.4) holds with $\epsilon = 2c - 1$. To compute $E_x T_K$, suppose $c \neq 1/2$, let $\epsilon = 2c - 1$, and let $u(x) = x^2/\epsilon$. $Lu = -1$ in $(1, b)$ so $u(X_t) + t$ is a local martingale until time $\tau_{(1,b)} = \inf\{t : X_t \notin (1, b)\}$. Using the optional stopping theorem at time $\tau_{(1,b)} \wedge t$ we have

$$\frac{x^2}{\epsilon} = \frac{E_x X_{\tau_{(1,b)} \wedge t}}{\epsilon} - E_x(\tau_{(1,b)} \wedge t)$$

Rearranging, then letting $t \to \infty$ and using the monotone and bounded convergence theorems, we have

$$E_x \tau_{(1,b)} = \frac{x^2}{\epsilon} - \frac{E_x X^2_{\tau_{(1,b)}}}{\epsilon}$$

To evaluate the right-hand side, we need the natural scale and doing this it is convenient to let start from 1 rather than 0:

$$\varphi(x) = \int_1^x \exp\left(\int_1^y \frac{2c}{z} dz\right)$$

$$= C \int_1^x y^{2c} dy = C'(y^{2c+1} - 1)$$

Using (3.2) in Chapter 6 with $a = 1$ it follows that

$$E_x \tau_{(1,b)} = \frac{x^2}{\epsilon} - \frac{b^{2c+1} - x^{2c+1}}{b^{2c+1} - 1} \cdot \frac{1}{\epsilon} - \frac{x^{2c+1} - 1}{b^{2c+1} - 1} \cdot \frac{b^2}{\epsilon}$$

The second term on the right always converges to $-1/\epsilon$ as $b \to \infty$. The third term on the right converges to 0 when $c > 1/2$ and to ∞ when $c < 1/2$ (recall that $\epsilon < 0$ in this case).

Exercise 5.1. Use (4.8) in Chapter 6 to show that $E_x T_K = \infty$ when $c \leq 1/2$.

Exercise 5.2. Consider $d = 1$. Suppose $a(x) = 1$ and $b(x) \le -\epsilon/x^\delta$ for $x \ge 1$ where $\epsilon > 0$ and $0 < \delta < 1$. Let $T_1 = \inf\{t : X_t = 1\}$ and show that for $x \ge 1$, $E_x T_1 \le (x^{1+\delta} - 1)/\epsilon$.

Example 5.2. When $d > 1$ and $a(x) = I$ the condition in (5.4) becomes
$$x \cdot b(x) \le -(d + \epsilon)/2$$
To see this is sharp, let $Y_t = |X_t|$, and use Itô's formula with $f(x) = |x|$ which has
$$D_i f = x_i/|x| \qquad D_{ii} f = 1/|x| - x_i^2/|x|^3$$
to get
$$Y_t - Y_0 = \int_0^t \frac{X_s}{|X_s|} \cdot b(X_s)\, ds + \int_0^t \frac{X_s}{|X_s|} \cdot dW_s + \frac{1}{2} \int_0^t \frac{d-1}{|X_s|}\, ds$$
We have written W here for a d dimensional Brownian motion, so that we can let
$$B_t = \int_0^t \frac{X_s}{|X_s|} \cdot dW_s$$
be a one dimensional Brownian motion B (it is a local martingale with $\langle B \rangle_t = t$) and write
$$dY_t = \frac{1}{|X_t|} \left(X_t \cdot b(X_t) + \frac{d-1}{2} \right) dt + dB_t$$
If $x \cdot b(x) = -(d-1+\epsilon)/2$ this reduces to the previous example and shows that the condition $x \cdot b(x) \le -(d+\epsilon)/2$ is sharp.

The last detail is to make the transition from $E_x T_K < \infty$ to $E_\alpha R < \infty$.

(5.5) Theorem. Let $K = \{x : |x| \le r\}$ and $H = \{x : |x| \le r+1\}$. If $\sup_{x \in H} E_x T_K < \infty$ then $E_\alpha R < \infty$.

Proof Let $U_0 = 0$, and for $m \ge 1$ let $V_m = \inf\{t > U_{m-1} : X_t \in K\}$, $U_m = \inf\{t > V_m : |X_t| \notin H \text{ or } t \in \mathbf{Z}\}$, and $M = \inf\{m \ge 1 : U_m \in \mathbf{Z}\}$. Since $X(U_m) \in A$, $R \le U_M$. To estimate $E_\alpha R$, we note that $X(U_{m-1}) \in A$ so
$$E\left(V_m - U_{m-1} \,\big|\, \mathcal{F}_{U_{m-1}}\right) \le C_0 = \sup_{x \in A} E_x T_K$$
To estimate M, we observe that if τ_H is the exit time from H then (3.6) implies
$$\inf_{x \in K} P_x(\tau_A > 1) \ge |K| \inf_{x,y \in K} p_1^{r+1}(x,y) = \epsilon_0 > 0$$
where $p_t^{r+1}(x,y)$ is the transition probability for the process killed when it leaves the ball of radius $r+1$. From this it follows that $P(M > m) < (1-\epsilon_0)^m$. Since $U_m - V_m \le 1$ we have $E U_M \le (1 + C_0)/\epsilon_0$ and the proof is complete. □

8 Weak Convergence

8.1. In Metric Spaces

We begin with a treatment of weak convergence on a general space S with a **metric** ρ, i.e., a function with (i) $\rho(x,x) = 0$, (ii) $\rho(x,y) = \rho(y,x)$, and (iii) $\rho(x,y)+\rho(y,z) \geq \rho(x,z)$. Open balls are defined by $\{y : \rho(x,y) < r\}$ with $r > 0$ and the **Borel sets** $\mathcal{S}$ are the σ-field generated by the open balls. Throughout this section we will suppose S is a metric space and $\mathcal{S}$ is the collection of Borel sets, even though several results are true in much greater generality. For later use, recall that S is said to be **separable** if there is a countable dense set, and **complete** if every Cauchy sequence converges.

A sequence of probability measures μ_n on $(S,\mathcal{S})$ is said to **converge weakly** if for each bounded continuous function $\int f(x)\mu_n(dx) \to \int f(x)\mu(dx)$. In this case we write $\mu_n \Rightarrow \mu$. In many situations it will be convenient to deal directly with random variables X_n rather than with the associated distributions $\mu_n(A) = P(X_n \in A)$. We say that X_n **converges weakly to** X and write $X_n \Rightarrow X$ if for any bounded continuous function f, $Ef(X_n) \to Ef(X)$.

Our first result, sometimes called the Portmanteau Theorem, gives five equivalent definitions of weak convergence.

(1.1) Theorem. The following statements are equivalent.

(i) $Ef(X_n) \to Ef(X)$ for any bounded continuous function f.

(ii) For all closed sets K, $\limsup_{n\to\infty} P(X_n \in K) \leq P(X \in K)$.

(iii) For all open sets G, $\liminf_{n\to\infty} P(X_n \in G) \geq P(X \in G)$.

(iv) For all sets A with $P(X \in \partial A) = 0$, $\limsup_{n\to\infty} P(X_n \in A) = P(X \in A)$.

(v) Let D_f = the set of discontinuities of f. For all bounded functions with $P(X \in D_f) = 0$, $Ef(X_n) \to Ef(X)$.

Remark. To help remember (ii) and (iii) think about what can happen when $P(X_n = x_n) = 1$, $x_n \to x$, and x lies on the boundary of the set. If K is closed we can have $x_n \notin K$ for all n but $x \in K$. If G is open we can have $x_n \in G$ for all n but $x \notin G$.

Proof We will go around the loop in roughly the order indicated. The last step, (v) implies (i), is trivial so we have four things to show.

(i) implies (ii) Let $\rho(x, K) = \inf\{\rho(x,y) : y \in K\}$, $\psi_j(r) = (1 - jr)^+$ where $z^+ = \max\{z, 0\}$ is the positive part, and let $f_j(x) = \psi_j(\rho(x, K))$. f_j is bounded, continuous, and $\geq 1_K$ so

$$\limsup_{n\to\infty} P(X_n \in K) \leq \lim_{n\to\infty} Ef_j(X_n) = Ef_j(X)$$

Now $f_j \downarrow 1_K$ as $j \uparrow \infty$, so letting $j \to \infty$ gives (ii).

(ii) is equivalent to (iii) This follows immediately from (a) $P(A)+P(A^c) = 1$, and (b) A is open if and only if A^c is closed.

(ii) and (iii) imply (iv) Let $K = \bar{A}$ = the closure of A, $G = A^o$ = the interior of A, and note that $\partial A = K - G$, so our assumptions imply

$$P(X \in K) = P(X \in A) = P(X \in G)$$

Using (ii) and (iii) now with the last equality we have

$$\limsup_{n\to\infty} P(X_n \in A) \leq \limsup_{n\to\infty} P(X_n \in K) \leq P(X \in K) = P(X \in A)$$
$$\liminf_{n\to\infty} P(X_n \in A) \geq \liminf_{n\to\infty} P(X_n \in G) \leq P(X \in G) = P(X \in A)$$

At this point we have shown

$$\liminf_{n\to\infty} P(X_n \in A) \geq P(X \in A) \geq \limsup_{n\to\infty} P(X_n \in A)$$

which with the triviality $\liminf \leq \limsup$ gives the desired result.

(iv) implies (v) Suppose $|f(x)| \leq K$ and pick $\alpha_0 < \alpha_1 < \ldots < \alpha_\ell$ with $\alpha_0 < -K$ and $\alpha_\ell > K$ so that $P(f(X) = \alpha_i) = 0$ for $0 \leq i \leq \ell$ and $\alpha_j - \alpha_{j-1} < \epsilon$ for $1 \leq j \leq \ell$. This is always possible since $\{\alpha : P(X = \alpha) > 0\}$ is a countable set. Let $A_i = \{x : \alpha_{i-1} < f(x) \leq \alpha_i\}$, $\partial A_i \subset \{x : f(x) \in \{\alpha_{i-1}, \alpha_i\}\} \cup D_f$, so $P(X \in \partial A_i) = 0$, and it follows from (iv) that

$$\sum_{i=1}^{\ell} \alpha_i P(X_n \in A_i) \to \sum_{i=1}^{\ell} \alpha_i P(X \in A_i)$$

The definition of the α_i implies

$$0 \le \sum_{i=1}^{\ell} \alpha_i P(X_n \in A_i) - Ef(X_n) \le \epsilon$$

and this inequality holds with X in place of X_n. Combining our conclusions we have

$$\limsup_{n\to\infty} |Ef(X_n) - Ef(X)| \le 2\epsilon$$

but ϵ is arbitrary so the proof is complete. □

An important consequence of the (1.1) is

(1.2) Continuous mapping theorem. Suppose μ_n on $(S, \mathcal{S})$ converge weakly to μ, and let $\varphi : S \to S'$ have a discontinuity set D_φ with $\mu(D_\varphi) = 0$. Then $\mu_n \circ \varphi^{-1} \Rightarrow \mu \circ \varphi^{-1}$.

Proof Let $f : S' \to \mathbf{R}$ be bounded and continuous. Since $D_{f\circ\varphi} \subset D_\varphi$, it follows from (v) in (1.1) that

$$\int \mu_n(dx) f(\varphi(x)) \to \int \mu(dx) f(\varphi(x))$$

Changing variables gives

$$\int \mu_n \circ \varphi^{-1}(dx) f(x) \to \int \mu \circ \varphi^{-1}(dx) f(x)$$

Since this holds for any bounded continuous f, the desired result follows. □

In checking convergence in distribution the following lemma is sometimes useful.

(1.3) Converging together lemma. Suppose $X_n \Rightarrow X$ and $\rho(X_n, Y_n) \to 0$ in probability then $Y_n \Rightarrow X$.

Proof Let K be a closed set and $K_\delta = \{y : \rho(y, K) \le \delta\}$, where $\rho(y, K) = \inf\{\rho(y, z) : z \in K\}$. It is easy to see that $z \notin K_\delta$ if and only if there is a $\gamma > \delta$ so that $B(z, \gamma) \cap K = \emptyset$, so $(K_\delta)^c$ is open and K_δ is closed. With this little detail out of the way the rest is easy.

$$P(Y_n \in K) \le P(X_n \in K_\delta) + P(\rho(X_n, Y_n) > \delta)$$

274 Chapter 8 Weak Convergence

Letting $n \to \infty$, noticing the second term on the right goes to 0 by assumption, and using (ii) in (1.1) we have

$$\limsup_{n\to\infty} P(Y_n \in K) \leq \limsup_{n\to\infty} P(X_n \in K_\delta) \leq P(X \in K_\delta)$$

Measure theory tells us that $P(X \in K_\delta) \downarrow P(X \in K)$ as $\delta \downarrow 0$ so we have shown

$$\limsup_{n\to\infty} P(Y_n \in K) \leq P(X \in K)$$

The desired conclusion now follows from (1.1). □

The next result, due to Skorokhod, is useful for proving results about a sequence that converges weakly, because it converts weak convergence into almost sure convergence.

(1.4) Skorokhod's representation theorem. Suppose S is a complete separable metric space. If $\mu_n \Rightarrow \mu_\infty$ then we can define random variables Y_n on $[0,1]$ (equipped with the Borel sets and Lebesgue measure λ) with distribution μ_n so that $Y_n \to Y_\infty$ almost surely.

Remark. When $S = \mathbf{R}$ this is easy. Let $F_n(x) = \mu_n(-\infty, x]$ be the distribution function, let

$$F_n^{-1}(y) = \inf\{x : F_n(x) \geq y\}$$

let $U(\omega) = \omega$ for $\omega \in [0,1]$ be a random variable that is uniform on $(0,1)$, and let $Y_n = F_n^{-1}(U)$. Then $Y_n \to Y_\infty$ almost surely. (For a proof, see (2.1) of Chapter 2 of Durrett (1995).) We invite the reader to contemplate the question: "Is there an easy way to do this when $S = \mathbf{R}^2$ (or even $S = [0,1]^2$)?" while we give the general proof. The details are somewhat messy and not important for the developments that follow, so if the reader finds herself asking "Who cares?", she can skip to the beginning of the next section.

Proof We will warm up for the real proof by treating the special case $\mu_n \equiv \mu$.

(1.5) Theorem. For each probability measure μ there is a random variable Y defined on $[0,1]$ with $P(Y \in A) = \mu(A)$.

Proof For each k construct a decomposition $\mathcal{A}_k = \{A_{k,1}, A_{k,2}, \ldots\}$ of S into disjoint sets of diameter less than $1/k$ and so that $\mathcal{A}_k$ refines $\mathcal{A}_{k-1}$, i.e., each set $A_{k-1,i}$ is a union of sets in $\mathcal{A}_k$. For each k construct a corresponding decomposition $\mathcal{I}_k$ of the unit interval so that $\lambda(I_{k,j}) = \mu(A_{k,j})$ and arrange things so that $A_{k-1,i} \supset A_{k,j}$ if and only if $I_{k-1,i} \supset I_{k,j}$.

Let $x_{k,j}$ be some point in $A_{k,j}$ and let $X_{k,j}(\omega) = x_{k,j}$ for $\omega \in I_{k,j}$. Since $X_k, X_{k+1}, \ldots$ is contained in some element of $\mathcal{A}_k$, which by definition has diameter $\leq 1/k$, it is a Cauchy sequence. So $X(\omega) = \lim_k X_k(\omega)$ exists and satisfies $\rho(X_k, X) \leq 1/k$. To show that X has distribution μ we let K be a closed set, let $S(k, K) = \{j : A_{k,j} \cap K \neq \emptyset\}$, and let $K_\delta = \{x : \rho(x, K) \leq \delta\}$ as in the proof of (1.3).

$$\lambda(X_k \in K) \leq \lambda(X_k(\omega) \in \bigcup_{j \in S(k,K)} A_{k,j})$$
$$\leq \sum_{j \in S(k,K)} \lambda(X_k(\omega) \in A_{k,j})$$
$$= \sum_{j \in S(k,K)} \lambda(I_{k,j}) = \sum_{j \in S(k,K)} \mu(A_{k,j}) \leq \mu(K_{1/k})$$

Letting $k \to \infty$ and noting $K_{1/k} \downarrow K$ it follows that

$$\limsup_k \lambda(X_k \in K) \leq \mu(K)$$

So (1.1) implies that $X_k \Rightarrow \mu$ and it follows that X has distribution μ. □

Proof of (1.4) Construct a decomposition $\mathcal{A}_k$ of the space S as in the preceding proof but this time require that each $A_{k,j}$ has $\mu_\infty(\partial A_{k,j}) = 0$. For each n construct decompositions $\mathcal{I}_{k,j}^n$ of $[0,1]$ so that $\lambda(I_{k,j}^n) = \mu_n(A_{k,j})$. To arrange the intervals in an appropriate way we introduce an ordering in which $[a,b] < [c,d]$ if $b \leq c$ and demand that $I_{k,i}^n < I_{k,j}^n$ if and only if $I_{k,i}^\infty < I_{k,j}^\infty$. Let $x_{k,j} \in A_{k,j}$ and define $X_k^n(\omega) = x_{k,j}$ for $\omega \in I_{k,j}^n$. As before, $X^n = \lim_{k \to \infty} X_k^n$ exists and has distribution μ_n.

Since $\sum_j \mu_\infty(A_{k,j}) - \mu_n(A_{k,j}) = 1 - 1 = 0$, we have

$$\sum_j |\lambda(I_{k,j}^\infty) - \lambda(I_{k,j}^n)| = \sum_j |\mu_\infty(A_{k,j}) - \mu_n(A_{k,j})|$$
$$= 2 \sum_j (\mu_\infty(A_{k,j}) - \mu_n(A_{k,j}))^+$$

where $y^+ = \max\{y, 0\}$ is the positive part of y. Since the $\mu_\infty(\partial A_{k,j}) = 0$, (1.1) implies $\mu_n(A_{k,j}) \to \mu_\infty(A_{k,j})$ and the dominated convergence theorem implies

(1.6) $$\sum_j |\lambda(I_{k,j}^\infty) - \lambda(I_{k,j}^n)| \to 0 \quad \text{as } n \to \infty$$

Fix k and j_0, let α_n be the left endpoint of I_{k,j_0}^n, and let $S(j_0) = \{j : I_{k,j}^n < I_{k,j_0}^n\}$ which by our construction does not depend on n. (1.6) implies that

$$\alpha_\infty = \sum_{j \in S(j_0)} |I_{k,j}^\infty| = \lim_{n \to \infty} \sum_{j \in S(j_0)} |I_{k,j}^n| = \lim_{n \to \infty} \alpha_n$$

Similarly if we let β_n be the right endpoint of I^n_{k,j_0} then $\beta_n \to \beta_\infty$ as $n \to \infty$. Hence if ω is in the interior of $I^\infty_{k,j}$ it lies in the interior of $I^n_{k,j}$ for large n, and it follows from the definition of the X_n that

$$\limsup_{n \to \infty} \rho(X_n, X_\infty) \leq 2/k$$

Letting $k \to \infty$ we see that if ω is not one of the countable set of points that is the end point of some $I^\infty_{k,j}$ we have $X_n \to X_\infty$, which proves (1.4). □

8.2. Prokhorov's Theorems

Let Π be a family of probability measures on a metric space (S, ρ). We call Π **relatively compact** if every sequence μ_n of elements of Π contains a subsequence μ_{n_k} that converges weakly to a limit μ, which may be $\notin \Pi$. We call Π **tight** if for each $\epsilon > 0$ there is a compact set K so that $\mu(K) \geq 1 - \epsilon$ for all $\mu \in \Pi$. This section is devoted to the proofs of

(2.1) Theorem. If Π is tight then it is relatively compact.

(2.2) Theorem. Suppose S is complete and separable. If Π is relatively compact, it is tight.

The first result is the one we want for applications, but the second one is comforting since it says that, in nice spaces, the two notions are equivalent.

Proof of (2.1) We shall prove the result successively for $\mathbf{R}^d$, $\mathbf{R}^\infty$, a countable union of compact sets, and finally, a general S. This approach is somewhat lengthy but to inspire the reader for the effort, we note that the first two special cases are important examples, and then the last two steps and the converse are remarkably easy.

Before we begin we will prove a lemma that will be useful for the first two examples.

(2.3) Lemma. Let $\mathcal{A}$ be a subclass of $\mathcal{S}$ that is closed under intersection, and so that each open set is a countable union of elements of $\mathcal{A}$. If $\mu_n(A) \to \mu(A)$ for all $A \in \mathcal{A}$ then $\mu_n \Rightarrow \mu$.

Proof The inclusion-exclusion formula says

$$P(\cup_{i=1}^m A_i) = \sum_{i=1}^m P(A_i) - \sum_{i<j} P(A_i \cap A_j) + \cdots + (-1)^{m+1} P(\cap_{i=1}^m A_i)$$

Combining this with the fact that $\mathcal{A}$ is closed under intersection it is easy to see that $\mu_n(\cup_{i=1}^k A_i) \to \mu(\cup_{i=1}^k A_i)$. Given an open set G and an $\epsilon > 0$, choose sets $A_1, \ldots A_k$ so that $\mu(\cup_{i=1}^k A_i) > \mu(G) - \epsilon$. Using the convergence just proved it follows that

$$\mu(G) - \epsilon < \mu(\cup_{i=1}^k A_i) \le \lim_{n \to \infty} \mu_n(\cup_{i=1}^k A_i) \le \liminf_{n \to \infty} \mu_n(G)$$

Since ϵ is arbitrary, (iii) in (1.1) follows and the proof is complete. □

Proof for $\mathbf{R}^d$ This is a fairly straightforward generalization of the argument for $d = 1$. (See e.g., (2.5) in Chapter 2 of Durrett (1995).) In this case we can rely on the distribution functions $F(x) = \mu(\{y \le x\})$, where for two vectors $y \le x$ means $y_i \le x_i$ for each i. Let $\mathbf{Q}^d$ be the points in $\mathbf{R}^d$ with rational coordinates. By enumerating the points in $\mathbf{Q}^d$ and then using a diagonal argument, we can find a subsequence F_{n_k} so that $F_{n_k}(q)$ converges for each $q \in \mathbf{Q}^d$. Call the limit $G(q)$ and define the proposed limiting distribution function by

$$F(x) = \inf\{G(q) : q \in \mathbf{Q}^d \text{ and } q > x\}$$

where $q > x$ means $q_i > x_i$ for each i.

To check that F is a distribution function we note that it is an immediate consequence of the definition that

(i) if $x \le y$ then $F(x) \le F(y)$

(ii) $\lim_{y \downarrow x} F(y) = F(x)$

where $y \downarrow x$ means $y_i \downarrow x_i$ for each i. The third condition for being a distribution function is easier to prove than it is to state:

(iii) let $A = (a_1, b_1] \times \cdots \times (a_d, b_d]$ be a rectangle with vertices $V = \{a_1, b_1\} \times \cdots \times \{a_d, b_d\}$, and let $\text{sgn}(v) = (-1)^{n(a,v)}$ where $n(a, v)$ is the number of a's that appear in v. The inclusion-exclusion formula implies that the probability of A is $\sum_{v \in V} \text{sgn}(v) F(v) \ge 0$.

The fact that (iii) holds for the F_n's implies that it holds for G and by taking limits we see that it holds for F.

Conditions (i)-(iii) imply that F is the distribution function of a measure on $\mathbf{R}^d$, which in our case will always have total mass ≤ 1. Define the ith marginal distribution function of F by $F^i(x) = \lim F(y)$ where the limit is taken through vectors y with $y_i = x$ and the other coordinates $y_j \to \infty$. F^i is a one dimensional distribution function and hence its discontinuities D^i are a countable set.

Call x a good point if $x_i \notin D^i$ for each i. We claim that F is continuous at good points. To see this, let $w < x < y$, let z_i be the vector which uses the first

i components from y and the last $d-i$ from w and note that the monotonicity properties that come from the interpretation of F as a measure imply

$$F(y) - F(w) = \sum_{i=1}^{d} F(z_i) - F(z_{i-1}) \leq \sum_{i=1}^{d} F_i(y_i) - F_i(w_i)$$

The continuity of F now follows from the monotonicity and the continuity of the F_i.

Our next claim is that for good x, $F_{n_k}(x) \to F(x)$. To prove this pick $p, q, r \in \mathbf{Q}^d$ with $p < q < x < r$ and

$$F(x) - \epsilon \leq F(p) \leq F(q) \leq F(x) \leq F(r) \leq F(x) + \epsilon$$

Since $u < v$ implies $F(u) \leq G(v) \leq F(v)$ it follows that

$$F(x) - \epsilon \leq G(q) \leq F(x) \leq G(r) \leq F(x) + \epsilon$$

Using $F_n(q) \leq F_n(x) \leq F_n(r)$ and the convergence $F_{n_k}(q) \to G(q)$ for $q \in \mathbf{Q}^d$ the claim follows easily.

Up to this point we have not used the tightness assumption. It enters into the proof only to guarantee that F is the distribution function of a probability measure. Given ϵ choose a compact set K so that $F_n(K) \geq 1 - \epsilon$ for all n, where $F_n(K)$ is short for the probability of K under the measure associated with the distribution function F_n. Now pick a good rectangle A, i.e., one with $a_i, b_i \notin D^i$ for all i, so that $K \subset A$. The formula in (iii) for the probability of A and the previous claim about convergence at good points imply

$$F(A) = \lim_{k \to \infty} F_{n_k}(A) \geq \liminf_{n \to \infty} F_n(K) \geq 1 - \epsilon$$

Finally, we have to prove that $F_{n_k} \Rightarrow F$. For this we use the following result which is of interest in its own right.

(2.4) Theorem. In $\mathbf{R}^d$ probability measure $\mu_n \Rightarrow \mu$ if and only if the associated distribution functions have $F_n(x) \to F(x)$ for all x that are good for F.

Proof If x is good for F, then $A = \{y : y \leq x\}$ has $\mu(\partial A) = 0$, so if $\mu_n \Rightarrow \mu$ then (iv) of (1.1) implies $F_n(x) \to F(x)$. To prove the converse we will apply (2.3) to the collection $\mathcal{A}$ of finite disjoint unions of good rectangles. (i.e., rectangles, all of whose vertices are good). We have already observed that $\mu_n(A) \to \mu(A)$ for all $A \in \mathcal{A}$. To complete the proof we note that $\mathcal{A}$ is closed under intersection, and any open set is a countable disjoint union of good rectangles. □

Proof for $\mathbf{R}^\infty$ Let $\mathbf{R}^\infty$ be the collection of all sequences $(x_1, x_2, \ldots)$ of real numbers. To define a metric on $\mathbf{R}^\infty$ we introduce the bounded metric $\rho_o(x,y) = |x-y|/(1+|x-y|)$ on $\mathbf{R}$ and let

$$\rho(x,y) = \sum_{i=1}^{\infty} 2^{-i}\rho_o(x_i, y_i)$$

It is easy to see that the induced topology is that of coordinatewise convergence, (i.e., $x^n \to x$ if and only if $x_i^n \to x_i$) for each i and hence this metric makes $\mathbf{R}^\infty$ complete and separable. A countable dense set is the collection of points with finitely many nonzero coordinates, all of which are rational numbers.

(2.5) Lemma. The Borel sets $\mathcal{S} = \mathcal{R}^\infty$, the product σ-field gnerated by the finite dimensional sets $\{x : x_i \in A_i \text{ for } 1 \le i \le k\}$.

Proof To argue that $\mathcal{S} \supset \mathcal{R}^\infty$ observe that if G_i are open then $\{x : x_i \in G_i \text{ for } 1 \le i \le k\}$ is open. For the other inclusion note that

$$\{y : \rho(x,y) \le \delta\} = \cap_{n=1}^{\infty} \left\{ y : \sum_{m=1}^{n} 2^{-m}\rho_o(x_m, y_m) \le \delta \right\} \in \mathcal{R}_\infty$$

so $\{y : \rho(x,y) < \gamma\} = \cup_{n=1}^{\infty}\{y : \rho(x,y) \le \gamma - 1/n\} \in \mathcal{R}^\infty$. □

Let $\pi_d : \mathbf{R}^\infty \to \mathbf{R}^d$ be the projection $\pi_d(x) = (x_1, \ldots, x_d)$. Our first claim is that if Π is a tight family on $(\mathbf{R}^\infty, \mathcal{R}^\infty)$ then $\{\mu \circ \pi_d^{-1} : \mu \in \Pi\}$ is a tight family on $(\mathbf{R}^d, \mathcal{R}^d)$. This follows from the following general result.

(2.6) Lemma. If Π is a tight family on $(S, \mathcal{S})$ and if h is a continuous map from S to S' then $\{\mu \circ h^{-1} : \mu \in \Pi\}$ is a tight family on $(S', \mathcal{S}')$.

Proof Given ϵ, choose in S a compact set K so that $\mu(K) \ge 1-\epsilon$ for all $\mu \in \Pi$. If $K' = h(K)$ then K' is compact and $h^{-1}(K') \supset K$ so $\mu \circ h^{-1}(K') \ge 1 - \epsilon$ for all $\mu \in \Pi$. □

Using (2.6), the result for $\mathbf{R}^d$, and a diagonal argument we can, for a given sequence μ_n, pick a subsequence μ_{n_k} so that for all d, $\mu_{n_k} \circ \pi_d$ converges to a limit ν_d. Since the measures ν_d obviously satisfy the consistency conditions of Kolmogorov's existence theorem, there is a probability measure ν on $(\mathbf{R}^\infty, \mathcal{R}^\infty)$ so that $\nu \circ \pi_d^{-1} = \nu_d$.

We now want to check that $\mu_{n_k} \Rightarrow \nu$. To do this we will apply (2.3) with $\mathcal{A} = $ the finite dimensional sets A with $\nu(\partial A) = 0$. Convergence of finite dimensional distributions and (1.1) imply that $\mu_{n_k}(A) \to \nu(A)$ for all $A \in \mathcal{A}$.

280   Chapter 8   Weak Convergence

The intersection of two finite dimensional sets is a finite dimensional set and $\partial(A \cap B) \subset \partial A \cup \partial B$ so $\mathcal{A}$ is closed under intersection.

To prove that any open set G is a countable union of sets in A, we observe that if $y \in G$ there is a $\delta > 0$ so that $B(y, \delta) = \{x : \rho(x, y) < \delta\} \subset G$. Pick δ so large that $B(y, 2\delta) \cap G^c \neq \emptyset$. ($G = \mathbf{R}^\infty \in \mathcal{A}$ so we can suppose $G^c \neq \emptyset$.) If we pick N so that $2^{-N} < \delta/2$ then it follows from the definition of the metric that for $r < \delta/2$ the finite dimensional set

$$A(y, r, N) = \{y : |x_i - y_i| < r \text{ for } 1 \le i \le N\} \subset B(y, \delta) \subset G$$

The boundaries of these sets are disjoint for different values of r so we can pick $r > \delta/4$ so that the boundary of $A(y, r, N)$ has ν measure 0.

As noted above $\mathbf{R}^\infty$ is separable. Let $y_1, y_2, \ldots$ be an enumeration of the members of the countable dense set that lie in G and let $A_i = A(y_i, r_i, N_i)$ be the sets chosen in the last paragraph. To prove that $\cup_i A_i = G$ suppose $x \in G - \cup_i A_i$, let $\gamma > 0$ so that $B(x, \gamma) \subset G$, and pick a point y_i so that $B(x, y_i) < \gamma/9$. We claim that $x \in A_i$. To see this, note that the triangle inequality implies $B(y_i, 8\gamma/9) \subset G$, so $\delta \ge 4\gamma/9$ and $r > \delta/4 = \gamma/9$, which completes the proof of $\mu_{n_k} \Rightarrow \nu$.   □

Before passing to the next case we would like to observe that we have shown

(2.7) Theorem. In $\mathbf{R}^\infty$, weak convergence is equivalent to convergence of finite dimensional distributions.

Proof for the σ-compact case   We will prove the result in this case by reducing it to the previous one. We start by observing

(2.8) Lemma. If S is σ-compact then S is separable.

Proof   Since a countable union of countable sets is countable, it suffices to prove this if S is compact. To do this cover S by balls of radius $1/n$, let x_m^n, $m \le m_n$ be the centers of the balls of a finite subcover, and check that the x_m^n are a countable dense set.   □

(2.9) Lemma. If S is a separable metric space then it can be embedded homeomorphically into $\mathbf{R}^\infty$.

Proof   Let $q_1, q_2, \ldots$ be a sequence of points dense in S and define a mapping from S into $\mathbf{R}^\infty$ by

$$h(x) = (\rho(x, q_1), \rho(x, q_2), \ldots)$$

Section 8.2 Prokhorov's Theorems   281

If the points $x_n \to x$ in S then $\lim \rho(x_n, q_i) = \rho(x, q_i)$ and hence $h(x_n) \to h(x)$.

Suppose $x \neq x'$ and let $\epsilon = \rho(x, x')$. If $\rho(x, q_i) < \epsilon/2$, which must be true for some i, then $\rho(x', q_i) > \epsilon/2$ or the triangle inequality would lead to the contradiction $\epsilon < \rho(x, x')$. This shows that h is 1-1.

Our final task is to show that h^{-1} is continuous. If x_n does not converge to x then $\limsup \rho(x_n, x) = \epsilon > 0$. If $\rho(x, q_i) < \epsilon/2$ which must be true for some q_i then $\limsup \rho(x_n, q_i) \geq \epsilon/2$, or again the triangle inequality would give a contradicition, and hence $h(x_n)$ does not converge to $h(x)$. This shows that if $h(x_n) \to h(x)$ then $x_n \to x$ so h^{-1} is continuous. □

It follows from (2.6) that if Π is tight on S then $\{\mu \circ h^{-1} : \mu \in \Pi\}$ is a tight family of measures on $\mathbf{R}^\infty$. Using the result for $\mathbf{R}^\infty$ we see that if μ_n is a sequence of measures in Π and we let $\nu_n = \mu_n \circ h^{-1}$ then there is a convergent subsequence ν_{n_k}. Applying the continuous mapping theorem, (1.2), now to the function $\varphi = h^{-1}$ it follows that $\mu_{n_k} = \nu_{n_k} \circ h$ converges weakly.

The general case Whatever S is, if Π is tight, and we let K_i be so that $\mu(K_i) \geq 1 - 1/i$ for all $\mu \in \Pi$ then all the measures are supported on the σ-compact set $S_0 = \cup_i K_i$ and this case reduces to the previous one. □

Proof of (2.2) We begin by introducing the intermediate statement:

(H) For each $\epsilon, \delta > 0$ there is a finite collection $A_1, \ldots, A_n$ of balls of radius δ so that $\mu(\cup_{i \leq n} A_i) \geq 1 - \epsilon$ for all $\mu \in \Pi$.

We will show that (i) if (H) holds then Π is tight and (ii) if (H) fails then Π is not relativley compact. Combining (ii) and (i) gives (2.2).

Proof of (i) Fix ϵ and choose for each k finitely many balls $A_1^k, \ldots, A_{n_k}^k$ of radius $1/k$ so that $\mu(\cup_{i \leq n_k} A_i^k) \geq 1 - \epsilon/2^k$ for all $\mu \in \Pi$. Let K be the closure of $\cap_{k=1}^\infty \cup_{i \leq n_k} A_i^k$. Clearly $\mu(K) \geq 1 - \epsilon$. K is totally bounded since if B_j^k is a ball with the same center as A_j^k and radius $2/k$ then B_j^k, $1 \leq j \leq n_k$ covers K. Being a closed and totally bounded subset of a complete space, K is compact and the proof is complete. □

Proof of (ii) Suppose (H) fails for some ϵ and δ. Enumerate the members of the countable dense set $q_1, q_2, \ldots$, let $A_i = B(q_i, \delta)$, and $G_n = \cup_{i=1}^n A_i$. For each n there is a measure $\mu_n \in \Pi$ so that $\mu_n(G_n) < 1 - \epsilon$. We claim that μ_n has no convergent subsequence. To prove this, suppose $\mu_{n_k} \Rightarrow \nu$. (iii) of (1.1) implies that for each m

$$\nu(G_m) \leq \limsup_{k \to \infty} \mu_{n_k}(G_m)$$
$$\leq \limsup_{k \to \infty} \mu_{n_k}(G_{n_k}) \leq 1 - \epsilon$$

However, $G_m \uparrow S$ as $m \uparrow \infty$ so we have $\nu(S) \leq 1 - \epsilon$, a contradiction. □

8.3. The Space C

Let $C = C([0,1], \mathbf{R}^d)$ be the space of continuous functions from $[0,1]$ to $\mathbf{R}^d$ equipped with the norm
$$\|\omega\| = \sup_{t \leq 1} |\omega(t)|$$
Let $\mathcal{B}$ be the collection of Borel subsets of C. Introduce the coordinate random variables $X_t(\omega) = \omega(t)$ and define the finite dimensional sets by
$$\{\omega : X_{t_i}(\omega) \in A_i \text{ for } 1 \leq i \leq k\}$$
where $0 \leq t_1 < t_2 < \ldots < t_k \leq 1$ and $A_i \in \mathcal{R}^d$, the Borel subsets of $\mathbf{R}^d$.

(3.1) Lemma. $\mathcal{B}$ is the same as the σ-field $\mathcal{C}$ generated by the finite dimensional sets.

Proof Observe that if ξ is a given continuous function
$$\{\omega : \|\omega - \xi\| \leq \epsilon - 1/n\} = \cap_q \{\omega : |\omega(q) - \xi(q)| \leq \epsilon - 1/n\}$$
where the intersection is over all rationals $q \in [0,1]$. Letting $n \to \infty$ shows $\{\omega : \|\omega - \xi\| < \epsilon\} \in \mathcal{C}$ and $\mathcal{B} \subset \mathcal{C}$. To prove the reverse inclusion observe that if the A_i are open the finite dimensional set $\{\omega : \omega(t_i) \in A_i\}$ is open so the $\pi - \lambda$ theorem implies $\mathcal{C} \subset \mathcal{B}$. □

Let $0 \leq t_1 < t_2 < \ldots < t_n \leq 1$ and $\pi_t : C \to (\mathbf{R}^d)^n$ be defined by
$$\pi_t(\omega) = (\omega(t_1), \ldots, \omega(t_n))$$
Given a measure μ on $(C, \mathcal{C})$, the measures $\mu \circ \pi_t^{-1}$, which give the distribution of the vectors $(X_{t_1}, \ldots, X_{t_n})$ under μ, are called the **finite dimensional distributions** or f.d.d.'s for short. On the basis of (3.1) one might hope that convergence of the f.d.d.'s might be enough for weak convergence of the μ_n. However, a simple example shows this is false.

Example 3.1. Let $a_n = 1/2 - 1/2n$, $b_n = 1/2 - 1/4n$, and let μ_n be the point mass on the function
$$f_n(t) = \begin{cases} 0 & x \in [0, a_n] \\ 4n(x - a_n) & x \in [a_n, b_n] \\ 4n(1/2 - x) & x \in [b_n, 1/2] \\ 0 & x \in [1/2, 1] \end{cases}$$

Section 8.3 The Space C   283

As $n \to \infty$, $f_n(t) \to f_\infty \equiv 0$ but not uniformly. To see that μ_n does not converge weakly to μ_∞, note that $h(\omega) = \sup_{0 \le t \le 1} \omega(t)$ is a continuous function but

$$\int h(\omega)\mu_n(d\omega) = 1 \not\to 0 = \int h(\omega)\mu_\infty(d\omega)$$

(3.2) Theorem. Let $\mu_n, 1 \le n \le \infty$ be probability measures on $\mathcal{C}$. If the finite dimensional distributions of μ_n converge to those of μ_∞ and if the μ_n are tight then $\mu_n \Rightarrow \mu_\infty$.

Proof If μ_n is tight then by (2.1) it is relatively compact and hence each subsequence μ_{n_m} has a further subsequence $\mu_{n'_m}$ that converges to a limit ν. If $f: \mathbf{R}^k \to \mathbf{R}$ is bounded and continuous then $f(X_{t_1}, \ldots X_{t_k})$ is bounded and continuous from $\mathcal{C}$ to $\mathbf{R}$ and hence

$$\int f(X_{t_1}, \ldots X_{t_k})\mu_{n'_m}(d\omega) \to \int f(X_{t_1}, \ldots X_{t_k})\nu(d\omega)$$

The last conclusion implies that $\mu_{n'_k} \circ \pi_t^{-1} \Rightarrow \nu \circ \pi_t^{-1}$, so from the assumed convergence of the f.d.d.'s we see that the f.d.d's of ν are determined and hence there is only one subsequential limit.

To see that this implies that the whole sequence converges to ν, we use the following result, which as the proof shows, is valid for weak convergence on any space. To prepare for the proof we ask the reader to do

Exercise 3.1. Let r_n be a sequence of real numbers. If each subsequence of r_n has a further subsequence that converges to r then $r_n \to r$.

(3.3) Lemma. If each subsequence of μ_n has a further subsequence that converges to ν then $\mu_n \Rightarrow \nu$.

Proof Note that if f is a bounded continuous function, the sequence of real numbers $\int f(\omega)\mu_n(d\omega)$ have the property that every subsequence has a further subsequence that converges to $\int f(\omega)\nu(d\omega)$. Exercise 3.1 implies that the whole sequence of real numbers converges to the indicated limit. Since this holds for any bounded continuous f the desired result follows. □

As Example 3.1 suggests, μ_n will not be tight if it concentrates on paths that oscillate too much. To find conditions that guarantee tightness, we introduce the **modulus of continuity**

$$w_\delta(\omega) = \sup\{|\omega(s) - \omega(t)| : |s - t| \le \delta\}$$

(3.4) Theorem. The sequence μ_n is tight if and only if for each $\epsilon > 0$ there are n_0, M and δ so that

(i) $\mu_n(|\omega(0)| > M) \le \epsilon$ for all $n \ge n_0$

(ii) $\mu_n(w_\delta > \epsilon) \le \epsilon$ for all $n \ge n_0$

Remark. Of course by increasing M and decreasing δ we can always check the condition with $n_0 = 1$ but the formulation in (3.4) eliminates the need for that final adjustment. Also by taking $\epsilon = \eta \wedge \zeta$ it follows that if μ_n is tight then there is a δ and an n_0 so that $\mu_n(w_\delta > \eta) \le \zeta$ for $n \ge n_0$.

Proof We begin by recalling (see e.g., Royden (1988), page 169)

(3.5) The Arzela-Ascoli Theorem. A subset A of C has compact closure if and only if $\sup_{\omega \in A} |\omega(0)| < \infty$ and $\lim_{\delta \to 0} \sup_{\omega \in A} w_\delta(\omega) = 0$.

To prove the necessity of (i) and (ii), we note that if μ_n is tight and $\epsilon > 0$ we can choose a compact set K so that $\mu_n(K) \ge 1 - \epsilon$ for all n. By (3.5), $K \subset \{X(0) \le M\}$ for large M and if $\epsilon > 0$ then $K \subset \{w_\delta \le \epsilon\}$ for small δ.

To prove the sufficiency of (i) and (ii), choose M so that $\mu_n(|X(0)| > M) \le \epsilon/2$ for all n and choose δ_k so that $\mu_n(w_{\delta_k} > 1/k) \le \epsilon/2^{k+1}$ for all n. If we let K be the closure of $\{X(0) \le M, w_{\delta_k} \le 1/k$ for all $k\}$ then (3.5) implies K is compact and $\mu_n(K) \ge 1 - \epsilon$ for all n. □

Condition (i) is usually easy to check. For example it is trivial when $X_n(0) = x_n$ and $x_n \to x$ as $n \to \infty$. The next result, (3.6), will be useful in checking condition (ii). Here, we formulate the result in terms of random variables X_n taking values in C, that is, to say random processes $\{X_n(t), 0 \le t \le 1\}$. However, one can easily rewrite the result in terms of their distributions $\mu_n(A) = P(X_n \in A)$. Here, $A \in \mathcal{C}$.

(3.6) Theorem. Suppose for some $\alpha, \beta > 0$

$$E|X_n(t_2) - X_n(t_1)|^\beta \le K|t_2 - t_1|^{1+\alpha}$$

then (ii) in (3.4) holds.

Remark. The condition should remind the reader of Kolmogorov's continuity criterion, (1.6) in Chapter 1. The key to our proof is the observation that the proof of (1.6) gives quantitative estimates on the modulus of continuity. For a much different approach to a slightly more general result, see Theorem 12.3 in Billingsley (1968).

Proof Let $\gamma < \alpha/\beta$, and pick $\eta > 0$ small enough so that

$$\lambda = (1-\eta)(1+\alpha-\beta\gamma) - (1+\eta) > 0$$

From (1.7) in Chapter 1 it follows that if $A = 3 \cdot 2^{(1-\eta)\gamma}/(1-2^{-\gamma})$ then with probability $\geq 1 - K2^{-N\lambda}/(1-2^{-\lambda})$ we have

$$|X_n(q) - X_n(r)| \leq A|q-r|^\gamma \quad \text{for } q,r \in Q_2 \cap [0,1] \text{ with } 2^{-(1-\eta)N}$$

Pick N so that $K2^{-N\lambda}/(1-2^{-\lambda}) \leq \epsilon$ then $\delta \leq 2^{-(1-\eta)N}$ so that $A\delta^\gamma \leq \epsilon$ and it follows that $P(w_\delta > \epsilon) \leq \epsilon$. □

8.4. Skorokhod's Existence Theorem for SDE

In this section, we will describe Skorokhod's approach to constructing solutions of stochastic differential equations. We will consider the special case

$$(\star) \qquad dX_t = \sigma(X_t)dB_t$$

where σ is bounded and continuous, since we can introduce the term $b(X_t)\,dt$ by change of measure. The new feature here is that σ is only assumed to be continuous, not Lipschitz or even Hölder continuous. Examples at the end of Section 5.3 show that we cannot hope to have uniqueness in this generality.

Skorokhod's idea for solving stochastic differential equations was to discretize time to get an equation that is trivial to solve, and then pass to the limit and extract subsequential limits to solve the original equation. For each n, define $X_n(t)$ by setting $X_n(0) = x$ and, for $m2^{-n} < t \leq (m+1)2^{-n}$,

$$X_n(t) = X_n(m2^{-n}) + \sigma(X_n(m2^{-n}))(B_t - B(m2^{-n}))$$

Since X_n is a stochastic integral with respect to Brownian motion, the formula for covariance of stochastic integrals implies

$$\langle X_n^i, X_n^j \rangle_t = \sum_k \int_0^t (\sigma_{ik}\sigma_{jk})(X_n([2^n s]/2^n))ds$$

$$= \int_0^t a_{ij}(X_n([2^n s]/2^n))ds$$

where as usual $a = \sigma\sigma^T$. If we suppose that $a_{ij}(x) \leq M$ for all i,j and x, it follows that if $s < t$, then

$$|\langle X_n^i, X_n^j \rangle_t - \langle X_n^i, X_n^j \rangle_s| \leq M(t-s)$$

so (5.1) in Chapter 3 implies

$$E \sup_{u\in[s,t]} |X_n^i(u) - X_n^i(s)|^p \leq C_p E|\langle X_n^i\rangle_t - \langle X_n^i\rangle_s|^{p/2} \leq C_p\{M(t-s)\}^{p/2}$$

Taking $p = 4$, we see that

$$E \sup_{u\in[s,t]} |X_n^i(u) - X_n^i(s)|^4 \leq CM^2(t-s)^2$$

Using (3.6) to check (ii) in (3.4) and then noting that $X_n(0) = x$ so (i) in (3.4) is trivial, it follows that the sequence X_n is tight. Invoking Prohorov's Theorem (2.1) now, we can conclude that there is a subsequence $X_{n(k)}$ that converges weakly to a limit X. We claim that X satisfies MP(0,a). To prove this, we will show

(4.1) Lemma. If $f \in C^2$ and $Lf = \frac{1}{2}\sum_{ij} a_{ij} D_{ij} f$ then

$$f(X_t) - f(X_0) - \int_0^t Lf(X_s)\,ds \quad \text{is a local martingale}$$

Once this is done the desired conclusion follows by applying the result to $f(x) = x_i$ and $f(x) = x_i x_j$.

Proof It suffices to show that if f, $D_i f$, and $D_{ij} f$ are bounded, then the process above is a martingale. Itô's formula implies that

$$f(X_n(t)) - f(X_n(s)) = \sum_i \int_s^t D_i f(X_n(r)) dX_n^i(r)$$
$$+ \frac{1}{2}\sum_{ij} \int_s^t D_{ij} f(X_n(r)) d\langle X_n^i, X_n^j\rangle_r$$

So it follows from the definition of X_n that

$$\langle X_n^i, X_n^j\rangle_r = \int_0^r a_{ij}(X_n([2^n u]2^{-n}))\,du$$

so if we let

$$L_n f(r) = \frac{1}{2}\sum_{ij} a_{ij}(X_n([2^n r]2^{-n})) D_{ij} f(X_r^n)$$

then $f(X_n(t)) - f(X_n(s)) - \int_s^t L_n f(r)\,dr$ is a local martingale.

Skorokhod's representation theorem, (1.4), implies that we can construct processes Y_k with the same distributions as the $X_{n(k)}$ on some probability space in such a way that with probability 1 as $k \to \infty$, $Y_k(t)$ converges to $Y(t)$ uniformly on $[0, T]$ for any $T < \infty$. If $s < t$ and $g : C \to R$ is a bounded continuous function that is measurable with respect to $\mathcal{F}_s$, then

$$E\left(g(Y) \cdot \left\{f(Y_t) - f(Y_s) - \int_s^t Lf(Y_r)\,dr\right\}\right)$$
$$= \lim E\left(g(Y^k) \cdot \left\{f(Y_t^k) - f(Y_s^k) - \int_s^t L_n f(Y_r^k)\,dr\right\}\right) = 0$$

Since this holds for any continuous g, an application of the monotone class theorem shows

$$E\left(f(Y_t) - f(Y_s) - \int_s^t Lf(Y_r)\,dr \,\Big|\, \mathcal{F}_s\right) = 0$$

which proves (4.1). □

8.5. Donsker's Theorem

Let $\xi_1, \xi_2, \ldots$ be i.i.d. with $E\xi_i = 0$ and $E\xi_i^2 = 1$. Let $S_n = \xi_1 + \cdots + \xi_n$ be the nth partial sum. The most natural way to turn S_m, $0 \le m \le n$ into a process indexed by $0 \le t \le 1$ let $\hat{B}_t^n = S_{[nt]}/\sqrt{n}$ where $[x]$ is the largest integer $\le x$. To have a continuous trajectory we will instead let

$$B_t^n = \begin{cases} S_m/\sqrt{n} & \text{if } t = m/n \\ \text{linear} & \text{if } t \in [m/n, (m+1)/n] \end{cases}$$

(5.1) Donsker's Theorem. As $n \to \infty$, $B^n \Rightarrow B$, where B is a standard Brownian motion.

Proof We will prove this result using (3.2). Thus there are two things to do:

Convergence of finite dimensional distributions. Since $S_{[nt]}$ is the sum of $[nt]$ independent random variables, it follows from the central limit theorem that if $0 < t_1 < \ldots t_n \le 1$ then

$$(\hat{B}_{t_1}^n, \hat{B}_{t_2}^n - \hat{B}_{t_1}^n, \ldots \hat{B}_{t_m}^n - \hat{B}_{t_{m-1}}^n) \Rightarrow (B_{t_1}, B_{t_2} - B_{t_1}, \ldots B_{t_m} - B_{t_{m-1}})$$

To extend the last conclusion from $\hat{B}^n$ to B^n, we begin by observing that if $r_n = nt - [nt]$ then

$$B_t^n = \hat{B}_t^n + r_n \xi_{[nt]+1}/\sqrt{n}$$

Since $0 \leq r_n < 1$, we have $r_n \xi_{[nt]+1}/\sqrt{n} \to 0$ in probability and the converging together lemma, (1.3), implies that the individual random variables $B_t^n \Rightarrow B_t$. To treat the vector we observe that

$$(B_{t_j}^n - B_{t_{j-1}}^n) - (\hat{B}_{t_j}^n - \hat{B}_{t_{j-1}}^n) = (B_{t_j}^n - \hat{B}_{t_j}^n) - (B_{t_{j-1}}^n - \hat{B}_{t_{j-1}}^n) \to 0$$

in probability, so it follows from (1.3) that

$$(B_{t_1}^n, B_{t_2}^n - B_{t_1}^n, \ldots, B_{t_m}^n - B_{t_{m-1}}^n) \Rightarrow (B_{t_1}, B_{t_2} - B_{t_1}, \ldots, B_{t_m} - B_{t_{m-1}})$$

Using the fact that $(x_1, x_2, \ldots, x_n) \to (x_1, x_1 + x_2, \ldots, x_1 + \cdots + x_m)$ is a continuous mapping and invoking (1.2) it follows that the finite dimensional distributions of B^n converge to those of B.

Tightness. The L^2 maximal inequality for martingales (see e.g., (4.3) in Chapter 4 of Durrett (1995)) implies

$$E\left(\max_{0 \leq j \leq \ell} S_j\right)^2 \leq 4ES_\ell^2$$

Taking $\ell = n/m$ and using Chebyshev's inequality it follows that

$$P\left(\max_{0 \leq j \leq n/m} |S_j| > \epsilon\sqrt{n}\right) \leq 4/m\epsilon^2$$

or writing things in terms of $\hat{B}_t^n$

$$P\left(\max_{s \leq t \leq s+1/m} |\hat{B}_t^n - \hat{B}_s^n| > \epsilon\right) \leq 4/m\epsilon^2$$

Since we need m intervals of length $1/m$ to cover $[0,1]$ the last estimate is not good enough to prove tightness. To improve this, we will truncate and compute fourth moments. To isolate these details we will formulate a general result.

(5.2) Lemma. Let $\xi_1, \xi_2, \ldots$ be i.i.d. with mean μ. Let $\bar{\xi}_i = \xi_i 1_{(|\xi| \leq M)}$, let $\bar{S}_\ell = \bar{\xi}_1 + \cdots + \bar{\xi}_\ell$, let $\alpha(M) = E(|\xi_i|^2; |\xi_i| > M)$, let $\bar{\mu}_M = E\bar{\xi}_i$, and $\bar{\nu}_M = E(\bar{\xi}_i^2)$. Then we have

(i) $P(\xi_i \neq \bar{\xi}_i) \leq M^{-2} E(|\xi_i|^2; |\xi| > M) = \alpha(M)/M^2$

(ii) $|\mu - \bar{\mu}_M| \leq E(|\xi_i|; |\xi| > M) \leq \alpha(M)/M$

(iii) If $M \geq 1$ and $\alpha(M) \leq 1$ then

$$E(\bar{S}_\ell - \ell\bar{\mu}_M)^4 \leq C_1 \ell M^2 + C_2 \ell^2$$

where $C_1 = 32\{\bar{\mu}_M^4 + \bar{\nu}_M\}$ and $C_2 = 24\{\bar{\mu}_M^2 + \bar{\nu}_M\}$.

Remark. We give the explicit values of C_1 and C_2 only to make it clear that they only depend on $\bar{\mu}_M$ and $\bar{\nu}_M$.

Proof Only (iii) needs to be proved. Since $E(\bar{\xi}_j - \bar{\mu}_M) = 0$ and the $\bar{\xi}_j$ are independent

$$E(\bar{S}_\ell - \ell\bar{\mu}_M)^4 = E\left(\sum_{j=1}^{\ell} \bar{\xi}_j - \bar{\mu}_M\right)^4$$

$$= \ell E(\bar{\xi}_j - \bar{\mu}_M)^4 + \binom{\ell}{2}\binom{4}{2}E(\bar{\xi}_j - \bar{\mu}_M)^2$$

If p is an even integer (we only care about $p = 2, 4$) then

$$E(\bar{\xi}_j - \bar{\mu}_M)^p \leq 2^p(\bar{\mu}_M^p + E\bar{\xi}_j^p)$$

From this, (ii), and our assumptions it follows that

$$E(\bar{\xi}_j - \bar{\mu}_M)^2 \leq 4(\bar{\mu}_M^2 + \bar{\nu}_M)$$

This gives the term $C_2\ell^2$. To estimate the fourth moment we notice that

$$E(\bar{\xi}_j^4) = \int_0^M 4x^3 P(|\bar{\xi}_j| > x)\,dx$$

$$\leq 2M^2 \int_0^M 2x P(|\bar{\xi}_j| > x)\,dx = 2M^2 \bar{\nu}_M$$

So using our inequality for the pth power again, we have

$$E(\bar{\xi}_j - \bar{\mu}_M)^4 \leq 16(\bar{\mu}_M^4 + 2M^2\bar{\nu}_M)$$

Since we have supposed $M \geq 1$, this gives the term $C_1\ell M^2$ and the proof is complete. □

We will apply (5.2) with $M = \delta\sqrt{n}$. Part (i) implies

(5.3) $$nP(\xi_i \neq \bar{\xi}_i) \leq n \cdot \frac{\alpha(\delta\sqrt{n})}{\delta^2 n} \to 0$$

by the dominated convergence theorem. Since $\mu = 0$ part (ii) implies

(5.4) $$n|\bar{\mu}_M| \leq n \cdot \frac{\alpha(\delta\sqrt{n})}{\delta\sqrt{n}} = o(\sqrt{n})$$

as $n \to \infty$. Using the L^4 maximal inequality for martingales (again see e.g., (4.3) in Chapter 4 of Durrett (1995)) and part (iii), it follows that if $\ell = n/m$ and n is large then (here and in what follows C will change from line to line)

$$E \sup_{0 \le k \le \ell} |\bar{S}_k - k\bar{\mu}_M|^4 \le CE|\bar{S}_\ell - \ell\bar{\mu}_M|^4$$

$$\le C(\ell M^2 + \ell^2) \le C\left(\frac{\delta^2}{m} + \frac{1}{m^2}\right) n^2$$

Using Chebyshev's inequality now it follows that

(5.5) $$P\left(\sup_{0 \le k \le \ell} |\bar{S}_k - k\bar{\mu}_M| \ge \epsilon\sqrt{n}/9\right) \le C\epsilon^{-4}\left(\frac{\delta^2}{m} + \frac{1}{m^2}\right)$$

Let $\bar{B}^n_t = \bar{S}_{[nt]}/\sqrt{n}$ and $I_{k,m} = [k/m, (k+1)/m]$. Adding up m of the probabilities in (5.5), it follows that

$$\limsup_{n \to \infty} P\left(\max_{0 \le k < m} \max_{s \in I_{k,m}} |\bar{B}^n_s - \bar{B}^n_{k/m}| > \epsilon/9\right) \le C\epsilon^{-4}\left(\delta^2 + \frac{1}{m}\right)$$

To turn this into an estimate of the modulus of continuity, we note that

$$w_{1/m}(f) \le 3 \max_{0 \le k < m} \max_{s \in I_{k,m}} |f(s) - f(k/m)|$$

since for example if $k/m \le s \le (k+1)/m \le t \le (k+2)/m$

$$|f(t) - f(s)| \le |f(t) - f((k+1)/m)|$$
$$+ |f((k+1)/m) - f(k/m)| + |f(k/m) - f(s)|$$

If we pick m and δ so that $C\epsilon^{-4}(\delta^2 + 1/m) \le \epsilon$ it follows that

(5.6) $$\limsup_{n \to \infty} P(w_{1/m}(\bar{B}^n) > \epsilon/3) \le \epsilon$$

To convert this into an estimate the modulus of continuity of $\hat{B}^n$ we begin by observing that (5.3) and (5.4) imply

$$P\left(\sup_{0 \le t \le 1} |\bar{B}^n_t - \hat{B}^n_t| > \epsilon/3\right) \to 0$$

as $n \to \infty$ so the triangle inequality implies

$$\limsup_{n \to \infty} P(w_{1/m}(\hat{B}^n) > \epsilon) \le \epsilon$$

Now the maximum oscillation of B^n over $[s,t]$ is smaller than the maximum oscillation of $\hat{B}^n$ over $[[ns]/n, ([nt]+1)/n]$ so

(5.7) $$w_\delta(B^n) \leq w_{\delta+2/n}(\hat{B}^n)$$

and it follows that

$$\limsup_{n\to\infty} P(w_{1/2m}(B^n) > \epsilon) \leq \epsilon$$

This verifies (ii) in (3.4) and completes the proof of Donsker's theorem. □

The main motivation for proving Donsker's theorem is that it gives as corollaries a number of interesting facts about random walks. The key to the vault is the continuous mapping theorem, (1.2), which with (5.1) implies:

(5.8) Theorem. If $\psi : C[0,1] \to \mathbf{R}$ has the property that it is continuous P_0-a.s. then $\psi(B_n) \Rightarrow \psi(B)$.

Example 5.1. Let $\psi(\omega) = \max\{\omega(t) : 0 \leq t \leq 1\}$. It is easy to see that $|\psi(\omega) - \psi(\xi)| \leq \|\omega - \psi\|$ so $\psi : C[0,1] \to \mathbf{R}$ is continuous, and (5.8) implies

$$\max_{0 \leq m \leq n} S_m/\sqrt{n} \Rightarrow M_1 \equiv \max_{0 \leq t \leq 1} B_t$$

To complete the picture, we observe that by (3.8) in Chapter 1 the distribution of the right-hand side is

$$P_0(M_1 \geq a) = P_0(T_a \leq 1) = 2P_0(B_1 \geq a)$$

Example 5.2. Let $\psi(\omega) = \sup\{t \leq 1 : \omega(t) = 0\}$. This time ψ is not continuous, for if ω_ϵ has $\omega_\epsilon(0) = 0$, $\omega_\epsilon(1/3) = 1$, $\omega_\epsilon(2/3) = \epsilon$, $\omega(1) = 2$, and linear on each interval $[j, (j+1)/3]$ then $\psi(\omega_0) = 2/3$ but $\psi(\omega_\epsilon) = 0$ for $\epsilon > 0$. It is easy to see that if $\psi(\omega) < 1$ and $\omega(t)$ has positive and negative values in each interval $(\psi(\omega) - \delta, \psi(\omega))$ then ψ is continuous at ω. By arguments in Example 3.1 of Chapter 1, the last set has P_0 measure 1. (If the zero at $\psi(\omega)$ was isolated on the left, it would not be isolated on the right.) Using (5.8) now

$$\sup\{m \leq n : S_{m-1} \cdot S_m \leq 0\}/n \Rightarrow L = \sup\{t \leq 1 : B_t = 0\}$$

The distribution of L, given in (4.2) in Chapter 1, is an arcsine law.

Example 5.3. Let $\psi(\omega) = |\{t \in [0,1] : \omega(t) > a\}|$. The point $\omega \equiv a$ shows that ψ is not continuous but it is easy to see that ψ is continuous at paths ω with $|\{t \in [0,1] : \omega(t) = a\}| = 0$. Fubini's theorem implies that

$$E_0|\{t \in [0,1] : B_t = a\}| = \int_0^1 P_0(B_t = a)\, dt = 0$$

so ψ is continuous P_0-a.s. With a little work (5.8) implies

$$|\{m \le n : S_m > a\sqrt{n}\}|/n \Rightarrow |\{t \in [0,1] : B_t > a\}|$$

Before doing that work we would like to observe that (9.2) in Chapter 4 shows that $|\{t \in [0,1] : B_t > 0\}|$ has an arcsine law under P_0.

Proof Application of (5.8) gives that for any a,

$$|\{t \in [0,1] : B_t^n > a\sqrt{n}\}| \Rightarrow |\{t \in [0,1] : B_t > a\}|$$

To convert this into a result about $|\{m \le n : S_m > a\sqrt{n}\}|$ we note that if $\epsilon > 0$ then

(5.9)
$$P(\max_{m \le n} |X_m| > \epsilon\sqrt{n}) \le nP(|X_m| > \epsilon\sqrt{n})$$
$$\le \epsilon^{-2} E(X_m^2 ; |X_m| > \epsilon\sqrt{n}) \to 0$$

by dominated convergence, and on $\{\max_{m \le n} |X_m| \le \epsilon\sqrt{n}\}$, we have

$$|\{t \in [0,1] : B_t^n > (a+\epsilon)\sqrt{n}\}| \le \frac{1}{n}|\{m \le n : S_m > a\sqrt{n}\}|$$
$$\le |\{t \in [0,1] : B_t^n > (a-\epsilon)\sqrt{n}\}|$$

Combining this with the first conclusion of the proof and using the fact that $b \to |\{t \in [0,1] : B_t > b\}|$ is continuous at $b = a$ with probability one, we arrive easily at the desired conclusion. $\square$

Example 5.4. Let $\psi(\omega) = \int_{[0,1]} \omega(t)^k dt$ where $k > 0$ is an integer. ψ is continuous so applying (5.8) gives

$$\int_0^1 (B_t^n/\sqrt{n})^k dt \Rightarrow \int_0^1 B_t^k dt$$

To convert this into a result about the original sequence, we begin by observing that if $x < y$ with $|x - y| \le \epsilon$ and $|x|, |y| \le M$ then

$$|x^k - y^k| \le \int_x^y \frac{|z|^{k+1}}{k+1} dz \le \frac{\epsilon M^{k+1}}{k+1}$$

From this it follows that on

$$G_n(M) = \left\{\max_{m \le n} |X_m| \le M^{-(k+2)}\sqrt{n}, \max_{m \le n} |S_m| \le M\sqrt{n}\right\}$$

we have
$$\left| \int_0^1 (S(nt)/\sqrt{n})^k \, dt - n^{-1-(k/2)} \sum_{m=1}^n S_m^k \right| \leq \frac{1}{(k+1)M}$$

For fixed M we have $P(\max_{m \leq n} |X_m| \leq M^{-(k+2)}\sqrt{n}) \to 0$ by (5.9) so using Example 5.1, and (1.1) it follows that

$$\liminf_{n \to \infty} P(G_n(M)) \geq P\left(\max_{0 \leq t \leq 1} |B_t| < M \right)$$

The right-hand side is close to 1 if M is large so

$$\int_0^1 (B_t^n/\sqrt{n})^k \, dt - n^{-1-(k/2)} \sum_{m=1}^n S_m^k \to 0$$

in probability and it follows from the converging together lemma (1.3) that

$$n^{-1-(k/2)} \sum_{m=1}^n S_m^k \Rightarrow \int_0^1 B_t^k \, dt$$

It is remarkable that the last result holds under the assumption that $EX_i = 0$ and $EX_i^2 = 1$, i.e., we do not need to assume that $E|X_i^k| < \infty$.

8.6. The Space D

In the previous section, forming the piecewise linear approximation was an annoying bookkeeping detail. In the next section when we consider processes that jump at random times, making them piecewise linear will be a genuine nuisance. To deal with that problem and to educate the reader, we will introduce the space $D([0,1], \mathbf{R}^d)$ of functions from $[0,1]$ into $\mathbf{R}^d$ that are right continuous and have left limits. Since we only need two simple results, (6.4) and (6.5) below, and one can find this material in Chapter 3 of Billingsley (1968), Chapter 3 of Ethier and Kurtz (1986), or Chapter VI of Jacod and Shiryaev (1987), we will content ourselves to simply state the results.

We begin by defining the **Skorokhod topology** on D. To motivate this consider

Example 6.1. For $1 \leq n \leq \infty$ let

$$f_n(t) = \begin{cases} 0 & t \in [0, (n+1)/2n) \\ 1 & t \in [(n+1)/2n, 1] \end{cases}$$

where $(n+1)/2n = 1/2$ for $n = \infty$. We certainly want $f_n \to f_\infty$ but $\|f_n - f_\infty\| = 1$ for all n.

Let Λ be the class of strictly increasing continuous mappings of $[0,1]$ onto itself. Such functions necessarily have $\lambda(0) = 0$ and $\lambda(1) = 1$. For $f, g \in D$ define $d(f,g)$ to be the infimum of those positive ϵ for which there is a $\lambda \in \Lambda$ so that
$$\sup_t |\lambda(t) - t| \le \epsilon \quad \text{and} \quad \sup_t |f(t) - g(\lambda(t))| \le \epsilon$$

It is easy to see that d is a metric. If we consider $f = f_n$ and $g = f_m$ in Example 6.1 then for $\epsilon < 1$ we must take $\lambda((n+1)/2n) = (m+1)/2m$ so
$$d(f_n, f_m) = \left|\frac{n+1}{2n} - \frac{m+1}{2m}\right| = \left|\frac{1}{2n} - \frac{1}{2m}\right|$$

When $m = \infty$ we have $d(f_n, f_\infty) = 1/2n$ so $f_n \to f_\infty$ in the metric d.

We will see in (6.2) that d defines the correct topology on D. However, in view of (2.2), it is unfortunate that the metric d is not complete.

Example 6.2. For $1 \le n < \infty$ let
$$g_n(t) = \begin{cases} 0 & t \in [0, 1/2) \\ 1 & t \in [1/2, (n+1)/2n) \\ 0 & t \in [(n+1)/2n, 1] \end{cases}$$

In order to have $\epsilon < 1$ in the definition of $d(g_n, g_m)$ we must have $\lambda(1/2) = 1/2$ and $\lambda((n+1)/2n) = (m+1)/2m$ so
$$d(g_n, g_m) = \left|\frac{1}{2n} - \frac{1}{2m}\right|$$

The pointwise limit of g_n is $g_\infty \equiv 0$ but $d(g_n, g_\infty) = 1$. We leave it to the reader to show

Exercise 6.1. If $h \in D$ then $\liminf_{n \to \infty} d(g_n, h) > 0$.

To fix the problem with completeness we require that λ be close to the identity in a more stringent sense: the slopes of all of its chords are close to 1. If $\lambda \in \Lambda$ let
$$\|\lambda\| = \sup_{s \ne t} \left|\log\left(\frac{\lambda(t) - \lambda(s)}{t - s}\right)\right|$$

For $f, g \in D$ define $d_0(x, y)$ to be the infimum of those positive ϵ for which there is a $\lambda \in \Lambda$ so that

$$||\lambda|| \leq \epsilon \quad \text{and} \quad \sup_t |f(t) - g(\lambda(t))| \leq \epsilon$$

It is easy to see that d_0 is a metric. The functions g_n in Example 6.2 have

$$d_0(g_n, g_m) = \min\{1, |\log(n/m)|\}$$

so they no longer form a Cauchy sequence. In fact, there are no more problems.

(6.1) Theorem. The space D is complete under the metric d_0.

For the reader who is curious why we discussed the simpler metric d we note:

(6.2) Theorem. The metrics d and d_0 are equivalent, i.e., they give rise to the same topology on D.

Our first step in studying weak convergence in D is to characterize tightness. Generalizing a definition from Section 7.3 we let

$$w_S(f) = \sup\{|f(t) - f(s)| : s, t \in S\}$$

for each $S \subset [0, 1]$. For $0 < \delta < 1$ put

$$w'_\delta(f) = \inf_{\{t_i\}} \max_{0 < i \leq r} w_{[t_i, t_{i+1})}(f)$$

where the infimum extends over the finite sets $\{t_i\}$ with

$$0 = t_0 < t_1 < \cdots < t_r \quad \text{and} \quad t_i - t_{i-1} > \delta \text{ for } 1 \leq i \leq r$$

The analogue of (3.4) in the current setting is

(6.3) Theorem. The sequence μ_n is tight if and only if for each $\epsilon > 0$ there are n_0, M, and δ so that

(i) $\mu_n(|\omega(0)| > M) \leq \epsilon$ for all $n \geq n_0$

(ii) $\mu_n(w'_\delta > \epsilon) \leq \epsilon$ for all $n \geq n_0$

Note that the only difference from (3.4) is the $'$ in (ii). If we remove the $'$ we get the main result we need.

(6.4) Theorem. If for each $\epsilon > 0$ there are n_0, M, and δ so that

(i) $\mu_n(|\omega(0)| > M) \le \epsilon$ for all $n \ge n_0$

(ii) $\mu_n(w'_\delta > \epsilon) \le \epsilon$ for all $n \ge n_0$

then μ_n is tight and every subsequential limit μ has $\mu(C) = 1$.

(6.1)-(6.4) are Theorems 14.2, 14.1, 15.2, and 15.5 in Billingsley (1968). We will also need the following, which is a consequence of the proof of (3.3).

(6.5) Theorem. If each subsequence of μ_n has a further subsequence that converges to μ then $\mu_n \Rightarrow \mu$.

8.7. Convergence to Diffusions

In this section we will prove a result, due to Stroock and Varadhan, about the convergence of Markov chains to diffusion processes. Our treatment here follows that in Chapter 11 of Stroock and Varadhan (1979) but we will discuss both discrete and continuous time in parallel. Our duplicity lengthens the proof somewhat, but is preferable we believe to the usual (somewhat dangerous) cop-out: "the same argument with minor changes handles the other case."

In discrete time, the basic data is a sequence of transition probabilities $\Pi_h(x, dy)$ for a Markov chain Y^h_{mh}, $m = 0, 1, 2, \ldots$, taking values in $S_h \subset \mathbf{R}^d$, i.e.,

$$P(Y^h_{(m+1)h} \in A | Y^h_{mh} = x) = \Pi_h(x, A) \quad \text{for } x \in S_h, A \in \mathcal{R}^d$$

In this case we define $X^h_t = Y^h_{h[t/h]}$, i.e., we make X^h_t constant on intervals $[mh, (m+1)h)$.

In continuous time, the basic data is a sequence of transition rates $Q_h(x, dy)$ for a Markov chain X^h_t, $t \ge 0$, taking values in $S_h \subset \mathbf{R}^d$, i.e.,

$$\frac{d}{dt} P(X^h_t \in A | X^h_0 = x) = Q_h(x, A) \quad \text{for } x \in S_h, A \in \mathcal{R}^d, \text{ with } x \notin A$$

Note that although $x \in A$ is excluded from the interpretation, we will allow $Q(x, \{x\}) > 0$ in which case jumps from x to x (which are invisible) occur at a positive rate. To have a well behaved Markov chain we will suppose

(B) For any compact set K, $\sup_{x \in K} Q_h(x, \mathbf{R}^d) < \infty$.

In words, our convergence theorem states that if the infinitesimal mean and covariance of the Markov chain converge to those of a diffusion for which the martingale problem is well posed, and we have a condition that rules out jumps in the limit then we have weak convergence. To make a precise statement we need some notation. To incorporate both cases we introduce a kernel

$$K_h(x, dy) = \begin{cases} \Pi_h(x, dy)/h & \text{in discrete time} \\ Q_h(x, dy) & \text{in continuous time} \end{cases}$$

and define

$$a_{ij}^h(x) = \int_{|y-x|\leq 1} (y_i - x_i)(y_j - x_j) K_h(x, dy)$$

$$b_i^h(x) = \int_{|y-x|\leq 1} (y_i - x_i) K_h(x, dy)$$

$$\Delta_\epsilon^h(x) = K_h(x, B(x, \epsilon)^c)$$

where $B(x, \epsilon) = \{y : |y - x| < \epsilon\}$. Suppose

(A) a^{ij} and b^i are continuous coefficients for which the martingale problem is well posed, i.e., for each x there is a unique measure P_x on $(C, \mathcal{C})$ so that the coordinate maps $X_t(\omega) = \omega(t)$ satisfy $P_x(X_0 = x) = 1$ and

$$X_t^i - \int_0^t b_i(X_s)\, ds \quad \text{and} \quad X_t^i X_t^j - \int_0^t a_{ij}(X_s)\, ds$$

are local martingales.

(7.1) Theorem. Suppose in continuous time that (B) holds, and in either case that (A) holds and for each i, j, $R < \infty$, and $\epsilon > 0$

(i) $\lim_{h\downarrow 0} \sup_{|x|\leq R} |a_{ij}^h(x) - a_{ij}(x)| = 0$

(ii) $\lim_{h\downarrow 0} \sup_{|x|\leq R} |b_i^h(x) - b_i(x)| = 0$

(iii) $\lim_{h\downarrow 0} \sup_{|x|\leq R} \Delta_\epsilon^h(x) = 0$

If $X_0^h = x_h \to x$ then we have $X_t^h \Rightarrow X_t$ the solution of the martingale problem with $X_0 = x$.

Remarks. Here $\Rightarrow$ denotes convergence in $D([0,1], \mathbf{R}^d)$ but the result can be trivially generalized to give convergence $D([0, T], \mathbf{R}^d)$ for all $T < \infty$.

In (i), (ii), and (iii) the sup is taken only over points $x \in S_h$. Condition (iii) cannot hold unless the points in S_h are getting closer together. However, there is no implicit assumption that S_h is becoming dense in all of $\mathbf{R}^d$.

Proof The rest of the section is devoted to the proof of this result. We will first prove the result under the stronger assumptions that for all i, j and $\epsilon > 0$ we have

(i′) $\lim_{h\downarrow 0}\sup_x |a^h_{ij}(x) - a_{ij}(x)| = 0$ (iv) $\sup_{h,x} |a^h_{ij}(x)| < \infty$
(ii′) $\lim_{h\downarrow 0}\sup_x |b^h_i(x) - b_i(x)| = 0$ (v) $\sup_{h,x} |b^h_i(x)| < \infty$
(iii′) $\lim_{h\downarrow 0}\sup_x \Delta^h_\epsilon(x) = 0$ (vi) $\sup_{h,x} \Delta^h_\epsilon(x) < \infty$

and in continuous time that (B′) $\sup_x Q(x, \mathbf{R}^d) < \infty$.

a. Tightness

If f is bounded and measurable, let

$$L^h f(x) = \int K_h(x, dy)(f(y) - f(x))$$

As the notation may suggest, we expect this to converge to

$$Lf(x) = \frac{1}{2}\sum_{i,j} a_{ij}(x) D_{ij} f(x) + \sum_i b_i(x) D_i f(x)$$

To explain the reason for this definition we note that if f is bounded and measurable then

(7.2a) In discrete time $f(X^h_{kh}) - \sum_{j=0}^{k-1} h L^h f(X^h_{jh})$ is a martingale.

(7.2b) In continuous time $f(X^h_t) - \int_0^t L^h f(X^h_s)$ is a martingale.

Proof (7.2a) follows easily from the definition of L^h and induction. Since we have (B′), (7.2b) follows from (2.1) and (1.6) in Chapter 7. □

The first step in the tightness proof is to see what happens when we apply L^h to $f_{\epsilon,y}(x) = f_\epsilon(x-y)$ where $f_\epsilon(x) = g(|x|^2/\epsilon^2)$, and

$$g(x) = \begin{cases} (1-x^2)(1-x)^3 & 0 \le x \le 1 \\ 0 & x \ge 1 \end{cases}$$

The exact formula for g is not important. All we care about is that $0 \le g(x) \le 1$ for $x \ge 0$, $g(0) = 1$, $g(x) = 0$ for $x \ge 1$, and $g \in C^2$ (since $g'(0) = g''(0) = 0$).

(7.3) **Lemma.** There is a C_ϵ which only depends on ϵ so that $|L^h f_{y,\epsilon}| \le C_\epsilon$.

Proof Applying Taylor's theorem to the function $h(t) = f(x + t(y-x))$ for $0 \le t \le 1$ we have for some $c_{x,y} \in [0,1]$

(7.4)
$$\begin{aligned} f(y) - f(x) &= h(1) - h(0) = h'(0) + h''(c_{x,y})/2! \\ &= \sum_i (y_i - x_i) D_i f(x) + \sum_{i,j}(y_i - x_i)(y_j - x_j) D_{ij} f(z_{x,y}) \end{aligned}$$

Section 8.7 Convergence to Diffusions   299

where $z_{x,y} = x + c_{x,y}(y-x)$. Integrating with respect to $K_h(x, dy)$ over the set $|y - x| \le 1$ and using the triangle inequality we have

$$L^h f(x) \le \left| \nabla f(x) \cdot \int_{|y-x| \le 1} (y - x) \, K_h(x, dy) \right|$$

$$+ \left| \int_{|y-x| \le 1} \sum_{i,j} (y_i - x_i)(y_j - x_j) D_{ij} f(z_{x,y}) K_h(x, dy) \right|$$

$$+ 2\|f\|_\infty K_h(x, B(x, 1)^c)$$

Let $A_\epsilon = \sup_x |\nabla f(x)|$, and $B_\epsilon = \sup_z \|D_{ij} f(z)\|$ where

$$\|m_{ij}\| = \sup \left\{ \left\| \sum_i u_i m_{ij} u_j \right\| : |u| = 1 \right\}$$

Noticing that the Cauchy-Schwarz inequality implies

(7.5) $$\left| \sum_i u_i v_i \right| \le |u||v|$$

and the definition of $\|m_{ij}\|$ gives

(7.6) $$\left| \sum_{i,j} (y_i - x_i)(y_j - x_j) D_{ij} f(z) \right| \le |y - x|^2 \|D_{ij} f(z)\|$$

it follows that

$$L^h f(x) \le A_\epsilon \left| \int_{|y-x| \le 1} (y - x) \, K_h(x, dy) \right|$$

$$+ B_\epsilon \int_{|y-x| \le 1} |y - x|^2 \, K_h(x, dy) + 2\|f\|_\infty K_h(x, B(x, 1)^c)$$

Now

(7.7) $$\sum_i a_{ii}^h(x) = \int_{|y-x| \le 1} |y - x|^2 \, K_h(x, dy)$$

so recalling the definitions of b_i^h and Δ_1^h, then using (iv)–(vi) the desired result follows. □

To estimate the oscillations of the sample paths, we will let $\tau_0 = 0$,

$$\tau_n = \inf\{t > \tau_{n-1} : |X_t^h - X_{\tau_{n-1}}^h| \ge \epsilon/4\}$$
$$N = \inf\{n \ge 0 : \tau_n > 1\}$$
$$\sigma = \min\{\tau_n - \tau_{n-1} : 1 \le n \le N\}$$
$$\theta = \max\{|X^h(t) - X^h(t-)| : 0 < t \le 1\}$$

Chapter 8 Weak Convergence

To relate these definitions to tightness, we will now prove

(7.8) Lemma. If $\sigma > \delta$ and $\theta \le \epsilon/4$ then $w_\delta(X^h) \le \epsilon$.

Proof To check this we note that if $\sigma > \delta$ and $0 < t - s < \delta$ there are two cases to consider

CASE 1. $\tau_n \le s < t < \tau_{n+1}$

$$|f(t) - f(s)| \le |f(t) - f(\tau_n)| + |f(\tau_n) - f(s)| \le 2\epsilon/4$$

CASE 2. $\tau_{n-1} \le s < \tau_n \le t < \tau_{n+1}$

$$|f(t) - f(s)| \le |f(t) - f(\tau_n)| + |f(\tau_n) - f(\tau_n-)|$$
$$+ |f(\tau_n-) - f(\tau_{n-1})| + |f(\tau_{n-1}) - f(s)| \le \epsilon \qquad \square$$

Tightness proof, discrete time. One of the probabilities in (7.8) is trivial to estimate

$$(7.9a) \qquad P_y(\theta \ge \epsilon/4) \le \frac{1}{h} \sup_x \Pi_h(x, B(x, \epsilon/4)^c) \le \sup_x \Delta^h_{\epsilon/4}(x) \to 0$$

by (iii'). The first step in estimating $P_y(\sigma > \delta)$ is to estimate $P_y(\tau_1 \le \delta)$. To do this we begin by observing that (7.2a) and (7.3) imply

$$f_{y,\epsilon/4}(X^h_{kh}) + C_{\epsilon/4}kh, \ k = 0, 1, 2, \ldots \text{ is a submartingale}$$

Using the optional stopping theorem at time $\tau \wedge \hat{\delta}$ where $\hat{\delta} = [\delta/h]h \le \delta$ and noticing $X_{\tau \wedge \hat{\delta}} = X_{\tau \wedge \delta}$ it follows that

$$E_y\{f_{y,\epsilon/4}(X_{\tau \wedge \delta}) + C_{\epsilon/4}(\tau \wedge \delta)\} \ge 1$$

or rearranging and using $\tau \wedge \delta \le \delta$

$$(7.10) \qquad C_{\epsilon/4}\delta \ge E_y\{1 - f_{y,\epsilon/4}(\hat{X}_{\tau \wedge \delta})\} \ge P_y(\tau_1 \le \delta)$$

Our next task is to estimate $P_y(N > k)$. To do this, we begin by observing

$$E_y(e^{-\tau_1}) \le P_y(\tau_1 \le \delta) + e^{-\delta} P_y(\tau_1 > \delta)$$
$$\le C_{\epsilon/4}\delta + e^{-\delta}(1 - C_{\epsilon/4}\delta) \equiv \lambda < 1$$

Iterating and using the strong Markov property it follows that $E_y(e^{-\tau_n}) \le \lambda^n$ and

(7.11) $\qquad P_y(N > k) = P_y(\tau_k \le 1) \le eE_y(e^{-\tau_k}) \le e\lambda^k$

Combining (7.10) and (7.11) we have

(7.12) $\qquad P_y(\sigma \le \delta) \le k \sup_x P_x(\tau \le \delta) + P_y(N > k) \le C_{\epsilon/4} k\delta + e\lambda^k$

If we pick k so that $e\lambda^k \le \epsilon/3$ and then pick δ so that $C_\epsilon k\delta \le \epsilon/3$ then (7.12), (7.9a), and (7.8) imply that for small h, $\sup_y P_y(w_\delta(X^h) > \epsilon) \le \epsilon$. This verifies (ii) in (6.4). Since $X_0^h = x_h$, (i) is trivial and the proof of tightness is complete in discrete time. □

Tightness proof, continuous time. One of the probabilities in (7.8) is trivial to estimate. Since jumps of size $> \epsilon/4$ occur at a rate smaller than $\sup_x \Pi_h(x, B(x, \epsilon/4)^c)$ and $e^{-z} \ge 1 - z$

(7.9b) $\quad P_y(\theta \ge \epsilon/4) \le 1 - \exp(-\sup_x Q_h(x, B(x, \epsilon/4)^c)) \le \sup_x \Delta^h_{\epsilon/4}(x) \to 0$

by (iii'). The first step in estimating $P_y(\sigma > \delta)$ is to estimate $P_y(\tau_1 \le \delta)$. To do this we begin by observing that (7.2b) and (7.3) imply

$$f_{y,\epsilon/4}(X^h_t) + C_{\epsilon/4} t, \ t \ge 0 \text{ is a submartingale}$$

Using the optional stopping theorem at time $\tau \wedge \delta$ it follows that

$$E_y\{f_{y,\epsilon/4}(X_{\tau \wedge \delta}) + C_{\epsilon/4}(\tau \wedge \delta)\} \ge 1$$

The remainder of the proof is identical to the one in discrete time. □

b. Convergence

Since we have assumed that the martingale problem has a unique solution, it suffices by (6.5) to prove that the limit of any convergent subsequence solves the martingale problem to conclude that the whole sequence converges. The first step is to show

(7.13) **Lemma.** If $f \in C_K^2$ then $\|L^h f - Lf\|_\infty \to 0$.

Proof Using (7.4) then integrating with respect to $K_h(x, dy)$ over the set $|y - x| \leq 1$ we have

(7.14)
$$L^h f(x) = \sum_i b_i^h(x) D_i f(x)$$
$$+ \int_{|y-x|\leq 1} \sum_{ij}(y_i - x_i)(y_j - x_j) D_{ij} f(z_{x,y}) K_h(x, dy)$$
$$+ \int_{|y-x|>1} \{f(y) - f(x)\} K_h(x, dy)$$

Recalling the definition of Δ_1^h and using (vi) it follows that the third term goes to 0 uniformly in x. To treat the first term we note that (7.5) and (v) imply

$$\left|\sum_i b_i^h(x) D_i f(x) - \sum_i b_i(x) D_i f(x)\right| \leq \sup_j |b_j^h(x) - b_j(x)| \sum_i |D_i f(x)| \to 0$$

uniformly in x.

Now the difference between the second term in (7.14) and $\sum_{ij} a_{ij} D_{ij} f(x)$ is smaller than

(7.15)
$$\left|\sum_{ij} a_{ij}(x) D_{ij} f(x) - \sum_{ij} a_{ij}^h(x) D_{ij} f(x)\right|$$
$$+ \left|\int_{|y-x|\leq 1} \sum_{ij}(y_i - x_i)(y_j - x_j) \{D_{ij} f(z_{x,y}) - D_{ij} f(x)\} K_h(x, dy)\right|$$

By (7.5) the first term is smaller than

$$\sup_{ij} |a_{ij}(x) - a_{ij}^h(x)| \sum_{ij} |D_{ij} f(x)|$$

which goes to 0 uniformly in x by (iv). To estimate the second term in (7.15) we note that for any ϵ the integral over $\{y : \epsilon < |y-x| \leq 1\}$ goes to 0 uniformly in x by (vi) and the fact that the integrand is uniformly bounded. Using (7.6), the last piece

(7.16)
$$\left|\int_{|y-x|\leq \epsilon} \sum_{ij}(y_i - x_i)(y_j - x_j) \{D_{ij} f(z_{x,y}) - D_{ij} f(x)\} K_h(x, dy)\right|$$
$$\leq \Gamma(\epsilon) \left|\int_{|y-x|\leq \epsilon} |y - x|^2 K_h(x, dy)\right|$$

where
$$\Gamma(\epsilon) = \sup_{|y-x|\le\epsilon} \|D_{ij}f(z_{x,y}) - D_{ij}f(x)\|$$

Now $z_{x,y}$ is on the line segment connecting x and y, $D_{ij}f$ is continuous and has compact support, so $\Gamma(\epsilon) \to 0$ as $\epsilon \to 0$. Using (7.7) and (iv) now the quantity in (7.16) converges to 0 and the proof of (7.13) is complete. □

Convergence proof, discrete time. Let $h_n \to 0$ so that X^{h_n} converges weakly to a limit X^0. (7.2a) implies

$$f(X_{kh_n}^{h_n}) - \sum_{j=0}^{k-1} h_n L^{h_n} f(X_{jh_n}^{h_n}), \quad k = 0, 1, 2, \ldots \text{ is a martingale}$$

To pass to the limit, we rewrite the martingale property by introducing a bounded continuous function $g : D \to \mathbf{R}$ which is measurable with respect to $\mathcal{F}_s$ and observe that if $k_n = [s/h_n] + 1$ and $\ell_n = [t/h_n] + 1$ the martingale property implies

$$E\left(g(X^{h_n})\left\{f(X_{\ell_n h_n}^{h_n}) - f(X_{k_n h_n}^{h_n}) - \sum_{j=k_n}^{\ell_n-1} h_n L^{h_n} f(X_{jh_n}^{h_n})\right\}\right) = 0$$

Let $h_\infty = 0$. Skorokhod's representation theorem (1.4) implies that we can construct processes Y^n, $1 \le n \le \infty$ with the same distribution as the X^{h_n} so that $Y^n \to Y^\infty$ almost surely. Combining this with (7.13) and using the bounded convergence theorem it follows that if $f \in C_K^2$ then

$$E\left(g(X^0)\left\{f(X_t^0) - f(X_s^0) - \int_s^t Lf(X_r^0)\,dr\right\}\right) = 0$$

Since this holds for all continuous $g : C \to \mathbf{R}$ measurable with respect to $\mathcal{F}_s$, it follows that if $f \in C_K^2$ then

$$f(X_t^0) - \int_0^t Lf(X_s^0)\,ds \quad \text{is a martingale}$$

Applying the last result to smoothly truncated versions of $f(x) = x_i$ and $f(x) = x_i x_j$ it follows that X^0 is a solution of the martingale problem. Since the martingale problem has a unique solution, the desired conclusion now follows from (6.5). □

Convergence proof, continuous time. Let $h_n \to 0$ so that X^{h_n} converges weakly to a limit X^0. (7.2b) implies

$$f(X_t^{h_n}) - \int_0^t L^{h_n} f(X_s^{h_n})\, ds \ \ t \geq 0 \ldots \text{ is a martingale}$$

and repeating the discrete time proof shows that X^0 is a solution of the martingale problem and the desired result follows. □

c. Localization

To extend the result to the original assumptions, we let φ_k be a C^∞ function that is 1 on $B(0,k)$ and 0 on $B(0,k+1)$ and define

$$K_{h,k}(x, dy) = \varphi_k(x) K_h(x, dy) + (1 - \varphi_k(x)) \delta_x(dy)$$

where δ_x is a point mass at x. It is easy to see that if K_h satisfies (i)-(iii) then $K_{h,k}$ satisfies (i')-(iii') and (iv)-(vi). From the first two parts of the proof it follows that for each k, $X^{h,k}$ is tight, and any subsequential limit solves the martingale problem with coefficients

$$a^k(x) = \varphi_k(x) a(x) \qquad b^k(x) = \varphi_k(x) b(x)$$

We have supposed that the martingale problem for a and b has a unique nonexplosive solution X^0. In view of (3.3), to prove (6.1) it suffices to show that given any sequence $h_n \to 0$ there is a subsequence h'_n so that $X^{h'_n} \Rightarrow X^0$. To prove this we note that since each sequence $X^{h_n,k}$ is tight, a diagonal argument shows that we can select a subsequence h'_n so that for each k, $X^{h'_n,k} \Rightarrow X^{0,k}$.

Let $G_k = \{\omega : \omega(t) \in B(0,k), 0 \leq t \leq 1\}$. G_k is an open set so if $X^{h'_n,k} \Rightarrow X^{0,k}$ and $H \subset C$ is open then (2.1) implies that

$$\liminf_{n \to \infty} P(X^{h'_n,k} \in G_k \cap H) \geq P(X^{0,k} \in G_k \cap H) = P(X^0 \in G_k \cap H)$$

since up until the exit from $B(0,k)$, $X^{0,k}$ is a solution of the martingale problem for b and a and hence its distribution agrees with that of X^0. Now up to the first exit from $B(0,k)$ $X^{h'_n}$ has the same distribution as that of $X^{h'_n,k}$ so we have

$$\liminf_{n \to \infty} P(X^{h'_n} \in H) \geq \liminf_{n \to \infty} P(X^{h'_n} \in G_k \cap H) \geq P(X^0 \in G_k \cap H)$$

Since there is no explosion the right-hand side increases to $P(X^0 \in H)$ as $k \uparrow \infty$ and the proof is complete.

8.8. Examples

In this section we will give applications of (7.1) beginning with the following very simple situation.

Example 8.1. Ehrenfest Chain. Here, the physical system is a box filled with air and divided in half by a plane with a small hole in it. We model this mathematically by two urns that contain a total of $2n$ balls, which we think of as the air molecules. At each time we pick a ball from the $2n$ in the two urns at random and move it to the other urn, which we think of as a molecule going through the hole. We expect the number of balls in the left urn to be about $n + C\sqrt{n}$, so we let Z_m^n be the number of balls at time m in the left urn minus n, and let $Y_{m/n}^{1/n} = Z_m^n/\sqrt{n}$. The state space $S_{1/n} = \{k/\sqrt{n} : -n \leq k \leq n\}$ while the transition probability is

$$\Pi_{1/n}(x, x + n^{-1/2}) = \frac{n - x\sqrt{n}}{2n} \qquad \Pi_{1/n}(x, x - n^{-1/2}) = \frac{n + x\sqrt{n}}{2n}$$

When $\epsilon > n^{-1/2}$, $\Delta_\epsilon^{1/n}(x) = 0$ for all x so (iii) holds. To check conditions (i) and (ii) we note that

$$b^{1/n}(x) = n\left\{\frac{1}{\sqrt{n}} \cdot \frac{n - x\sqrt{n}}{2n} - \frac{1}{\sqrt{n}} \cdot \frac{n - x\sqrt{n}}{2n}\right\} = -x$$

$$a^{1/n}(x) = n\left\{\frac{1}{n} \cdot \frac{n - x\sqrt{n}}{2n} + \frac{1}{n} \cdot \frac{n - x\sqrt{n}}{2n}\right\} = 1$$

The limiting coefficients $b(x) = -x$ and $a(x) = 1$ are Lipschitz continuous, so the martingale problem is well posed and (A) holds. Letting $X_t^{1/n} = Y_{[nt]/n}^{1/n}$ and applying (7.1) now it follows that

(8.1) Theorem. As $n \to \infty$, $X_t^{1/n}$ converges weakly to an Ornstein-Uhlenbeck process X_t, i.e., the solution of

$$dX_t = -X_t\,dt + dB_t$$

Next we state a lemma that will help us check the hypotheses in the next two examples. The point here is to replace the truncated moments by ordinary moments that are easier to compute.

$$\hat{a}_{ij}^h(x) = \int (y_i - x_i)(y_j - x_j) K_h(x, dy)$$

$$\hat{b}_i^h(x) = \int (y_i - x_i) K_h(x, dy)$$

$$\gamma_p^h(x) = \int |y - x|^p K_h(x, dy)$$

(8.2) **Lemma.** If $p \geq 2$ and for all $R < \infty$ we have

(a) $\lim_{h \downarrow 0} \sup_{|x| \leq R} |\hat{a}_{ij}^h(x) - a_{ij}(x)| = 0$

(b) $\lim_{h \downarrow 0} \sup_{|x| \leq R} |\hat{b}_i^h(x) - b_i(x)| = 0$

(c) $\lim_{h \downarrow 0} \sup_{|x| \leq R} \gamma_p^h(x) = 0$

then (i), (ii), and (iii) of (7.1) hold.

Proof Taking the conditions in reverse order, we note $\Delta_\epsilon^h(x) \leq \epsilon^{-p} \gamma_p^h(x)$

$$|\hat{b}_i^h(x) - b_i^h(x)| \leq \int_{|y-x|>1} |y - x| K_h(x, dy) \leq \gamma_p^h(x)$$

Using the Cauchy-Schwarz inequality with the triviality $(y_k - x_k)^2 \leq |y - x|^2$

$$|\hat{a}_{ij}^h(x) - a_{ij}^h(x)| \leq \int_{|y-x|>1} |(y_i - x_i)(y_j - x_j)| K_h(x, dy)$$

$$\leq \int_{|y-x|>1} |y - x|^2 K_h(x, dy) \leq \gamma_p^h(x)$$

The desired conclusion follows easily from the last three observations. □

Our first two examples are the last two in Section 5.1, and date back to the beginnings of the subject, see Feller (1951).

Example 8.2. Branching Processes. Consider a sequence of branching processes $\{Z_m^n, m \geq 0\}$ in which the probability of k children p_k^n has mean $1 + (\beta_n/n)$ and variance σ_n^2. Suppose that

(A1) $\beta_n \to \beta \in (-\infty, \infty)$,

(A2) $\sigma_n \to \sigma \in (0, \infty)$,

(A3) for any $\delta > 0$, $\sum_{k > \delta n} k^2 p_k^n \to 0$

Following the motivation in Example 1.6 of Chapter 5, we let $X_t^{1/n} = Z_{[nt]}^n/n$. Our goal is to show

(8.3) **Theorem.** If (A1), (A2), and (A3) hold then $X_t^{1/n}$ converges weakly to Feller's branching diffusion X_t, i.e., the solution of

$$dX_t = \beta X_t \, dt + \sigma \sqrt{X_t} \, dB_t$$

References. For a classical generating function approach and the history of the result, see Sections 3.3–3.4 of Jagers (1975). For an approach based on semigroups and generators see Section 9.1 of Ethier and Kurtz (1986).

Proof We first prove the result under the assumption

(A3') for any $\delta > 0$, $\sum_{k > \delta n} k^2 p_k^n = 0$ for large n

Then we will argue that if n is large then with high probability we never see any families of size larger than $n\delta$.

The limiting coefficients satisfy (3.3) in Chapter 5, so using (4.1) there, we see that the martingale problem is well posed. Calculations in Example 1.6 of Chapter 5 show that (a) and (b) hold. To check (c) with $p = 4$, we begin by noting that when $Z_0^n = \ell$, Z_1^n is the sum of ℓ independent random variables ξ_i^n with mean $1 + \beta_n/n \geq 0$ and variance σ_n^2. Let $\delta > 0$. Under (A3') we have in addition $|\xi_i^n| \leq n\delta$ for large n. Using $(a+b)^4 \leq 2^4 a^4 + 2^4 b^4$ we have

$$\gamma_4^{1/n}(x) = nE(n^{-4}\{Z_1^n - nx\}^4 | Z_0^n = nx)$$
$$\leq 16n^{-3}(x\beta_n)^4 + 16n^{-3}E(\{Z_1^n - nx(1 + \beta_n/n)\}^4 | Z_0^n = nx)$$

To bound the second term we note that

$$E(\{Z_1^n - nx(1 + \beta_n/n)\}^4 | Z_0^n = nx) = nx E(\xi_i^n - (1 + \beta_n/n))^4$$
$$+ 6\binom{nx}{2} E(\xi_i^n - (1 + \beta_n/n))^2$$

To bound $E(\xi_i^n - (1 + \beta_n/n))^4$, we note that if $1 + \beta_n/n \leq n\delta$

$$E(\xi_i^n - (1 + \beta_n/n))^4 = \int_0^{n\delta} 4x^3 P(|\xi_i^n - (1 + \beta_n/n)| > x)\, dx$$
$$\leq 2(n\delta)^2 \int_0^{n\delta} 2x\, P(|\xi_i^n - (1 + \beta_n/n)| > x)\, dx \leq 2\delta^2 \sigma_n^2 n^2$$

Combining the last three estimates we see that if $|x| \leq R$ then

$$\gamma_4^{1/n}(x) \leq 32\delta^2 \sigma_n^2 R + 96\sigma_n^2 R^2 n^{-1} + 16(x\beta_n/n)^4$$

Since $\delta > 0$ is arbitrary we have established (c) and the desired conclusion follows from (8.2) and (7.1).

To replace (A3') by (A3) now, we observe that the convergence for the special case and use of the continuous mapping theorem as in Example 5.4 imply

$$n^{-2} \sum_{m=0}^{n-1} Z_m^n \Rightarrow \int_0^1 X_s\, ds$$

(A3) implies that the probability of a family of size larger than $n\delta$ is $o(n^{-2})$. Thus if we truncate the original sequence by not allowing more than $\delta_n n$ children where $\delta_n \to 0$ slowly, then (A1), (A2) and (A3') will hold and the probability of a difference between the two systems will converge to 0. □

Example 8.3. Wright Fisher Diffusion. Recall the setup of Example 1.7 in Chapter 5. We have an urn with n letters in it which may be A or a. To build up the urn at time $m+1$ we sample with replacement from the urn at time n but with probability α/n we ignore the draw and place an a in, and with probability β/n we ignore the draw and place an A in. Let Z_m^n be the number of A's in the urn at time n, and let $X_t^{1/n} = Z_{[nt]}^n/n$. Our goal is to show

(8.4) Theorem. As $n \to \infty$, $X_t^{1/n}$ converges weakly to the Wright-Fisher diffusion X_t, i.e., the solution of

$$dX_t = (-\alpha X_t + \beta(1-X_t))\,dt + \sqrt{X_t(1-X_t)}\,dB_t$$

References. Again the first results are due to Feller (1951). Chapter 10 of Ethier and Kurtz (1986) treats this problem and generalizations allowing more than 2 alleles.

Proof The limiting coefficients satisfy (3.3) in Chapter 5 so using (4.1) there we see that the martingale problem is well posed. Calculations in Example 1.7 in Chapter 5 show that (a) and (b) hold. To check condition (c) with $p = 4$ now, we note that if $Z_0^n = nx$ then the distribution of Z_1^n is the same as that of $S_n = \xi_1 + \cdots + \xi_n$ where $\xi_1, \ldots, \xi_n \in \{0,1\}$ are i.i.d. with $P(\xi_j = 1) = x(1-\alpha/n) + (1-x)\beta/n$.

$$E(S_n - np)^4 = E\left(\sum_{j=1}^n \xi_j - p\right)^4$$

$$= nE(\xi_j - p)^4 + 6\binom{n}{2}E(\xi_j - p)^2 \leq Cn^2$$

since $E(\xi_j - p)^m \leq 1$ for all $m \geq 0$. From this and the inequality $(a+b)^4 \leq 2^4 a^4 + 2^4 b^4$ it follows that

$$nE(\{Z_1^n/n - x\}^4 | Z_0^n = x) \leq n(-\alpha x + \beta(1-x))/n^4 + nCn^2/n^4 \to 0$$

uniformly for $x \in [0,1]$ and the proof is complete. □

Our final example has a deterministic limit. In such situations the following lemma is useful.

(8.5) Lemma. If for all $R < \infty$ we have

(a) $\lim_{h \downarrow 0} \sup_{|x| \leq R} |\hat{a}_{ij}^h(x)| = 0$

(b) $\lim_{h \downarrow 0} \sup_{|x| \leq R} |\hat{b}_i^h(x) - b_i(x)| = 0$

then (i), (ii), and (iii) of (7.1) hold with $a_{ij}(x) \equiv 0$.

Proof The new (a) and (b) imply the ones in (8.2). To check (c) of (8.2) with $p = 2$ we note that

$$\sum_i \hat{a}_{ii}^h(x) = \int |y - x|^2 K_h(x, dy) = \gamma_2^h(x) \qquad \square$$

Example 8.4. Epidemics. Consider a population with a fixed size of n individuals and think of a disease like measles so that recovered individuals are immune to further infection. Let S_t^n and I_t^n be the number of susceptible and infected individuals at time n, so that $n - (S_t^n + I_t^n)$ is the number of recovered individuals. We suppose that (S_t^n, I_t^n) makes transitions at the following rates:

$(s, i) \to (s + 1, i)$ $\alpha(n - s - i)$
$(s, i) \to (s - 1, i + 1)$ $\beta si/n$
$(s, i) \to (s, i - 1)$ γi

In words, each immune individual dies at rate α and is replaced by a new susceptible. Each susceptible individual becomes infected at rate β times the fraction of the population that is infected, i/n. Finally, each infected individual recovers at rate γ and enters the immune class.

Let $X_t^{1/n} = (S_t^n/n, I_t^n/n)$. To check (a) in (8.5) we note that

$$a_{11}^{1/n}(x_1, x_2) = \frac{1}{n^2} \cdot \{\alpha n(1 - x_1 - x_2) + \beta n x_1 x_2\}$$

$$a_{12}^{1/n}(x_1, x_2) = \frac{1}{n^2} \cdot \beta n x_1 x_2$$

$$a_{22}^{1/n}(x_1, x_2) = \frac{1}{n^2} \cdot \{\beta n x_1 x_2 + \gamma n x_1 x_2\}$$

and all three terms converge to 0 uniformly in $\Gamma = \{(x_1, x_2) : x_1, x_2 \geq 0, x_1 + x_2 \leq 1\}$. To check (b) in (8.5) we observe

$$b_1^{1/n}(x_1, x_2) = \alpha n(1 - x_1 - x_2) \cdot \frac{1}{n} - \beta n x_1 x_2 \cdot \frac{1}{n}$$

$$= \alpha(1 - x_1 - x_2) - \beta x_1 x_2 \equiv b_1(x)$$

$$b_2^{1/n}(x_1, x_2) = \beta n x_1 x_2 \cdot \frac{1}{n} - \gamma n x_2 \cdot \frac{1}{n}$$

$$= \beta x_1 x_2 - \gamma x_2 \equiv b_2(x)$$

310   *Chapter 8   Weak Convergence*

The limiting coefficients are Lipschitz continuous on Γ. Their values outside Γ are irrelevant so we extend them to be Lipschitz continuous. It follows that the martingale problem is well posed and we have

(8.6) Theorem. As $n \to \infty$, $X_t^{1/n}$ converges weakly to X_t, the solution of the ordinary differential equation

$$dX_t = b(X_t)\,dt$$

Remark. If we set $\alpha = 0$ then our epidemic model reduces to one considered in Sections 11.1-11.3 of Ethier and Kurtz (1986). In addition to (8.6) they prove a central limit theorem for $\sqrt{n}(X_t^{1/n} - X_t)$.

Solutions to Exercises

Chapter 1

1.1. Let $\mathcal{A} = \{A = \{\omega : (\omega(t_1), \omega(t_2), \ldots) \in B\} : B \in \mathcal{R}^{\{1,2,\ldots\}}\}$. Clearly, any $A \in \mathcal{A}$ is in the σ-field generated by the finite dimensional sets. To complete the proof, we only have to check that $\mathcal{A}$ is a σ-field. The first and easier step is to note if $A = \{\omega : (\omega(t_1), \omega(t_2), \ldots) \in B\}$ then $A^c = \{\omega : (\omega(t_1), \omega(t_2), \ldots) \in B^c\} \in \mathcal{A}$. To check that $\mathcal{A}$ is closed under countable unions, let $A_n = \{\omega : (\omega(t_1^n), \omega(t_2^n), \ldots) \in B_n\}$, let $t_1, t_2, \ldots$ be an ordering of $\{t_m^n : n, m \geq 1\}$ and note that we can write $A_n = \{\omega : (\omega(t_1), \omega(t_2), \ldots) \in E_n\}$ so $\cup_n A_n = \{\omega : (\omega(t_1), \omega(t_2), \ldots) \in \cup_n E_n\} \in \mathcal{A}$.

1.2. Let $A_n = \{\omega :$ there is an $s \in [0,1]$ so that $|B_t - B_s| \leq C|t-s|^\gamma$ when $|t - s| \leq k/n\}$. For $1 \leq i \leq n - k + 1$ let

$$Y_{i,n} = \max\left\{\left|B\left(\frac{i+j}{n}\right) - B\left(\frac{i+j-1}{n}\right)\right| : j = 0, 1, \ldots k-1\right\}$$

$B_n = \{$ at least one $Y_{i,n}$ is $\leq (2k-1)C/n^\gamma\}$

Again $A_n \subset B_n$ but this time if $\gamma > 1/2 + 1/k$, i.e., $k(1/2 - \gamma) < -1$, then

$$P(B_n) \leq nP(|B(1/n)| \leq (2k-1)C/n^\gamma)^k$$
$$\leq nP(|B(1)| \leq (2k-1)Cn^{1/2-\gamma})^k$$
$$\leq C'n^{k(1/2-\gamma)+1} \to 0$$

1.3. The first step is to observe that the scaling relationship (1.3) implies

$(\star)$ $\qquad\qquad\qquad\qquad \Delta_{m,n} \stackrel{d}{=} 2^{-n/2}\Delta_{1,0}$

while the definition of Brownian motion shows $E\Delta_{1,0}^2 = t$, and $E(\Delta_{1,0}^2 - t)^2 = C < \infty$. Using $(\star)$ and the definition of Brownian motion, it follows that if $k \neq m$ then $\Delta_{k,n}^2 - t2^{-n}$ and $\Delta_{m,n}^2 - t2^{-n}$ are independent and have mean 0 so

$$E\left(\sum_{1 \leq m \leq 2^n}(\Delta_{m,n}^2 - t2^{-n})\right)^2 = \sum_{1 \leq m \leq 2^n} E\left(\Delta_{m,n}^2 - t2^{-n}\right)^2 = 2^n C 2^{-2n}$$

where in the last equality we have used (*) again. The last result and Chebyshev's inequality imply

$$P\left(\left|\sum_{1\le m\le 2^n} \Delta_{m,n}^2 - t\right| \ge 1/n\right) \le n^2 \cdot C 2^{-n}$$

The right-hand side is summable so the Borel-Cantelli lemma implies

$$P\left(\left|\sum_{m\le 2^n} \Delta_{m,n}^2 - t\right| \ge 1/n \text{ infinitely often}\right) = 0$$

2.1. Take $Y(\omega) = f(\omega_{t-s})$.

2.2. When $f(x) = x_i$, 2.1 becomes $E_x(B_t^i|\mathcal{F}_s) = E_{B(s)}B_{t-s}^i = B_s^i$ since under P_x, B_{t-s}^i is normal with mean x_i. When $f(x) = x_i x_j$ with $i \ne j$ 2.1 becomes

$$E_x(B_t^i B_t^j|\mathcal{F}_s) = E_{B_s}(B_{t-s}^i B_{t-s}^j) = B_s^i B_s^j$$

since under P_x, B_{t-s}^i and B_{t-s}^j are independent normals with means x_i and x_j.

2.3. Let $Y = 1_{(T_0 > t)}$ and note that $T_0 \circ \theta_1 = R - 1$ so $Y \circ \theta_1 = 1_{(R > 1+t)}$. Using the Markov property gives

$$P_x(R > 1 + t|\mathcal{F}_1) = P_{B_1}(T_0 > t)$$

Taking expected value now and recalling $P_x(B_1 = y) = p_1(x, y)$ gives

$$P_x(R > 1 + t) = \int p_1(x, y) P_y(T_0 > t)\, dy$$

2.4. Let $Y = 1_{(T_0 > 1-t)}$ and note that $Y \circ \theta_t = 1_{(L \le t)}$. Using the Markov property gives

$$P_0(L \le t|\mathcal{F}_t) = P_{B_t}(T_0 > 1 - t)$$

Taking expected value now and recalling $P_0(B_t = y) = p_t(0, y)$ gives

$$P_0(L \le t) = \int p_t(0, y) P_y(T_0 > 1 - t)\, dy$$

2.5. We will prove the result by induction on n. By assumption it holds for $n = 1$. Suppose now it is true for n. Let $Y(\omega) = 1_{(T > 1, \omega_1 \in K)}$. Applying the Markov property at time n gives

$$E_x(Y \circ \theta_n|\mathcal{F}_n) = P_{B_n}(T > 1, B_1 \in K)$$

Integrating both sides over $A_n = \{T > n, B_n \in K\}$, recalling the definition of conditional expectation, and using our assumption shows

$$E_x(Y \circ \theta_n; A_n) \geq \alpha P(A_n)$$

The right-hand side is smaller than $P(A_{n+1})$ and the desired result follows.

2.6. Since $\int_0^s h(r, B_r)\, dr \in \mathcal{F}_s$, we have

$$E_x\left(\int_0^t h(r, B_r)\, dr \bigg| \mathcal{F}_s\right) = \int_0^s h(r, B_r)\, dr$$
$$+ E_x\left(\int_s^t h(r, B_r)\, dr \bigg| \mathcal{F}_s\right)$$

To evaluate the second term we let $Y(\omega) = \int_0^{t-s} h(s+u, \omega_u)\, du$ and note

$$Y \circ \theta_s = \int_0^{t-s} h(s+u, B_{s+u})\, du$$

so the Markov property implies

$$E_x\left(\int_s^t h(r, B_r)\, dr \bigg| \mathcal{F}_s\right) = E_{B_s}\left(\int_0^{t-s} h(s+u, B_u)\, du\right)$$

2.7. Since $\int_0^s c(B_r)\, dr \in \mathcal{F}_s$, we have

$$E_x\left(f(B_t)\exp(c_t)| \mathcal{F}_s\right) = \exp(c_s) E_x\left(f(B_t)\exp\left(\int_s^t c(B_r)\, dr\right) \bigg| \mathcal{F}_s\right)$$

To evaluate the second term we let $Y(\omega) = f(B_{t-s})\exp(\int_0^{t-s} c(\omega_u)\, du)$ and note $Y \circ \theta_s = f(B_t)\exp(\int_0^{t-s} c(B_{s+u})\, du)$ so the Markov property implies

$$E_x\left(f(B_t)\exp(\int_s^t c(B_r)\, dr)\bigg| \mathcal{F}_s\right) = E_{B(s)}\left(f(B_{t-s})\exp\int_0^{t-s} c(\omega_u)\, du\right)$$

2.8. (2.8), (2.9), and the Markov property imply that with probability one there are times $s_1 > t_1 > s_2 > t_2 \ldots$ that converge to a so that $B(s_n) = B(a)$ and $B(t_n) > B(a)$. In each interval (s_{n+1}, s_n) there is a local maximum.

2.9. $Z \equiv \limsup_{t \downarrow 0} B(t)/f(t) \in \mathcal{F}_0^+$ so (2.7) implies $P_0(Z > c) \in \{0, 1\}$ for each c, which implies that Z is constant almost surely.

2.10. Let $C < \infty$, $t_n \downarrow 0$, $A_N = \{B(t_n) \geq C\sqrt{t_n} \text{ for some } n \geq N\}$ and $A = \cap_N A_N$. A trivial inequality and the scaling relation (1.3) imply

$$P_0(A_N) \geq P_0(B(t_N) \geq C\sqrt{t_N}) = P_0(B(1) \geq C) > 0$$

Letting $N \to \infty$ and noting $A_N \downarrow A$ we have $P_0(A) \geq P_0(B_1 \geq C) > 0$. Since $A \in \mathcal{F}_0^+$ it follows from (2.7) that $P_0(A) = 1$, that is, $\limsup_{t \to 0} B(t)/\sqrt{t} \geq C$ with probability one. Since C is arbitrary the proof is complete.

2.11. Since coordinates are independent it suffices to prove the result in one dimension. Continuity implies that if δ is small

$$P_0\left(\sup_{0 \leq s \leq \delta} |B_s| < \epsilon/2\right) \equiv 2\alpha > 0$$

so it follows from symmetry that

$$P_0\left(\sup_{0 \leq s \leq \delta} |B_s| < \epsilon/2, B_\delta < 0\right)$$
$$= P_0\left(\sup_{0 \leq s \leq \delta} |B_s| < \epsilon/2, B_\delta > 0\right) = \alpha > 0$$

Using the top result for $x \geq 0$ and the bottom one for $x \leq 0$ shows that if $x \in [-\epsilon/2, \epsilon/2]$

$$P_x\left(\sup_{0 \leq t \leq \delta} |B_t| < \epsilon, B_\delta \in [-\epsilon/2, \epsilon/2]\right) \geq \alpha > 0$$

Iterating and using Exercise 2.5 gives that if $x \in [-\epsilon/2, \epsilon/2]$

$$P_x\left(\sup_{0 \leq t \leq n\delta} |B_t| < \epsilon, B_{n\delta} \in [-\epsilon/2, \epsilon/2]\right) \geq \alpha^n > 0$$

3.1. Suppose $A = \cup_{n=1}^\infty K_n$ where K_n are closed. Let $H_n = \cup_{m=1}^n K_m$. H_n is closed so (3.4) implies $T_n = \inf\{t : B_t \in H_n\}$ is a stopping time. In view of (3.2) we can complete the proof now by showing that $T_n \downarrow T_A$ as $n \uparrow \infty$. Clearly, $T_A \leq \lim T_n$ for all n. To see that $<$ cannot hold let t be a time at which $B_t \in A$. Then $B_t \in K_n$ for some n and hence $\lim T_n \leq t$. Taking the inf over t with $B_t \in A$ we have $\lim T_n \leq T_A$.

3.2. If $m2^{-n} < t \leq (m+1)2^{-n}$ then $\{S_n < t\} = \{S < m2^{-n}\} \in \mathcal{F}_{m2^{-n}} \subset \mathcal{F}_t$.

3.3. Since constant times are stopping times the last three statements follow from the first three.
$$\{S \wedge T \leq t\} = \{S \leq t\} \cup \{T \leq t\} \in \mathcal{F}_t.$$
$$\{S \vee T \leq t\} = \{S \leq t\} \cap \{T \leq t\} \in \mathcal{F}_t$$
$$\{S + T < t\} = \cup_{q,r \in \mathbb{Q}: q+r<t} \{S < q\} \cap \{T < r\} \in \mathcal{F}_t$$

3.4. Define R_n by $R_1 = T_1$, $R_n = R_{n-1} \vee T_n$. Repeated use of Exercise 3.3 shows that R_n is a stopping time. As $n \uparrow \infty$ $R_n \uparrow \sup_n T_n$ so the first conclusion follows from (3.3). Define S_n by $S_1 = T_1$, $S_n = S_{n-1} \wedge T_n$. Repeated use of Exercise 3.3 shows that S_n is a stopping time. As $n \uparrow \infty$, $S_n \downarrow \inf_n T_n$ so the second conclusion follows from (3.2). The last two conclusions follow from the first two since

$$\limsup_n T_n = \inf_n \sup_{m \geq n} T_m \quad \text{and} \quad \liminf_n T_n = \sup_n \inf_{m \geq n} T_m$$

3.5. First if $A \in \mathcal{F}_S$ then

$$A \cap \{S < t\} = \cup_n (A \cap \{S \leq t - 1/n\}) \in \mathcal{F}_t$$

On the other hand if $A \cap \{S < t\} \in \mathcal{F}_t$ and the filtration is right continuous then $A \cap \{S \leq t\} = \cap_n (A \cap \{S < t + 1/n\}) \in \cap_n \mathcal{F}_{t+1/n} = \mathcal{F}_t$.

3.6. $\{R \leq t\} = \{S \leq t\} \cap A \in \mathcal{F}_t$ since $A \in \mathcal{F}_S$

3.7. (i) Let $r = s \wedge t$.

$$\{S < t\} \cap \{S < s\} = \{S < r\} \in \mathcal{F}_r \subset \mathcal{F}_s$$
$$\{S \leq t\} \cap \{S \leq s\} = \{S \leq r\} \in \mathcal{F}_r \subset \mathcal{F}_s$$

This shows $\{S < t\}$ and $\{S \leq t\}$ are in $\mathcal{F}_S$. Taking complements and intersections we get $\{S \geq t\}$, $\{S > t\}$, and $\{S = t\}$ are in $\mathcal{F}_S$.

(ii) $\{S < T\} \cap \{S < t\} = \cup_{q<t} \{S < q\} \cap \{T > q\} \in \mathcal{F}_t$ by (i), so $\{S < T\} \in \mathcal{F}_S$. $\{S < T\} \cap \{T < t\} = \cup_{q<t} \{S < q\} \cap \{q < T < t\} \in \mathcal{F}_t$ by (i), so $\{S < T\} \in \mathcal{F}_T$. Here the unions were taken over rational q. Interchanging the roles of S and T we have $\{S > T\}$ in $\mathcal{F}_S \cap \mathcal{F}_T$. Taking complements and intersections we get $\{S \geq T\}$, $\{S \leq T\}$, and $\{S = T\}$ are in $\mathcal{F}_S \cap \mathcal{F}_T$.

3.8. If $A \in \mathcal{R}$ then

$$\{B(S_n) \in A\} \cap \{S_n \leq t\} = \cup_{0 \leq m \leq 2^n t} \{S_n = m/2^n\} \cap \{B(m/2^n) \in A\} \in \mathcal{F}_t$$

by (i) in Exercise 3.7. This shows $\{B(S_n) \in A\} \in \mathcal{F}_{S_n}$ so $B(S_n) \in \mathcal{F}_{S_n}$. Letting $n \to \infty$ we have $B_S = \lim_n B(S_n) \in \cap_n \mathcal{F}_{S_n} = \mathcal{F}_S$.

3.9. (3.5) implies $\mathcal{F}_{S \wedge T_n} \subset \mathcal{F}_S$. To argue the other inclusion let $A \in \mathcal{F}_S$. Since

$$A = \cup_n (A \cap \{T_n > S\})$$

it suffices to show that $A \cap \{T_n > S\} \in \mathcal{F}_{S \wedge T_n}$. To do this we observe

$$A \cap \{T_n > S\} \cap \{T_n \wedge S < t\} = A \cap \{T_n > S\} \cap \{S < t\}$$
$$= \cup_{q < t} (A \cap \{S < q\}) \cap \{T_n > q\} \in \mathcal{F}_t$$

since $A \in \mathcal{F}_S$ implies $A \cap \{S < q\} \in \mathcal{F}_q$ and the fact that T_n is a stopping time implies $\{T_n > q\} \in \mathcal{F}_q$.

3.10. Since $\int_0^S g(B_s)\,ds \in \mathcal{F}_S$ we have

$$E_x\left(\int_0^S g(B_s)\,ds \,\bigg|\, \mathcal{F}_S\right) = \int_0^S g(B_s)\,ds$$
$$+ E_x\left(\int_S^T g(B_s)\,ds \,\bigg|\, \mathcal{F}_S\right)$$

Applying the strong Markov property to $Y(\omega) = \int_0^{T(\omega)} g(\omega_s)\,ds$ we have

$$E_x\left(\int_S^T g(B_s)\,ds \,\bigg|\, \mathcal{F}_S\right) = u(B_S)$$

3.11. Let $Y_s(\omega) = 1$ if $s < t$ and $u < \omega(t - s) < v$, 0 otherwise. Let

$$\bar{Y}_s(\omega) = \begin{cases} 1 & \text{if } s < t,\, 2a - v < \omega(t-s) < 2a - u \\ 0 & \text{otherwise} \end{cases}$$

Symmetry of the normal distribution implies $E_a Y_s = E_a \bar{Y}_s$, so if we let $S = \inf\{s < t : B_s = a\}$ and apply the strong Markov property then on $\{S < \infty\}$

$$E_x(Y_S \circ \theta_S | \mathcal{F}_S) = E_a Y_S = E_a \bar{Y}_S = E_x(\bar{Y}_S \circ \theta_S | \mathcal{F}_S)$$

Taking expected values now gives the desired result.

4.1. We begin by noting symmetry and (2.5) imply

$$P_0(R \le 1+t) = 2\int_0^\infty p_1(0,y) \int_0^t P_y(T_0 = s)\, ds\, dy$$
$$= \int_0^t 2 \int_0^\infty p_1(0,y) \int_0^t P_y(T_0 = s)\, dy\, ds$$

by Fubini's theorem, so the integrand gives the density $P_0(R = 1+t)$. Since $P_y(T_0 = t) = P_0(T_y = t)$, (4.1) gives

$$P_0(R = 1+t) = 2\int_0^\infty \frac{1}{\sqrt{2\pi}} e^{-y^2/2} \frac{1}{\sqrt{2\pi t^3}} y e^{-y^2/2t}\, dy$$
$$= \frac{1}{2\pi t^{3/2}} \int_0^\infty y e^{-y^2(1+t)/2t}\, dy = \frac{1}{2\pi t^{3/2}} \frac{t}{(1+t)}$$

4.2. Translation invariance implies that if $r < s$ then

$$E_{(x,r)} f(B_{\tau_s} - B_0) = E_{(0,0)} f(B_{\tau_{s-r}})$$

Applying the strong Markov property to $Y(\omega) = f(\omega(\tau_s) - \omega_0)$ at time τ_r, and using the last identity we have

$$E_{(0,0)}(f(B_{\tau_s} - B_{\tau_r})|\mathcal{F}_{\tau_r}) = E_{(0,0)} f(B_{\tau_{s-r}})$$

The rest of the argument is identical to that for (4.4).

4.3. As $s \to 0$, $\varphi(s) \to 1$, so for some $\delta > 0$ we have $\varphi_\theta(s) > 0$ for all $s \le \delta$. Using the equation $\varphi(s)\varphi(t) = \varphi(s+t)$ we can now conclude by induction that for all n, $\varphi(s) > 0$ for $s \le n\delta$, and we can take logarithms without feeling guilty. The rest is easy: $\psi(2^{-n}) + \psi(2^{-n}) = \psi(2^{-(n-1)})$ and induction tells us that $\psi(2^{-n}) = 2^{-n}\psi(1)$, while $\psi((m-1)2^{-n}) + \psi(2^{-n}) = \psi(m 2^{-n})$ and induction implies $\psi(m 2^{-n}) = m 2^{-n} \psi(1)$. Since $\varphi(s)$ is continuous and does not vanish, $\psi(s) = \varphi(s)$ is continuous and the desired result follows.

4.4. Let $\varphi_\lambda(a) = E_0(\exp(-\lambda T_a))$. (4.4) and (4.5) imply

$$\varphi_\lambda(a)\varphi_\lambda(b) = \varphi_\lambda(a+b) \qquad \varphi_\lambda(a) = \varphi_{\lambda a^2}(1)$$

As before the first equation implies $\varphi_\lambda(a) = \exp(c_\lambda a)$ while the second with $\lambda = 1$ and $a = \sqrt{b}$ gives $c_b = \sqrt{b} c_1$ and this implies $c_\lambda = -\kappa\sqrt{\lambda}$.

4.5. Let $M_1 = \max_{0 \le s \le 1} B_s$. Exercise 3.10 implies that with probability 1, $M_1 > B_1$ and hence $a \to T_a$ is discontinuous at $a = M_1$. This shows $P_0(a \to T_a$ is discontinuous in $[0,n]) \to 1$ as $n \to \infty$ and the result follows from the hint.

4.6. Let $M_1 = \max_{0 \le s \le 1} B_s^2$ and $T = \inf\{t > 1 : B_s^2 > M_1\}$. Since $B_1^2 < M_1$ the strong Markov property and (4.3) imply that with probability one $B_T^2 \ne B_{M_1}^2$ so $s \to C_s$ is discontinuous at $s = M_1$. This shows $P_0(s \to C_s$ is discontinuous in $[0, n]) \to 1$ as $n \to \infty$ and the result follows from the hint.

4.7. If we let $\bar{f}(x) = f(\bar{x})$, then it follows from the strong Markov property and symmetry of Brownian motion that

$$E_x(f(B_t); \tau \le t) = E_x[E_0 f(B_{t-\tau}); \tau \le t]$$
$$= E_x[E_0 \bar{f}(B_{t-\tau}); \tau \le t]$$
$$= E_x[\bar{f}(B_t); \tau \le t] = E_x(\bar{f}(B_t))$$

The last equality holds since $\bar{f}(y) = 0$ when $y_d \ge 0$, and the desired formula follows as in (4.8).

Chapter 2

1.1. If $A \in \mathcal{F}_a$ and $a < b$ then $H_n(s, \omega) = 1_{[a+1/n, b+1/n)} 1_A$ is optional and converges to $1_{(a,b]}(s) 1_A(\omega)$ as $n \to \infty$.

2.1. The martingale property w.r.t. a filtration $\mathcal{G}_t$ is that

$$E(X_t^S; A) = E(X_s^S; A)$$

for all $A \in \mathcal{G}_s$. (3.5) in Chapter 1 implies that $\mathcal{F}_{S \wedge s} \subset \mathcal{F}_s$ so if X_s^S is a martingale with respect to $\mathcal{F}_s$ it is a martingale with respect to $\mathcal{F}_{S \wedge s}$. To argue the other direction let $A \in \mathcal{F}_s$. We claim that $A \cap \{S > s\} \in \mathcal{F}_{S \wedge s}$. To prove this we have to check that for all r

$$(A \cap \{S > s\}) \cap \{S \wedge s \le r\} \in \mathcal{F}_r$$

but this is $\emptyset$ for $r < s$ and true for $r \ge s$. If X_s^S is a martingale with respect to $\mathcal{F}_{S \wedge s}$ it follows that if $A \in \mathcal{F}_s$ then

$$E(X_t^S; A \cap \{S > s\}) = E(X_s^S; A \cap \{S > s\})$$

On the other hand

$$E(X_t^S; A \cap \{S \le s\}) = E(X_S 1_{(S>0)}; A \cap \{S \le s\}) = E(X_s^S; A \cap \{S \le s\})$$

Adding the last two equations gives the desired conclusion.

2.2. To check integrability we note that

$$E_x|X_t|^p = \int (2\pi t)^{-1} e^{-|y-x|^2/2t} |\log |y||^p \, dy < \infty$$

since $|\log |y||$ is integrable near 0 and $e^{-|y-x|^2/2t}$ takes care of the behavior for large y. To show that $E_x X_t \to \infty$ we observe that for any $R < \infty$

$$E_x X_t \geq (2\pi t)^{-d/2} \int_{|y| \leq 1} \log |y| \, dy + (\log R) P_x(|B_t| > R)$$

The integral is convergent and as we showed in Example 2.1, $P_x(|B_t| > R) \to 1$ as $t \to \infty$ so the desired result follows from the fact that R is arbitrary.

2.3. Let $T_n = \inf\{t : |X_t| > n\}$. By (2.3) this sequence will reduce X_t. Note that $X_t^{T_n} \leq n$. Jensen's inequality for conditional expectation (see e.g., (1.1.d) in Chapter 4 of Durrett (1995)) implies

$$E(\varphi(X_t^{T_n})|\mathcal{F}_{T_n \wedge s}) \geq \varphi(E(X_t^{T_n}|\mathcal{F}_{T_n \wedge s})) = \varphi(X_s^{T_n})$$

2.4. Let $T_n \leq n$ be a sequence which reduces X, and $S_n = (T_n - R)^+$. If $s < t$ then definitions involved (note $S_n = T_n - R$ on $\{S_n > 0\} = \{T_n > R\}$), the optional stopping theorem, and the fact $1_{(T_n > R)} \in \mathcal{F}_R \subset \mathcal{F}_{R+s}$ imply

$$E(Y_{t \wedge S_n} 1_{(S_n > 0)}|\mathcal{G}_s) = E(X_{(R+t) \wedge T_n} 1_{(T_n > R)}|\mathcal{F}_{R+s})$$
$$= 1_{(T_n > R)} E(X_{(R+t) \wedge T_n}|\mathcal{F}_{R+s})$$
$$= X_{(R+s) \wedge T_n} 1_{(T_n > R)} = Y_{s \wedge S_n} 1_{(S_n > 0)}$$

2.5. Let T_n be a sequence that reduces X. The martingale property shows that if $A \in \mathcal{F}_0$ then

$$E(X_{t \wedge T_n} 1_{(T_n > 0)}; A) = E(X_0 1_{(T_n > 0)}; A)$$

Letting $n \to \infty$ and using Fatou's lemma and the dominated convergence theorem gives

$$E(X_t; A) \leq \liminf_{n \to \infty} E(X_{t \wedge T_n} 1_{(T_n > 0)}; A)$$
$$\leq \lim_{n \to \infty} E(X_0 1_{(T_n > 0)}; A) = E(X_0; A)$$

Taking $A = \Omega$ we see that $EX_t \leq EX_0 < \infty$. Since this holds for all $A \in \mathcal{F}_0$ we get $E(X_t|\mathcal{F}_0) \leq X_0$. To replace 0 by s apply the last conclusion to $Y_t = X_{s+t}$, which by Exercise 2.4 is a local martingale.

3.1. Since $t \to V_t$ is increasing, we only have to rule out jump discontinuities. Suppose there is a t and an $\epsilon > 0$ so that $s < t < u$ then $V(u) - V(s) > 3\epsilon$ for all n. It is easy to see that $V(u) - V(s)$ is the variation of X over $[s, u]$. Let $s_0 < t < u_0$. We will now derive a contradiction that the variation over $[s_0, u_0]$ is infinite. Pick $\delta > 0$ so that if $|r - t| < \delta$ then $|X_t - X_r| < \epsilon$. Assuming s_n and u_n have been defined, pick a partition of $[s_n, u_n]$ not containing t with mesh $< \delta$ and variation $> 2\epsilon$. Let s_{n+1} be the largest point in the partition $< t$ and u_{n+1} be the smallest point in the partition $> t$. Our construction shows that the variation of X over $[s_n, u_n] - [s_{n+1}, t_{n+1}]$ is always $> \epsilon$, so the total variation over $[s_0, u_0]$ is infinite, a contradiction which implies $t \to V_t$ is continuous.

3.2. $(X + Y)_t Z_t - \langle X, Z \rangle_t - \langle Y, Z \rangle_t$ is a local martingale.

3.3. $-X_0 Z_t$ is a local martingale, so if $Y_t = -X_0$ then $\langle Y, Z \rangle_t \equiv 0$ and the desired result follows from Exercise 3.2.

3.4. $(aX_t)(bY_t) - ab\langle X, Y \rangle_t = ab(X_t Y_t - \langle X, Y \rangle_t)$ is a local martingale.

3.5. If T is such that X^T is a bounded martingale it follows from (3.7) and the definition of $\langle X \rangle$ that $\langle X^T \rangle = \langle X \rangle^T$. For a general stopping time T, let $T_n = \inf\{t : |X_t| > n\}$, note that the last result implies $\langle X^{T \wedge T_n} \rangle = \langle X \rangle^{T \wedge T_n}$ and then let $n \to \infty$. The result for the covariance process follows immediately from the definition and the result for the variance process.

3.6. Since $X_t^2 - \langle X \rangle_t$ is a martingale and $|X_t| \leq M$ for all t

$$E\langle X \rangle_t = EX_t^2 - EX_0^2 \leq M^2$$

so letting $t \to \infty$ we have $E\langle X \rangle_\infty \leq M^2$. This shows that $X_t^2 - \langle X \rangle_t$ is dominated by an integrable random variable and hence is uniformly integrable.

3.7. The optional stopping theorem implies

$$E(X_{S+t}|\mathcal{F}_{S+s}) = X_{S+s}$$

and $X_S \in \mathcal{F}_S \subset \mathcal{F}_{S+s}$ so

$$E(X_{S+t} - X_S|\mathcal{F}_{S+s}) = X_{S+s} - X_S$$

i.e., Y_s is a martingale. To prepare for the proof of the second result we note that imitating the proof of (2.4)

$$E((X_{S+t} - X_S)^2|\mathcal{F}_{S+s}) = E((X_{S+t} - X_{S+s})^2|\mathcal{F}_{S+s}) + (X_{S+s} - X_S)^2$$
$$= E(X_{S+t}^2 - X_{S+s}^2|\mathcal{F}_{S+s}) + (X_{S+s} - X_S)^2$$

Let $Z_t = (X_{S+t} - X_S)^2 - \{\langle X \rangle_{S+t} - \langle X \rangle_S\}$. Using the last equality we get

$$\begin{aligned} E(Z_t | \mathcal{F}_{S+s}) &= E(X_{S+t}^2 - X_{S+s}^2 - \{\langle X \rangle_{S+t} - \langle X \rangle_S\} | \mathcal{F}_{S+s}) + (X_{S+s} - X_s)^2 \\ &= E(X_{S+t}^2 - \langle X \rangle_{S+t} | \mathcal{F}_{S+s}) - X_{S+s}^2 + \langle X \rangle_S + (X_{S+s} - X_s)^2 \\ &= -\langle X \rangle_{S+s} + \langle X \rangle_S + (X_{S+s} - X_s)^2 = Z_s \end{aligned}$$

where the last equality follows from the optional stopping theorem and Exercise 3.6. This shows that Z_t is a martingale, so $\langle Y \rangle_t = \langle X \rangle_{S+t} - \langle X \rangle_S$.

3.8. By stopping at $T_n = \inf\{t : |X_t| > n\}$ and using Exercise 3.5 it suffices to prove the result when X is a bounded martingale. Using Exercise 3.7, we can suppose without loss of generality that $S = 0$. Using the L^2 maximal inequality on the martingale $X_{t \wedge T}$, and the optional stopping theorem on the martingale $X_t^2 - \langle X \rangle_t$ it follows that

$$E\left(\sup_{t \le n} X_{t \wedge T}^2\right) \le 4 E(X_{T \wedge n}^2) = E(\langle X \rangle_{t \wedge T}) = 0$$

Letting $n \to \infty$ now gives the desired conclusion.

3.9. This is an immediate consequence of (3.8).

4.1. It is simply a matter of patiently checking all the cases.
1. $s < t \le a$. $Y_t = Y_0 \in \mathcal{F}_0$ so

$$E(Y_t | \mathcal{F}_s) = E(Y_0 | \mathcal{F}_s) = Y_0 = Y_s$$

2. $s < a \le t \le b$

$$\begin{aligned} E(Y_t | \mathcal{F}_s) &= E(E(Y_t | \mathcal{F}_a) \mathcal{F}_s) = E(Y_a | \mathcal{F}_s) \\ &= E(Y_0 | \mathcal{F}_s) = Y_0 = Y_s \end{aligned}$$

3. $s < b$, $t > b$. $Y_t = Y_b$ so this reduces to case 2 if $s < a$ or to our assumption if $a \le s < b$. 4. $b \le s < t$. $E(Y_t | \mathcal{F}_s) = E(Y_b | \mathcal{F}_s) = Y_b = Y_s$.

4.2. If we fix ω then $t \to \langle X \rangle_t$ defines a measure. The triangle inequality for L^2 of that measure implies

$$\left(\int H_s^2 \, d\langle X \rangle_t \right)^{1/2} + \left(\int K_s^2 \, d\langle X \rangle_t \right)^{1/2} \le \left(\int (H_s + K_s)^2 \, d\langle X \rangle_t \right)^{1/2}$$

Now take expected value.

4.3. $X \in \mathcal{M}^2$ implies $E \sup_t X_t^2 < \infty$ and hence $EX_0^2 < \infty$. To check the other condition let T_n be a sequence of stopping times $\uparrow \infty$ so that X^{T_n} is bounded and $\langle X \rangle_{T_n} \leq n$. The optional stopping theorem implies

$$E\left(X_{T_n}^2 - \langle X \rangle_{T_n}\right) 1_{(T_n > 0)} = EX_0^2 1_{(T_n > 0)}$$

Letting $n \to \infty$ it follows that $E(X_\infty^2 - \langle X \rangle_\infty) = EX_0^2$. To prove the converse rearrange the displayed equation to conclude

$$E\left(X_{T_n}^2 1_{(T_n > 0)}\right) \leq EX_0^2 + E\langle X \rangle_\infty$$

so X is L^2 bounded.

4.4. The triangle inequality implies $\|x\| \leq \|x_n\| + \|x - x_n\|$ so

$$\|x\| \leq \liminf \|x_n\|$$

For the other inequality note that $\|x_n\| \leq \|x\| + \|x_n - x\|$ so

$$\|x\| \geq \limsup \|x_n\|$$

4.5. Let $H^n \in b\Pi_1$ with $\|H^n - H\|_X \to 0$. Since $\|(H^n \cdot X) - (H \cdot X)\|_2 \to 0$, using Exercise 4.4, the isometry property for $b\Pi_1$, and Exercise 4.4 again

$$\|H \cdot X\|_2 = \lim \|H^n \cdot X\|_2 = \lim \|H^n\|_X = \|H\|_X$$

5.1. If $L, L' \in \mathcal{M}^2$ have the desired properties then taking $N = L - L'$ we have $\langle L, L - L' \rangle = \langle L', L - L' \rangle$ so $\langle L - L' \rangle \equiv 0$. Using Exercise 3.8 now it follows that $L - L'$ is constant and hence must be $\equiv 0$.

5.2. If we let $H_s \equiv 1$ and $K_s = 1_{(s \leq T)}$ then $X = H \cdot X$ and $Y^T = K \cdot Y$. (It is for this reason we have assumed $X_0 = Y_0 = 0$.) Using (5.4) now it follows that

$$\langle X, Y^T \rangle_t = \int_0^t 1_{(s \leq T)} d\langle X, Y \rangle_s = \langle X, Y \rangle_t^T$$

5.3. If we let $H_s = 1_{(s \leq T)}$ and $K_s = 1_{(s > T)}$ then $X^T = H \cdot X$ and $Y - Y^T = K \cdot Y$. Using (5.4) now it follows that $\langle X^T, Y - Y^T \rangle_t \equiv 0$ so the desired result follows from (3.11).

6.1. Write $H_s = H_s^1 + H_s^2$ and $K_s = K_s^1 + K_s^2$ where

$$H_s^1 = K_s^1 = H_s 1_{(S \leq s \leq T)} = K_s 1_{(S \leq s \leq T)}$$

Clearly, $(H^1 \cdot X)_t = (K^1 \cdot Y)_t$ for all t. Since $H_s^2 = K_s^2 = 0$ for $S \leq s \leq T$, (5.4) implies
$$\langle H^2 \cdot X \rangle_s = \langle K^2 \cdot X \rangle_s = 0 \quad \text{for } S \leq s \leq T$$
and it follows from Exercise 3.8 that $(H^2 \cdot X)_t$ and $(K^2 \cdot X)_t$ are constant on $[S, T]$. Combining this with the first result and using (4.3.b) gives the desired conclusion.

6.2. Stop X at $T_n = \inf\{t : |X_t| > n \text{ or } \int_0^t H_s^2 \, d\langle X \rangle_s\}$ to get a bounded martingale and an integrand in $\Pi_2(X)$. Exercise 4.5 implies
$$\|H^{T_n} \cdot X^{T_n}\|_2 = E \int_0^{T_n} H_t^2 \, d\langle X \rangle_t$$

Using the L^2 maximal inequality it follows that
$$E\left(\sup_{t \leq T_n} (H \cdot X)_t^2\right) \leq 4E \int_0^{T_n} H_t^2 \, d\langle X \rangle_t$$

Letting $n \to \infty$ now we can conclude
$$E\left(\sup_t (H \cdot X)_t^2\right) < \infty$$

With this in hand we can start with
$$E(H \cdot X)_{T_n}^2 = E \int_0^{T_n} H_t^2 \, d\langle X \rangle_t$$
and let $n \to \infty$ to get the desired result.

6.3. By stopping we can reduce to the case in which X is a bounded continuous martingale and $\langle X \rangle_t \leq N$ for all t, which implies $H \in \Pi_2(X)$. If we replace S and T by S_n and T_n which stop at the next dyadic rational and let $H_s^n = C$ for $S_n < s \leq T_n$ then $H_s^n \in \Pi_1$ and it follows easily from the definition of the integral in Step 2 in Section 2.4 that
$$\int H_s^n \, dX_s = C(X_{T_n} - X_{S_n})$$

To complete the proof now we observe that if $|C(\omega)| \leq K$ then
$$\|H_n - H\|_X \leq K^2 \{E(\langle X \rangle_{T_n} - \langle X \rangle_T) + E(\langle X \rangle_{S_n} - \langle X \rangle_S)\} \to 0$$

as $n \to \infty$ by the bounded convergence theorem. Using Exercise 4.4 and (4.3.b) it follows that $\|(H_n \cdot X) - (H \cdot X)\|_2 \to 0$ and hence $\int H_s^n \, dX_s \to \int H_s \, dX_s$. Clearly, $C(X_{T_n} - X_{S_n}) \to C(X_T - X_S)$ and the desired result follows.

6.4. $X_t^2 - X_0^2 = \sum_i X^2(t_{i+1}^n) - X^2(t_i^n)$

$$= \sum_i \{X(t_{i+1}^n) - X(t_i^n)\}^2 + \sum_i 2X(t_i^n)\{X(t_{i+1}^n) - X(t_i^n)\}$$

(3.8) implies that the first term converges in probability to $\langle X \rangle_t$. The previous exercise shows the second term converges in probability to $\int_0^t 2X_s \, dX_s$.

6.5. The difference between evaluating at the right and the left end points is

$$2\sum_i \{X(t_{i+1}^n) - X(t_i^n)\}^2 \to 2\langle X \rangle_t$$

6.6. Let $C_{k,n} = B((k+1/2)2^{-n}t) - B(k2^{-n}t)$ and $D_{k,n} = B((k+1)2^{-n}t) - B((k+1/2)2^{-n}t)$. In view of (6.7) it suffices to show that as $n \to \infty$

$$S_n = \sum_k C_{k,n}(C_{k,n} + D_{k,n}) \to t/2$$

in probability. $EC_{k,n}^2 = 2^{-n}t/2$ and $EC_{k,n}D_{k,n} = 0$ so $ES_n = t/2$. To complete the proof now we note that the terms in the sum are independent so

$$E(S_n - t/2)^2 = E\left(\sum_k C_{k,n}^2 - 2^{-n}t/2 + C_{k,n}D_{k,n}\right)^2$$

$$= E\sum_k (C_{k,n}^2 - 2^{-n}t/2)^2 + C_{k,n}^2 D_{k,n}^2 \leq 2^n Ct2^{-2n} \to 0$$

and then use Chebyshev's inequality.

6.7. (6.7) implies that if t_i^n is a sequence of partitions of $[0,t]$ with mesh $\to 0$ as $n \to \infty$ we have

$$\sum_i h_{t_i^n}(B_{t_{i+1}^n} - B_{t_i^n}) \to \int_0^t h_s \, dB_s$$

in probability. The left-hand side has a normal distribution with mean 0 and variance

$$\sum_i h_{t_i^n}^2 (t_{i+1}^n - t_i^n) \to \int_0^t h_s^2 \, ds$$

Answers for Chapter 3   325

as $n \to \infty$ by the convergence of the Riemann approximating sums for the integral. Convergence of the means and variances for a sequence of normal random variables is sufficent for them to converge in distribution (consider the characteristic functions) and the desired result follows.

7.1. $\sup_t |((K^n - K) \cdot A)_t| \le M \sup_s |K^n_s - K_s| \to 0$

7.2. By stopping it suffices to prove the result when $|A|_t \le M$. If we let $G^\delta_s = f'(c(A_{t_i}, A_{t_{i+1}}))$ then as in (7.4) we get

$$f(A_t) - f(A_0) = \int_0^t G^\delta_s \, dA_s$$

As $\delta \to 0$, $G^\delta_s \to f'(A_s)$ uniformly on $[0,t]$ so the desired conclusion from Exercise 7.1.

8.1. Clearly $M_t + M'_t$ is a continuous local martingale and $A_t + A'_t$ is continuous adapted, locally of bounded variation and has $A_0 + A'_0 = 0$.

Chapter 3

1.1. The optional stopping theorem implies that

$$x = E_x B_T = a P_x(B_T = a) + b(1 - P_x(B_T = a))$$

Solving gives $P_x(B_T = a) = (b-x)/(b-a)$.

1.2. From (1.6) it follows that for any s and n $P(\sup_{t \ge s} B_t \ge n) = 1$. This implies $P(\sup_{t \ge s} B_t = \infty) = 1$ for all s and hence $\limsup_{s \to \infty} B_s = \infty$ a.s.

1.3. If $S_r(\omega) \uparrow t(\omega)$ which is $< \infty$ with positive probability then $B_{t(\omega)}(\omega) = 0$ and we have a contradiction.

3.1. Using the optional stopping theorem at time $T \wedge t$ we have $E_0 B^2_{T \wedge t} = E_0(T \wedge t)$. Letting $t \to \infty$ and using the bounded and monotone convergence theorems with (1.4) we have

$$E_0 T = E_0 B^2_T = a^2 \cdot \frac{b}{b+a} + b^2 \cdot \frac{a}{b+a} = ab$$

3.2. Let $f(x,t) = x^6 - ax^4 t + bx^2 t^2 - ct^3$. Differentiating gives

$$D_t f = -ax^4 + 2bx^2 t - 3ct^2$$
$$(1/2) D_{xx} f = 15x^4 - 6ax^2 t + bt^2$$

Setting $a = 15$, $2b = 6a$, and $b = 3c$, that is $a = 15$, $b = 45$, $c = 15$, we have $D_t f + (1/2)D_{xx}f = 0$, so $f(B_t, t)$ is a local martingale. Using the optional stopping theorem at time $\tau_a \wedge t$ we have

$$E_0\{B^6_{\tau_a \wedge t} - 15 B^4_{\tau_a \wedge t}(\tau_a \wedge t) + 45 B^2_{\tau_a \wedge t}(\tau_a \wedge t)^2\} = 15 E_0 (\tau_a \wedge t)^3$$

Letting $t \to \infty$ and using the bounded and monotone convergence theorems with the results in (3.3) we have

$$15 E_0 \tau_a^3 = a^6 - 15 a^4 E\tau_a + 45 a^2 E\tau_a^2 = a^6(1 - 15 + 75)$$

so $E_0 \tau_a^3 = 61/15$.

3.3. Using the optional stopping theorem at time $T_{-a} \wedge n$ we have

$$1 = E_0 \exp(-(2\mu/\sigma^2) Z_{T_{-a} \wedge n})$$

Letting $n \to \infty$ and noting that on $\{T_{-a} = \infty\}$ we have $Z_t/t = \sigma B_t/t + \mu \to \mu$ almost surely and hence $Z_t \to \infty$ a.s., the desired result follows.

4.1. Let $\theta : [0, \infty) \to \mathbf{R}^d$ be piecewise constant and have $|\theta_s| = 1$ for all s. Let $Y_t = \sum_i \int_0^t \theta^i_s \, dX^i_s$. The formula for the covariance of stochastic integrals shows Y_t is a local martingale with $\langle Y \rangle_t = t$, so (4.1) implies that Y_t is a Brownian motion. Letting $0 = t_0 < t_1 < \ldots < t_n$ and taking $\theta_s = v_j$ for $s \in (t_{j-1}, t_j]$ for $1 \le j \le n$ shows that $X_{t_1} - X_{t_0}, \ldots, X_{t_n} - X_{t_{n-1}}$ are independent multivariate normals with mean 0 and covariance matrices $(t_1 - t_0)I, \ldots, (t_n - t_{n-1})I$, and it follows that X_t is a d-dimensional Brownian motion.

4.2. X_s is a local martingale with $\langle X \rangle_s = \int_0^s h_r^2 \, dr$. By modifying h_s after time t we can suppose without loss of generality that $\langle X \rangle_\infty = \infty$. Using (4.4) now we see that if $\gamma(u) = \inf\{s : \langle X \rangle_s > u\}$ then $X_{\gamma(u)}$ is a Brownian motion. Since $\langle X \rangle_s$ and hence the time change $\gamma(u)$ are deterministic, the desired result follows.

Chapter 4

4.1. Let $\tau = \inf\{t > 0 : B_t \notin G\}$, let $y \in \partial G$, and let $T_y = \inf\{t > 0 : B_t = y\}$. (2.9) in Chapter 1 implies $P_y(T_y = 0) = 1$. Since $\tau \le T_y$ it follows that y is regular.

4.2. $P_y(\tau = 0) < 1$ and Blumenthal's 0-1 law imply $P_y(\tau = 0) = 0$. Since we are in $d \ge 2$, it follows that $P_y(B_\tau = y) = 0$ and hence $\epsilon \equiv 1 - E_y f(B_\tau) > 0$. Let $\sigma_n = \inf\{t > 0 : B_t \notin D(y, 1/n)\}$. $\sigma_n \downarrow 0$ as $n \uparrow \infty$ so $P_y(\sigma_n < \tau) \to 1$.

$$1 - \epsilon = E_y f(B_\tau) = E_y(f(B_\tau); \tau \le \sigma_n) + E_y(v(B_{\sigma_n}); \tau > \sigma_n)$$

As $n \to \infty$ the first term on the right $\to 0$, so

$$E_y(v(B_{\sigma_n}); \tau > \sigma_n) \to 1 - \epsilon$$

and it follows that for large n, $\inf_{x \in \partial D(y, 1/n)} v(x) \le 1 - \frac{\epsilon}{2}$.

4.3. Let $y \in \partial G$ and consider $U = V(y, \nabla g(y), 1)$. Calculus gives us

$$g(z) - g(y) = \int_0^1 \nabla g(y + \theta(z-y)) \cdot \theta(z-y)\, d\theta$$

Continuity of ∇g implies that if $|z - y| < r$ and $z \in U$ then $\nabla g(y + \theta(z - y)) \cdot \theta(z - y) \ge 0$ so $g(z) \ge 0$. This shows that for small r the truncated cone $U \cap D(y, r) \subset G^c$ so the regularity of y follows from (4.5c).

4.4. Let $\tau = \inf\{t > 0 : B_t \in \bar{V}(y, v, a)\}$. By translation and rotation we can suppose $y = 0$, $v = (1, 0, \ldots, 0)$, and the $d-1$ dimensional hyperplane is $z_d = 0$. Let $T_0 = \inf\{t > 0 : B_t^d = 0\}$. (2.9) in Chapter 1 implies that $P_0(T_0 = 0) = 1$. If $a > 0$ then for some $k < \infty$ the hyperplane can be covered by k rotations of $\bar{V}$ so $P_0(\tau = 0) \ge 1/k$ and it follows from the 0-1 law that $P_0(\tau = 0) = 1$.

7.1. (a) Ignoring C_d and differentiating gives

$$D_{x_i} h_\theta = -\frac{d}{2} \cdot \frac{2(x_i - \theta_i)y}{(|x - \theta|^2 + y^2)^{(d+2)/2}}$$

$$D_{x_i x_i} h_\theta = -d \cdot \frac{y}{(|x - \theta|^2 + y^2)^{(d+2)/2}} + \frac{d(d+2)(x_i - \theta)^2 y}{(|x - \theta|^2 + y^2)^{(d+4)/2}}$$

$$D_y h_\theta = \frac{1}{(|x - \theta|^2 + y^2)^{d/2}} - d \cdot \frac{y^2}{(|x - \theta|^2 + y^2)^{(d+2)/2}}$$

$$D_{yy} h_\theta = -d \cdot \frac{y + 2y}{(|x - \theta|^2 + y^2)^{(d+2)/2}} + \frac{d(d+2)y^3}{(|x - \theta|^2 + y^2)^{(d+4)/2}}$$

Adding up we see that

$$\sum_{i=1}^{d-1} D_{x_i x_i} h_\theta + D_{yy} h_\theta = \frac{\{(-d)(d-1) + (-d) \cdot 3\}y}{(|x - \theta|^2 + y^2)^{(d+2)/2}}$$
$$+ \frac{d(d+2)y(|x - \theta|^2 + y^2)}{(|x - \theta|^2 + y^2)^{(d+4)/2}} = 0$$

The fact that $\Delta u(x) = 0$ follows from (1.7) as in the proof of (7.1).

(b) Clearly $\int d\theta\, h_\theta(x, y)$ is independent of x. To show that it is independent of y, let $x = 0$ and change variables $\theta_i = y\varphi_i$ for $1 \leq i \leq d-1$ to get

$$\int d\theta\, h_\theta(0, y) = \int \frac{C_d y}{(y^2|\varphi|^2 + y^2)^{d/2}} \cdot y^{d-1} d\varphi = \int d\varphi\, h_\varphi(0, 1) = 1$$

(c) Changing variables $\theta_i = x_i + r_i y$ and using dominated convergence

$$\int_{D(x,\epsilon)^c} d\theta\, h_\theta(x, y) = \int_{D(0,\epsilon/y)^c} dr\, h_r(0, 1) \to 0$$

(d) Since $P_x(\tau < \infty) = 1$ for all $x \in H$, this follows from (4.3).

9.1. $c(x) \equiv -\beta < 0$ so $w(x) = E_x e^{-\beta\tau} \leq 1$ and it follows from (6.3) that $v(x) = E_x e^{-\beta\tau}$ is the unique solution of

$$\frac{1}{2}u'' - \beta u = 0 \qquad u(-a) = u(a) = 1$$

Guessing $u(x) = B\cosh(bx)$ with $b > 0$ we find

$$\frac{1}{2}u'' - \beta u = \left(\frac{Bb^2}{2} - \beta B\right)\cosh(bx) = 0$$

if $b = \sqrt{2\beta}$. Then we take $B = 1/\cosh(a\sqrt{2\beta})$ to satisfy the boundary condition.

Chapter 5

3.1. We first prove that h is Lipschitz continuous with constant $C_2 = 2C_1 + R^{-1}|h(0)|$ on $D_2 = \{R \leq |x| \leq 2R\}$. Let $f(x) = (2R - |x|)/R$ and $g(x) = h(Rx/|x|)$. If $x, y \in D_2$ then

$$|f(x) - f(y)| = \frac{||y| - |x||}{R} \leq \frac{1}{R}|x - y|$$

Since h is Lipschitz continuous on $D_1 = \{|x| \leq R\}$, and $Rx/|x|$ is the projection of x onto D_1 we have

$$|g(x) - g(y)| = |h(Rx/|x|) - h(Ry/|y|)|$$
$$\leq C_1 \left|\frac{Rx}{|x|} - \frac{Ry}{|y|}\right| \leq C_1|x - y|$$

Combining the last two results, introducing $\|k\|_{\infty,i} = \sup\{|k(x)| : x \in D_i\}$, and using the triangle inequality:

$$|h(x) - h(y)| \leq \|f\|_{\infty,2}|g(x) - g(y)| + |f(x) - f(y)|\|g\|_{\infty,2}$$
$$\leq 1 \cdot C_1|x - y| + \frac{1}{R}\|h\|_{\infty,1}$$
$$\leq (2C_1 + |h(0)|/R)|x - y| = C_2|x - y|$$

since for $x \in D_1$, $|h(x)| \leq |h(0)| + C_1 R$.

To extend the last result to Lipschitz continuity on $\mathbf{R}^d$ we begin by observing that if $x \in D_1$, $y \in D_2$, and z is the point of ∂D_1 on the line segment between x and y then

$$|h(x) - h(y)| \leq |h(x) - h(z)| + |h(z) - h(y)|$$
$$\leq C_1|x - z| + C_2|z - y| \leq C_2|x - y|$$

since $C_1 \leq C_2$ and $|x - z| + |z - y| = |x - y|$. This shows that h is Lipschitz continuous with constant C_2 on $|z| \leq 2R$. Repeating the last argument taking $|x| \leq 2R$ and $y > 2R$ completes the proof.

3.2. (a) Differentiating we have

$$\kappa(x) = -Cx \log x$$
$$\kappa'(x) = -C \log x - C > 0 \quad \text{if } 0 < x < e^{-1}$$
$$\kappa''(x) = -C/x < 0 \quad \text{if } x > 0$$

so κ is strictly increasing and concave on $[0, e^{-2}]$. Since $\kappa(e^{-2}) = 2Ce^{-2}$ and $\kappa'(e^{-2}) = C$, κ is strictly increasing and concave on $[0, \infty)$. When $\epsilon \leq e^{-2}$

$$\int_0^\epsilon \frac{-1}{Cx \log x} dx = -C^{-1} \log\log x \Big|_0^\epsilon = \infty$$

(b) If $g(t) = \exp(-1/t^p)$ with $p > 0$ then

$$g'(t) = \frac{p}{t^{p+1}} \exp(-1/t^p) = pg(t)\{\log(1/g(t))\}^{(p+1)/p}$$

5.1. Under Q the coordinate maps $X_t(\omega)$ satisfy MP$(\beta + b, \sigma)$. Let

$$\hat{X}_t = X_t - \int_0^t \beta(X_s) + b(X_s)\, ds$$
$$= \tilde{X}_t - \int_0^t b(X_s)\, ds$$

Since we are interested in adding drift $-b$ we let $c = a^{-1}b$ as before, but let

$$\hat{Y}_t = -\int_0^t c(X_s) \cdot d\hat{X}_s$$
$$= -Y_t + \int_0^t c(X_s) \cdot b(X_s)\, ds$$
$$= -Y_t + \int_0^t ba^{-1}b(X_s)\, ds$$

Since Y_t and $\hat{Y}_t$ differ by a process of bounded variation we have

$$\langle \hat{Y} \rangle_t = \langle Y \rangle_t = \int_0^t ba^{-1}b(X_s)\, ds$$

The change of measure in this case is given by

$$\hat{\alpha}_t = \exp\left(\hat{Y}_t - \frac{1}{2}\langle \hat{Y} \rangle_t\right)$$
$$= \exp\left(-Y_t + \frac{1}{2}\langle Y \rangle_t\right) = 1/\alpha_t$$

6.1. It suffices to show that P_0 a.s.

$$\limsup_{n \to \infty} \int_0^{T_1} |B_s|^{-1} 1_{(|B_s| \leq 2^{-n})}\, ds > 0$$

To prove the last result, let $R_0 = 0$, $S_n = \inf\{t > R_n : |B_t| = 2^{-n}\}$, $R_{n+1} = \inf\{t > S_n : B_t = 0\}$ for $n \geq 0$, and $N = \sup\{n : R_n < T_1\}$. Now

$$\int_0^{T_1} |B_s|^{-1} 1_{(|B_s| \leq 2^{-n})}\, ds \geq \sum_{m=0}^{N} 2^n(S_m - R_m)$$

where the number of terms, $N+1$, has a geometric distribution with $E(N+1) = 2^n$, and the $S_m - R_m$ are i.i.d. with $E(S_m - R_m) = 2^{-2n}$ and $E(S_m - R_m)^2 = (5/3)2^{-4n}$. Using the formula for the variance of a random sum or noticing that $P(N \geq 2^n) \to e^{-1}$ as $n \to \infty$ one concludes easily that $\limsup \geq C > 0$ a.s.

Chapter 6

3.1. (i) If $b(z) \leq 0$ then

$$\exp\left(-\int_0^y 2b(z)/a(z)\, dz\right) \geq 1$$

so $\varphi(x) \geq x$, $\varphi(\infty) = \infty$ and recurrence follows from (3.3).

(ii) If $b(z)/a(z) \geq \epsilon$ then

$$\exp\left(-\int_0^y 2b(z)/a(z)\,dz\right) \leq e^{-2\epsilon y}$$

so $\varphi(\infty) < \infty$ and transience follows from (3.3).

3.2. The condition is $\int_0^1 -2b(y)/a(y)\,dy = 0$. Let $I = \int_0^1 -2b(y)/a(y)\,dy$. $\varphi'(n+y) = e^{nI}\varphi'(y)$ so if $I > 0$ $\varphi(-\infty) > -\infty$ and if $I < 0$ then $\varphi(\infty) < \infty$. If $I = 0$ then $\varphi'(y)$ is periodic so $0 < c < \varphi'(y) < C < \infty$, and it follows that $\varphi(\infty) = \infty$ and $\varphi(-\infty) = -\infty$.

5.1. (a) If $\beta = 0$ then $\varphi(x) = x$ and $m(x) = 1/\sigma^2 x$ so

$$\int_0^1 m(x)(\varphi(x) - \varphi(0))\,dx = \int_0^1 \sigma^{-2}\,dx < \infty$$

$M(0) = -\infty$ so $J = \infty$ and it follows that 0 is absorbing. To deal with ∞ note that $\varphi(\infty) = M(\infty) = \infty$ so $I = J = \infty$.

(b) When $\beta < 0$, $\varphi'(x) = e^{2|\beta|x/\sigma^2}$, so

$$\varphi(x) = \frac{\sigma^2}{2|\beta|}\left(e^{2|\beta|x/\sigma^2} - 1\right)$$

$$m(x) = e^{-2|\beta|x/\sigma^2}/\sigma^2 x$$

To evaluate I for the boundary at 0, we note

$$\int_0^1 m(x)(\varphi(x) - \varphi(0))\,dx = \int_0^1 \frac{1 - e^{-2|\beta|x/\sigma^2}}{2|\beta|x}\,dx < \infty$$

since the integrand converges to $1/\sigma^2$ as $x \to 0$. Again $M(0) = -\infty$ so $J = \infty$ and it follows that 0 is absorbing. To deal with ∞ we note that $\varphi(\infty) = \infty$ so $I = \infty$. To evaluate J we begin by noting that

$$\int_x^\infty e^{-2|\beta|y/\sigma^2}\,dy = \frac{\sigma^2}{2|\beta|}e^{-2|\beta|x/\sigma^2}$$

Since most of the integral comes from $x \leq y \leq x + \sqrt{x}$ it follows easily that

$$M(\infty) - M(x) \sim \frac{\sigma^2}{|\beta|x}e^{-2|\beta|x/\sigma^2}$$

Since $\varphi'(x) = e^{-2|\beta|x/\sigma^2}$ it follows that $J = \infty$ and ∞ is a natural boundary.

5.2. The computations in Example 5.3 show that (a) $I < \infty$ when $\gamma < 1$ and (b) for $\gamma < -1$ we have $M(0) = -\infty$ and hence $J = \infty$. Consulting the table of definitions we see that 0 is an absorbing boundary for $\gamma < -1$.

5.3. (a) $H'_\delta \leq T_0 < \infty$.

(b) (2.9) in Chapter 3 implies

$$E_1 H_\delta = \int_0^1 2yy^{-2\delta}\, dy < \infty$$

since $\delta < 1$ implies $1 - 2\delta > -1$.

(c) If $\delta > 1$ then $H_\delta > H_1$ so it suffices to prove the result when $\delta = 1$. To do this, we note that if $B_0 = x$ then $x^{-1} B_{x^2 s}$ is a Brownian motion starting at 1, so the distribution of

$$\int_0^{T_{x/2} \wedge T_{2x}} B_s^{-2}\, ds \qquad \text{under } P_x$$

does not depend on x. Letting $S_x = (T_{x/2} \wedge T_{2x}) \circ \theta_{T_x}$ and using the strong Markov property it follows that the distribution of

$$I_x = \int_{T_x}^{S_x} B_s^{-2}\, ds \qquad \text{under } P_1$$

does not depend on x. Pick $\delta > 0$ so that $P_1(I_x \geq \delta) > 0$. The $I_{2^{-n}}$ are independent so the second Borel-Cantelli lemma implies that $P_1(I_{2^{-n}} \geq \delta \text{ i.o.}) = 1$ but this is inconsistent with $P_1(H_1 < \infty) > 0$.

5.4. $\varphi(x) = x$ so $\varphi(\infty) = \infty$ and $I = \infty$. When $\delta \leq 1/2$, $M(\infty) = \infty$ and hence $J = \infty$. When $\delta > 1/2$

$$J = \int_1^\infty \frac{z^{1-2\delta} - 1}{2\delta - 1}\, dz \begin{cases} < \infty & \text{if } \delta > 1 \\ = \infty & \text{if } \delta \leq 1 \end{cases}$$

Combining the results for I and J and consulting the table we have the indicated result.

5.5. When $\alpha = 0$

$$\varphi'(x) = \exp\left(\int_{1/2}^y -\frac{2\beta}{z}\, dz\right) = Cy^{-2\beta}$$

Here and in what follows C is a constant which will change from line to line. If $\beta \geq 1/2$ then $\varphi(0) = -\infty$ and $I = \infty$. If $\beta < 1/2$ then $\varphi(z) - \varphi(0) = Cz^{1-2\beta}$ and

$$m(y) = \frac{1}{\varphi'(y)a(y)} = \frac{C}{y^{-2\beta} \cdot y(1-y)} \sim Cy^{2\beta-1}$$

as $y \to 0$. From this we see that $I < \infty$ if $\beta < 1/2$. When $\beta = 0$, $M(0) = -\infty$ so $J = \infty$. When $\beta > 0$, $M(z) - M(0) \sim Cz^{2\beta}$ as $z \to 0$ so $J < \infty$ if $\beta > 0$. Comparing the possibilities for I and J gives the desired result.

Chapter 7

5.1. Letting $b \to \infty$ in (4.7) of Chapter 6 we have

$$E_x T_{(1,\infty)} = 2(\varphi(x) - \varphi(1)) \int_x^\infty m(z)\, dz + 2 \int_1^x (\varphi(z) - \varphi(1))\, m(z)\, dz$$

The result now follows from $m(z) = 1/\varphi'(z)a(z) = z^{-2c}$.

5.2. We apply (5.3) to $v(x) = x^{1+\delta} - 1$ and note that

$$Lu = \frac{(1+\delta)\delta}{2} \cdot \frac{x^{\delta-1}}{\epsilon} + b(x)(1+\delta)\frac{x^\delta}{\epsilon} \leq -1$$

since $b(x) \leq -\epsilon/x^\delta$, $(1+\delta)/2 \leq 1$, and $x^{\delta-1} \leq 1$.

Chapter 8

3.1. Suppose $r_n \not\to r$. Then there is an $\epsilon > 0$ and a subsequence r_{n_k} so that $|r_{n_k} - r| > \epsilon$ for all k. However it is impossible for a subsequence of r_{n_k}, so we have a contradiction which proves $r_n \to r$.

6.1. If $d(g_n, h) \to 0$ then h can only take the values 0 and 1. In Example 6.2 we showed that h cannot be $\equiv 0$ so it must take the value 1 at some point a. Since h is right continuous at a there must be some point $b \neq 1/2$ where $h(b) = 1$ but in this case it is impossible for $d(g_n, h) \to 0$.

References

M. Aizenman and B. Simon (1982) Brownian motion and a Harnack inequality for Schrödinger operators. *Comm. Pure Appl. Math.* **35**, 209–273

P. Billingsley (1968) *Convergence of Probability Measures.* John Wiley and Sons, New York

R.M. Blumenthal and R.K. Getoor (1968) *Markov Processes and their Potential Theory.* Academic Press, New York

D. Burkholder (1973) Distribution function inequalities for martingales. *Ann. Probab.* **1**, 19–42

D. Burkholder, R. Gundy, and M. Silverstein (1971) A maximal function characterization of the class H^p. *Trans A.M.S.* **157**, 137–153

K.L. Chung and R.J. Williams (1990) *Introduction to Stochastic Integration.* Second edition. Birkhauser, Boston

K.L. Chung and Z. Zhao (1995) *From Brownian Motion to Schrödinger's Equation.* Springer, New York

Z. Ciesielski and S.J. Taylor (1962) First passage times and sojurn times for Brownian motion in space and exact Hausdorff measure. *Trans. A.M.S.* **103**, 434–450

E. Cinlar, J. Jacod, P. Protter, and M. Sharpe (1980) Semimartingales and Markov processes. *Z. fur. Wahr.* **54**, 161–220

B. Davis (1983) On Brownian slow points. *Z. fur Wahr.* **64**, 359–367

C. Dellacherie and P.A. Meyer (1978) *Probabilities and Potentials.* North Holland, Amsterdam

C. Dellacherie and P.A. Meyer (1982) *Probabilities and Potentials B. Theory of Martingales.* North Holland, Amsterdam

C. Doléans-Dade (1970) Quelques applications de la formule de changement de variables pour les semimartingales. *Z. fur Wahr.* **16**, 181–194

R. Durrett (1995) *Probability: Theory and Examples.* Second edition. Duxbury Press, Belmont, CA

References

A. Dvoretsky and P. Erdös (1951) Some problems on random walk in space. Pages 353–367 in *Proc. 2nd Berkeley Symp.*, U. of California Press

E.B. Dynkin (1965) *Markov Processes.* Springer, New York

P. Echeverria (1982) A criterion for invariant measures of Markov processes. *Z. fur Wahr.* **61**, 1–16

S. Ethier and T. Kurtz (1986) *Markov Processes: Characterization and Convergence.* John Wiley and Sons, New York

W. Feller (1951) Diffusion processes in genetics. Pages 227–246 in *Proc. 2nd Berkeley Symp.*, U. of California Press

G.B. Folland (1976) *An Introduction to Partial Differential Equations.* Princeton U. Press, Princeton, NJ

D. Freedman (1970) *Brownian Motion and Diffusion.* Holden-Day, San Francisco, CA

A. Friedman (1964) *Partial Differential Equations of Parabolic Type.* Prentice-Hall, Englewood Cliffs, NJ

A. Friedman (1975) *Stochastic Differential Equations and Applications, Volume 1.* Academic Press, New York

R.K. Getoor and M. Sharpe (1979) Excursions of Brownian motion and Bessel processes. *Z. fur. Wahr.* **47**, 83–106

D. Gilbarg and J. Serrin (1956) On isolated singularities of second order elliptic differential equations. *J. d'Analyse Math.* **4**, 309–340

D. Gilbarg and N.S. Trudinger (1977) *Elliptic Partial Differential Equations of Second Order.* Springer, New York

E. Hewitt and K. Stromberg (1969) *Real and Abstract Analysis.* Springer, New York

N. Ikeda and S. Watanabe (1981) *Stochastic Differential Equations and Diffusion Processes.* North Holland, Amsterdam

J. Jacod and A.N. Shiryaev (1987) *Limit Theorems for Stochastic Processes.* Springer, New York

P. Jagers (1975) *Branching Processes with Biological Applications.* John Wiley and Sons, New York

F. John (1982) *Partial Differential Equations.* Fourth edition. Springer-Verlag, New York

M. Kac (1951) On some connections between probability theory and differential and integral equations. Pages 189–215 in *Proc. 2nd Berkeley Symposium*, U. of California Press

I. Karatzas and S. Shreve (1991) *Brownian Motion and Stochastic Calculus*. Second edition. Springer, New York

S. Karlin and H. Taylor (1981) *A Second Course in Stochastic Processes*. Academic Press, New York

T. Kato (1973) Schrödinger operators with singular potentials. *Israel J. Math.* **13**, 135–148

R.Z. Khasminskii (1960) Ergodic properties of recurrent diffusion processes and stabilization of the solution to the Cauchy problem for parabolic equations. *Theor. Prob. Appl.* **5**, 179–196

F. Knight (1981) *Essentials of Brownian Motion and Diffusion*. American Math. Society, Providence, RI

H.P. McKean (1969) *Stochastic Integrals*. Academic Press, New York

P.A. Meyer (1976) Un cours sur les intégrals stochastiques. Pages 246–400 in Lecture Notes in Math. 511, Springer, New York

N. Meyers and J. Serrin (1960) The exterior Dirichlet problem for second order elliptic partial differential equations. *J. Math. Mech.* **9**, 513–538

B. Øksendal (1992) *Stochastic Differential Equations*. Third edition. Springer, New York

S. Port and C. Stone (1978) *Brownian Motion and Classical Potential Theory*. Academic Press, New York

P. Protter (1990) *Stochastic Integration and Differential Equations*. Springer, New York

D. Revuz and M. Yor (1991) *Continuous Martingales and Brownian Motion*. Springer, New York

L.C.G. Rogers and D. Williams (1987) *Diffusions, Markov Processes, and Martingales. Volume 2: Itô Calculus*. John Wiley and Sons, New York

H. Royden (1988) *Real Analysis*. Third edition. MacMillan, New York

B. Simon (1982) Schrödinger semigroups. *Bulletin A.M.S.* **7**, 447–526

D.W. Stroock and S.R.S. Varadhan (1979) *Multidimensional Diffusion Processes*. Springer, New York

H. Tanaka (1963) Note on continuous additive functionals of one-dimensional Brownian motion. *Z. fur Wahr.* **1**, 251–257

J. Walsh (1978) A diffusion with a discontinuous local time. *Asterique* **52–53**, 37–45

T. Yamada and S. Watanabe (1971) On the uniqueness of solutions to stochastic differential equations. *J. Math. Kyoto* **11**, I, 155–167; II, 553–563

Index

adapted 36
arcsine law 173, 291
Arzela-Ascoli theorem 284
associative law 76
averaging property 148

basic predictable process 52
Bessel process 179, 232
Bichteler-Dellacherie theorem 71
Blumenthal's 0-1 law 14
boundary behavior of diffusions 231
Brownian motion
    continuity of paths 4
    definition 1
    Hölder continuity 6
    Lévy's characterization of 111
    nondifferentiability 6
    quadratic variation 7
    writes your name 15
    zeros of 23
Burkholder Davis Gundy theorem 116

C, the space 4, 282
Cauchy process 29
change of measure in SDE 202
change of time in SDE 207
Ciesielski-Taylor theorem 171
cone condition 147
continuous mapping theorem 273
contraction semigroup 245
converging together lemma 273
covariance process 50

D, the space 293
differentiating under an integral 129
diffusion process 177
Dini function 239
Dirichlet problem 143
distributive laws 72
Doléans measure 56
Donsker's theorem 287
Dvoretsky-Erdös theorem 99

effective dimension 239
exit distributions
    ball 164
    half space 166
exponential Brownian motion 178
exponential martingale 108

Feller semigroup 246
Feller's branching diffusion 181, 231
Feller's test 214, 227
Feynman-Kac formula 137
filtration 8
finite dimensional sets 3, 10
Fubini's theorem 86
fundamental solution 251

generator of a semigroup 246
Girsanov's formula 91
Green's functions
    Brownian motion 104
    one dimensional diffusions 222
Gronwall's inequality 188

harmonic function 95
   averaging property 148
   Dirichlet problem 143
Harris chains 259
heat equation 126
   inhomogeneous 130
hitting times 19, 26

infinitesimal drift 178
infinitesimal generator 246
infinitesimal variance 179
integration by parts 76
isometry property 56
Itô's formula 68
   multivariate 77

Kalman-Bucy filter 180
Khasminskii's lemma 141
Khasminskii's test 237
Kolmogorov's continuity theorem 5
Kunita-Watanabe inequality 59

law of the iterated logarithm 115
Lebesgue's thorn 146
Lévy's theorem 111
local martingale 37, 113
local time 83
   continuity of 88
locally equivalent measures 90

Markov property 9
martingale problem 198
   well posed 210
Meyer-Tanaka formula 84
modulus of continuity 283
monotone class theorem 12

natural scale 212
nonnegative definite matrices 198

occupation times
   ball 167

half space 31
optional process 36
   in discrete time 34
optional stopping theorem 39
Ornstein-Uhlenbeck process 180

pathwise uniqueness for SDE 184
$\pi - \lambda$ theorem 11
Poisson's equation 151
potential kernel 102
predictable process 36
   basic 52
   in discrete time 34
   simple 53
predictable σ-field 35
progressivley measurable process 36
Prokhorov's theorems 276
punctured disc 146

quadratic variation 50

recurrence and transience
   Brownian motion 17, 98
   one dimensional diffusions 220
   Harris chains 262
reflecting boundary 229
reflection principle 25
regular point 144
resolvent operator 249
relatively compact 276
right continuous filtration

scaling relation 2
Schrödinger equation 156
semigroup property 245
semimartingale 70
shift transformations 9
simple predictable process 53
skew Brownian motion 89
Skorokhod representation 274
Skorokhod topology on D 293
speed measure 224

stopping time 18
strong Markov property
    for Brownian motion 21
    for diffusions 201
strong solution to SDE 183

tail σ-field 17
tight sequence of measures 276
transience, see recurrence
transition density 257
transition kernels 250

uniqueness in distribution for SDE 197

variance process 42

weak convergence 271
Wright-Fisher diffusion 182, 234, 308

Yamada-Watanabe theorem 193